Neuropeptides in Neuroprotection and Neuroregeneration

Neuropeptides in Neuroprotection and Neuroregeneration

EDITED BY

FRED J. NYBERG

CRC Press
Taylor & Francis Group
Boca Raton London New York

CRC Press is an imprint of the
Taylor & Francis Group, an **informa** business

CRC Press
Taylor & Francis Group
6000 Broken Sound Parkway NW, Suite 300
Boca Raton, FL 33487-2742

First issued in paperback 2019

© 2012 by Taylor & Francis Group, LLC
CRC Press is an imprint of Taylor & Francis Group, an Informa business

No claim to original U.S. Government works

ISBN-13: 978-1-4398-3062-8 (hbk)
ISBN-13: 978-0-367-38123-3 (pbk)

Library of Congress Cataloging-in-Publication Data

Neuropeptides in neuroprotection and neuroregeneration / editor, Fred Nyberg.
    p. ; cm.
  Includes bibliographical references and index.
  ISBN 978-1-4398-3062-8 (hardback : alk. paper)
  I. Nyberg, F.
  [DNLM: 1. Neuropeptides. 2. Neuroprotective Agents--therapeutic use. WL 104]

572'.65--dc23   2012014549

**Visit the Taylor & Francis Web site at
http://www.taylorandfrancis.com**

**and the CRC Press Web site at
http://www.crcpress.com**

Contents

Preface

Over the past four decades, the number of peptides identified as neurotransmitters or neuromodulators in the central and peripheral nervous systems has significantly increased. These compounds, known as neuropeptides, have been recognized for playing an important role in the communication between cells in a variety of neuronal networks. Although a considerable number of neuropeptides have been characterized so far, their quantity is limited compared to the number of precursor proteins that are actually found to be expressed in the cells of the nervous system. Numerous studies have confirmed that neuropeptides are involved in a number of biological activities. These include modulation of brain reward, pain processing, and immune response, as well as neuroendocrine regulations, control of neurovegetative functions, and trophic effects.

In recent years, it has become evident that the role of neuropeptides as fast-acting neurotransmitters is challenged by the observation that many of them may act as growth factors by stimulating cell proliferation and slow-acting mitogenesis. It has thus been demonstrated that a number of neuroactive peptides, such as pituitary adenylate cyclase–activating polypeptide (PACAP), adrenocorticotrophin (ACTH), opioid peptides, somatostatin, and substance P, may modulate proliferation and cell viability. Some peptides act as stimulatory factors on brain circuits involved in cognition, whereas others may act as inhibitory factors. Accordingly, many neuropeptide systems appear as important targets for neuroprotective drugs and drugs that promote neuroregeneration. The recently discovered nuclear protein or polypeptide prothymosin alpha (ProTα), which inhibits neuronal necrosis, suggested to be of clinical use for stroke, is also a neuroprotective agent of interest in this regard. This volume aims to describe recent aspects on the impact of neuropeptides on processes involved in neuroprotection and neuroregeneration. It is by no means exhaustive, but intends to highlight some examples of neuropeptides, neuroactive polypeptides, and growth factors that typify the actual area of research.

This book begins with chapters describing important features of the endogenous neuropeptide systems with regard to their formation, receptor signaling, and inactivation. However, unlike many books dealing with this topic, it also includes chapters focused on the design and development of peptide-like drugs (peptidomimetics). It further includes a contribution highlighting neuropeptide circuits as targets of cognitive enhancers. Most of the authors who have contributed to this volume belong to the core of international top scientists in their particular area of research.

Finally, it is interesting to note that it has been possible to get contributors to this volume who are such a good representation of geographical districts around the world.

With the current advances in neuropeptide research, this volume presents a timely book that underlines an important aspect of interest to a diverse group of people, from scientists and medical practitioners to basic clinical investigators and students.

I am grateful to all my colleagues who have contributed such excellent chapters to this volume. As editor, I would like to point out that all these chapters were submitted within a time period of 15 months and therefore they present the most recent knowledge in the respective fields covered within the volume.

Fred J. Nyberg
Uppsala University, Sweden

List of Contributors

Csaba Adori
Department of Neuroscience
Karolinska Institutet
Stockholm, Sweden

Georgy Bakalkin
Division of Biological Research on
    Drug Dependence
Department of Pharmaceutical
    Biosciences
Uppsala University
Uppsala, Sweden

Swapnali Barde
Department of Neuroscience
Karolinska Institutet
Stockholm, Sweden

Margery C. Beinfeld
Department of Molecular Physiology
    and Pharmacology
Tufts University School of Medicine
Boston, Massachusetts

Jennifer K. Callaway
Department of Pharmacology
University of Melbourne
Victoria, Australia

Gabriele Campana
Department of Pharmacology
University of Bologna
Bologna, Italy

Lee E. Eiden
Section on Molecular Neuroscience
Laboratory of Cellular and Molecular
    Regulation
National Institute of Mental Health
National Institute of Health
Bethesda, Maryland

Daniel Förster
Medical Faculty
Institute for Neurobiochemistry
Otto-von-Guericke University of
    Magdeburg
Magdeburg, Germany

Rebecca Fransson
Department of Medicinal Chemistry
Uppsala University
Uppsala, Sweden

Luca Gentilucci
Department of Chemistry
    "G. Ciamician"
University of Bologna
Bologna, Italy

Sebok K. Halder
Division of Molecular Pharmacology
    and Neuroscience
Nagasaki University Graduate School of
    Biomedical Sciences
Nagasaki, Japan

Mathias Hallberg
Department of Pharmaceutical
    Biosciences
Uppsala University
Uppsala, Sweden

Kurt F. Hauser
Department of Pharmacology and
    Toxicology
Institute for Drug and Alcohol
    Studies
Virginia Commonwealth University
    School of Medicine
Richmond, Virginia

Tomas Hökfelt
Department of Neuroscience
Karolinska Institutet
Stockholm, Sweden

Pamela E. Knapp
Department of Anatomy and Neurobiology
and
Department of Pharmacology and
    Toxicology
and
Institute for Drug and Alcohol Studies
Virginia Commonwealth University
    School of Medicine
Richmond, Virginia

Takaaki Komatsu
Department of Pharmacology and
    Pharmacodynamics
Medical University
Lublin, Poland

Jolanta Kotlinska
Department of Pharmacology and
    Pharmacodynamics
Medical University
Lublin, Poland

Alberto Loizzo
Department of Therapeutic Research
    and Medicines Evaluation
Istituto Superiore di Sanità
Roma, Italy

Stefano Loizzo
Department of Therapeutic Research
    and Medicines Evaluation
Istituto Superiore di Sanità
Roma, Italy

Kristina Magnusson
Department of Public Health and
    Caring Science
Uppsala University
Uppsala, Sweden

Hayato Matsunaga
Division of Molecular Pharmacology
    and Neuroscience
Nagasaki University Graduate School of
    Biomedical Sciences
Nagasaki, Japan

Claudia A. McCarthy
Department of Pharmacology
Monash University Clayton
Victoria, Australia

Aleksandra Misicka
Faculty of Chemistry
University of Warsaw
Warsaw, Poland

Hirokazu Mizoguchi
Department of Physiology and Anatomy
Tohoku Pharmaceutical University
Sendai, Japan

Fred J. Nyberg
Department of Pharmaceutical
    Biosciences
Uppsala University
Uppsala, Sweden

Georg Reiser
Medical Faculty
Institute for Neurobiochemistry
Otto-von-Guericke University of
    Magdeburg
Magdeburg, Germany

Chikai Sakurada
Laboratory of Molecular Pathophysiology
Department of Pharmaceutical Health
    Care and Life Sciences
Nihon Pharmaceutical University
Saitama, Japan

Shinobu Sakurada
Department of Physiology and Anatomy
Tohoku Pharmaceutical University
Sendai, Japan

Tsukasa Sakurada
Department of Pharmacology
Daiichi College of Pharmaceutical
   Sciences
Fukuoka, Japan

Anja Sandström
Department of Medicinal Chemistry
Uppsala University
Uppsala, Sweden

Tiejun Shi
Department of Neuroscience
Karolinska Institutet
Stockholm, Sweden

Jerzy Silberring
Department of Biochemistry and
   Neurobiology
AGH University of Science and
   Technology
Krakow, Poland

and

Centre of Polymer and Carbon
   Materials
Polish Academy of Sciences
Zabrze, Poland

Santi Spampinato
Department of Pharmacology
University of Bologna
Bologna, Italy

Ulrike Muscha Steckelings
Center for Cardiovascular Research
Charité University of Medicine
Berlin, Germany

Hiroshi Ueda
Division of Molecular Pharmacology
   and Neuroscience
Nagasaki University Graduate School of
   Biomedical Sciences
Nagasaki, Japan

Dineke S. Verbeek
Department of Genetics
University Medical Center
   Groningen
University of Groningen
Groningen, the Netherlands

Robert E. Widdop
Department of Pharmacology
Monash University Clayton
Victoria, Australia

Zhi-Qing David Xu
Department of Neuroscience
Karolinska Institutet
Stockholm, Sweden

Tatiana Yakovleva
Division of Biological Research on
   Drug Dependence
Department of Pharmaceutical
   Biosciences
Uppsala University
Uppsala, Sweden

Mingdong Zhang
Department of Neuroscience
Karolinska Institutet
Stockholm, Sweden

Kang Zheng
Department of Neuroscience
Karolinska Institutet
Stockholm, Sweden

Gregor Zündorf
Medical Faculty
Institute for Neurobiochemistry
Otto-von-Guericke University of
   Magdeburg
Magdeburg, Germany

1 Neuropeptide Systems—Some Basic Concepts

Tomas Hökfelt, Zhi-Qing David Xu,
Tiejun Shi, Csaba Adori, Kang Zheng,
Swapnali Barde, and Mingdong Zhang

CONTENTS

1.1 NEUROPEPTIDE RESEARCH—THE BEGINNINGS

Early in the 1970s, an intense research on neuropeptides was initiated, fueled by the following seminal discoveries: (1) most of the hypothalamic-releasing and -inhibiting hormones were shown to be peptides by Roger Guillemin and Andrew Schally's groups; (2) substance P, though discovered in 1931 by Ulf von Euler and John Gaddum, was chemically identified as an 11-amino acid peptide by Susan Leeman and coworkers only in 1971; (3) the first endogenous ligands for the morphine receptors were identified as two pentapeptides, leucine- and methionine-enkephalin

1

by John Hughes, Hans Kosterlitz, and coworkers; (4) the gut peptide cholecystokinin (CCK), discovered by Viktor Mutt and Erik Jorpes, and the vasodilatory peptide vasoactive intestinal polypeptide (VIP), isolated by Sami Said and Viktor Mutt, were found in the brain; and last but not least (5) David de Wied and coworkers' pioneering behavioral work, showing interesting central effects of peripherally administered peptides like vasopressin and oxytocin, opening up the idea that these posterior pituitary hormones can bypass the blood–brain barrier (BBB) and act on receptors in the brain, further corroborated by Abba Kastin and coworkers. Subsequently, the rate of discovery of neuropeptides accelerated, and during the first decade of the third millennium new members of the family have been added. Thus, we now know of several hundred neuropeptides, many of them belonging to chemically, or functionally, distinct families. Clearly, neuropeptides have emerged by far as the largest group of messenger molecules in the nervous system, as recently reviewed by Burbach (2010) (Table 1.1).

1.2   NEUROPEPTIDES—TRANSMITTERS OR TROPHIC FACTORS, OR BOTH?

Neuropeptides were initially considered as transmitter-like substances, in any case in our laboratory, perhaps with a slow onset and certainly with a long duration of action. As discussed later in this chapter, it was at that time unclear on which type of receptor neuropeptides act. However, as time passed by, more and more reports described trophic, regenerative, neuroprotective, and developmental effects. One important peptide in this context was alpha-melanocyte-stimulating hormone (α-MSH) that, in addition to many other effects, influenced nerve regeneration and was neuroprotective (Strand 1999). Other important neuropeptides in this category are VIP (Brenneman 2007; Gozes et al. 1999; Waschek 1995), pituitary adenylyl cyclase-activating peptide (PACAP) (Brenneman 2007; Gozes et al. 1999; Waschek 1995) (see Chapter 11 by Dr. Eiden in this book), and neuropeptide Y (NPY) (Hansel et al. 2001; Hökfelt et al. 2008; Xapelli et al. 2006). In this chapter, as an example for the aforementioned activities, we will describe some of galanin's actions, both as a transmitter and a trophic factor. It should be mentioned that, surprisingly, there are also reports indicating the contrary, for example, neurotoxic actions of somatostatin (Gaumann et al. 1990; Mollenholt et al. 1988) and dynorphin (Gaumann et al. 1990; Skilling et al. 1992; Stewart and Issac 1989) (see Chapter 6 by Dr. Hauser in this book). Taken together, neuropeptides display a wide range of biological effects, making it difficult to arrange them into a single category. Having said this, equally interesting, classic transmitters have neurotrophic effects (Schwartz 1992).

1.3   HOW NEUROPEPTIDES HAVE BEEN DISCOVERED

The methods of discovery have developed over time, as a consequence of the breathtaking generation of new tools during the past 50 years. The aforementioned hypothalamic-releasing and -inhibiting factors as well as most gut hormones were extracted from huge amounts of tissues from appropriate organs/tissues, and the purity was monitored by bioassays. Tatemoto and Mutt (1980) realized

TABLE 1.1 (from Burbach 2010)
Neuropeptide Gene Families: Classical Neuropeptides

1. Opioid Peptide Family

Prepro enkephalin: Leu-enkephalin, met-enkephalin, amidorphin, adrenorphin, peptide B, peptide E, peptide F, BAM22P

Proopiomelanocortin (POMC): α-melanocyte-stimulating hormone (α-MSH), γ-melanocyte-stimulating hormone (γ-MSH), β-melanocyte-stimulating hormone (β-MSH), adrenocorticotropic hormone (ACTH), β-endorphin, α-endorphin, γ-endorphin, β-lipoprotein (β-LPH), γ-lipoprotein (γ-LPH), corticotropin-like intermediate peptide (CLIP)

Prepro dynorphin: Dynorphin A, dynorphin B, α-neo-endorphin, β-neo-endorphin, dynorphin-32, leu-morphin

Prepro nociceptin, prepro-orphanin: Nociceptin (orphanin FQ), neuropeptide 1, neuropeptide 2

2. Vasopressin/Oxytocin Gene Family

Prepro vasopressin-neurophysin II: Vasopressin (VP), neurophysin II (NP II), C-terminal glycopeptide CPP

Prepro oxytocin-neurophysin I: Oxytocin (OT), neurophysin I (NP 1)

3. CCK/Gastrin Gene Family

Prepro gastrin: Gastrin-34, gastrin-17, gastrin-4

Prepro cholecystokinin (CCK): CCK-8, CCK-33, CCK-58

4. Somatostatin Gene Family

Prepro somatostatin (SST): SS-12, SS-14, SS-28, antrin

Prepro cortistatin: Cortistatin-29, cortistatin-17

5. F- and Y-Amide Family

Prepro neuropeptide RF: QRF-amide (neuropeptide RF-amide, gonadotropin inhibitory hormone [GnIH], p518, RF-related peptide-2), RF-related peptide-1, RF-related peptide-3, neuropeptide VF

Prepro neuropeptide FF (NPFF): Neuropeptide FF, neuropeptide AF, neuropeptide SF

Prepro neuropeptide Y (NPY): C-flanking peptide of NPY (CPON)

Prepro pancreatic polypeptide Y (PPY): PPY

Prepro peptide YY (PYY): PYY, PYY-(3-36)

Prepro prolactin-releasing peptide (PrLH): PrRP-31, PrRP-20

6. Calcitonin Gene Family

Prepro calcitonin: Calcitonin, katacalcin

Prepro CGRP-α: Calcitonin gene-related peptide I (α-CGRP)

Prepro CGRP-β: Calcitonin gene-related peptide II (β-CGRP)

Prepro islet amyloid polypeptide (IAPP): Amylin: IAPP (amylin, amyloid polypeptide)

Prepro adrenomedullin: Adrenomedullin, AM, PAMP

Prepro adrenomedullin-2: Adrenomedullin-2, intermedin-long (IMDL), intermedin-short (IMDS)

7. Natriuretic Factor Gene Family

Preproatrial natriuretic factor: Atrial natriuretic factor (natriuretic peptide A, ANF, ANP, natriodilatine, cardiodilatine-related peptide)

Prepro brain natriuretic factor: Brain natriuretic factor (natriuretic peptide B, BNF, BNP)

Prepro natriuretic peptide precursor C: C-type natriuretic peptide (CNP-23), CNP-29, CNP-53

(Continued)

TABLE 1.1 (CONTINUED)
Neuropeptide Gene Families: Classical Neuropeptides

8. Bombesin-Like Peptide Gene Family

Prepro gastrin-releasing peptide 1 (GRP-1): GRP-27, GRP-14, GRP-10 (neuromedin C)
Prepro gastrin-releasing peptide 2 (GRP-2): GRP-27, GRP-14, GRP-10 (neuromedin C)
Prepro gastrin-releasing peptide 3 (GRP-3): GRP-27, GRP-14, GRP-10 (neuromedin C)
Prepro neuromedin B1: Neuromedin B (ranatensin-like peptide, RLP)
Prepro neuromedin B2: Neuromedin B (ranatensin-like peptide, RLP)

9. Endothelin Gene Family

Prepro endothelin 1 (PPET1): Endothelin 1 (ET-1)
Prepro endothelin 2 (PPET2): Endothelin 2 (ET-2)
Prepro endothelin 3 (PPET3): Endothelin 3 (ET-3)

10. Glucagon/Secretin Gene Family

Prepro glucagon: Glicentin, glicentin-related polypeptide (GRPP), oxyntomodulin (OXY) (OXM), glucagon, glucagon-like peptide 1 (GLP-1), glucagon-like peptide 1(7-37) [GLP1 (7-37)], glucagon-like peptide 1(7-36) [GLP-1(7-36)], glucagon-like peptide 2 (GLP-2)
Prepro secretin (SCT): Secretin
Prepro vasoactive intestinal peptide (VIP-1): VIP, PHM-27/PHI-27, PHV-42
Prepro vasoactive intestinal peptide (VIP-2): VIP, PHM-27/PHI-27, PHV-42
Prepro pituitary adenylcyclase-activated peptide: PACAP-38, PACAP-27, prolactin-relating peptide (PRP)-48
Prepro growth hormone-releasing hormone (GHRH): GHRH (somatoliberin, GRF, somatocrinin, somatorelin, sermorelin)
Prepro gastric inhibitory peptide (GIP): GIP (gastric inhibitory peptide, glucose-dependent insulinotropic polypeptide)

11. CRH-Related Gene Family

Prepro corticotropin-releasing hormone (CRH): CRH
Prepro urocortin-I: UNC I
Prepro urocortin-II: UNC II, stresscopin-related peptide
Prepro urocortin-III: UNC III, stresscopin
Prepro urotensin-2, isoform a: Urotensin-2
Prepro urotensin-2, isoform b: Urotensin-2
Prepro urotensin-2B: Urotensin-2-related peptide, urotensin-2B

12. Kinin and Tensin Gene Family

α-Prepro tachykinin A (α-PPTA): Substance P, neurokinin A (NKA, substance K, neuromedin L), neuropeptide K, neuropeptide γ
β-Prepro tachykinin A (β-PPTA): Substance P, neuropeptide K, neurokinin A
γ-Prepro tachykinin A (γ-PPTA): Substance P, neurokinin A, neuropeptide γ
δ-Prepro tachykinin A (δ-PPTA): Substance P, neuropeptide K, neurokinin A
Prepro tachykinin B (PPTB), isoform 1: Neuromedin K, neurokinin B
Prepro tachykinin B (PPTB), isoform 2: Neuromedin K, neurokinin B

13. Neuromedins

Prepro neuromedin S: Neuromedin S (NMS)
Prepro neuromedin U, multiple isoforms Neuromedin U: (NMU)

TABLE 1.1 (CONTINUED)
Neuropeptide Gene Families: Classical Neuropeptides

14. Tensins and Kinins

Kininogen-1 precursor, isoform 1: Bradykinin, kallidin, LMW-K-kinin, HMW-K-kinin

Kininogen-1 precursor, isoform 2: Bradykinin, kallidin, LMW-K-kinin, HMW-K-kinin

Angiotensinogen preprotein: Angiotensin I, angiotensin II, angiotensin (1-7)

Prepro neurotensin: Neurotensin (NT), neuromedin N

15. Motilin Gene Family

Prepro motilin isoform 1: Motilin, motilin-associated peptide

Prepro motilin isoform 2: Motilin, motilin-associated peptide

Prepro ghrelin isoform 1: Ghrelin, obestatin

Prepro ghrelin isoform 2: Ghrelin, obestatin

Prepro ghrelin isoform 3: Obestatin

Prepro ghrelin isoform 4: Obestatin

Prepro ghrelin isoform 5: Obestatin

16. Galanin Gene Family

Prepro galanin: Galanin, galanin message-associated peptide (GMAP)

Galanin-like peptide precursor: Galanin-like peptide (GALP)

17. GnRH Gene Family

Prepro gonadotropin-releasing hormone 1: GnRH (LHRH, gonadoliberin)

Prepro gonadotropin-releasing hormone 2, isoform-a: GnRH2 (LHRH II, gonadoliberin II)

Prepro gonadotropin-releasing hormone 2, isoform-b: GnRH2 (LHRH II, gonadoliberin II)

Prepro gonadotropin-releasing hormone, isoform-c: GnRH2 (LHRH II, gonadoliberin II)

18. Neuropeptide B/W Gene Family

Prepro neuropeptide B, PPL7: Neuropeptide B-23 (peptide L7), neuropeptide B-29

Prepro neuropeptide W, PPL8: Neuropeptide W-23 (peptide L8), neuropeptide W-30

Prepro neuropeptide S: Neuropeptide S

19. Insulin/Relaxins

Prepro relaxin-1: Relaxin-1

Prepro relaxin-2, isoform 1: Relaxin-2

Prepro relaxin-2, isoform 2: Relaxin-2

Prepro relaxin-3: Relaxin-3

20. No-Family Neuropeptides

Prepro thyrotropin-releasing hormone: TRH (thyroliberin)

Prepro parathyroid hormone-like hormone, isoform CRA_a: PTHrP (1-36), PTHrP (38-94), PTHrP (107-139) (osteostatin)

Prepro parathyroid hormone-like hormone, isoform CRA_b

Prepro melanin-concentrating hormone: MCH, neuropeptide Glu-Ile (NEI), neuropeptide Gly-Glu (NGE)

Prepro hypocretin: Hypocretin-1 (orexin A), hypocretin-2 (orexin B)

Prepro cocaine- and amphetamine-regulated transcript (CART): CART (1-39), CART (42-89)

(*Continued*)

TABLE 1.1 (CONTINUED)
Neuropeptide Gene Families: Classical Neuropeptides

Agouti-related protein precursor isoform 1: AGRP

Agouti-related protein precursor isoform 2: AGRP

Prolactin precursor: Prolactin (PRL)

Prepro apelin: Apelin-13, apelin-17, apelin-36 (APJ ligand, AGTRL1 ligand)

Metastasis-suppressor KISS 1: Metastin (kisspeptin-54), (Golgi transport 1 homolog A, golt1a), kisspeptin-14, kisspeptin-13, kisspeptin-10

Diazepam-binding inhibitor isoform 1: Diazepam-binding inhibitory peptide

Diazepam-binding inhibitor isoform 2: Diazepam-binding inhibitory peptide

Diazepam-binding inhibitor isoform 3: Diazepam-binding inhibitory peptide

Prokineticin-I precursor: Prokineticin-1, (PK1) endocrine gland-derived EGF (EGVEGF)

Prokineticin-2 precursor isoform 1: Prokineticin-2 (PK2)

Prokineticin-2 precursor isoform 2: Prokineticin-2 (PK2)

21. Cerebellins

Cerebellin-1 precursor: Cerebellin-1 (Cbln1)

Cerebellin-2 precursor: Cerebellin-2 (Cbln2)

Cerebellin-3 precursor: Cerebellin-3 (Cbln3)

Cerebellin-4 precursor: Cerebellin-4 (Cbln4, cerebellin-like glycoprotein-1)

22. Granins

Chromogranin A precursor: Chromogranin A, β-granin, vasostatin

Chromogranin B precursor: Chromogranin B (secretogranin I), CCB peptide, GAWK peptide

Secretogranin II precursor, chromogranin C precursor: Secretogranin II (chromogranin C), EM66, secretoneurin

Secretogranin III precursor: Secretogranin III

Secretory granule neuroendocrine precursor: Secretory granule neuroendocrine protein-1 (7B2, secretogranin 5)

VGF nerve growth factor inducible protein precursor: VGF (NGF-inducible protein, neurosecretory protein), TLPQ-62, TLPQ-21, AQEE-30, LQEQ-19

23. Neuroexophilins

Neuroexophilin-1 precursor: Neuroexophilin-1

Neuroexophilin-2 precursor: Neuroexophilin-2

Neuroexophilin-3 precursor: Neuroexophilin-3

Neuroexophilin-4 precursor: Neuroexophilin-4

24. Adipose Neuropeptides

Prepro-leptin: Leptin (obesin)

Adiponectin precursor: Adiponectin (Acpr30, adipocyte complement-related protein, adipocyte, C1Q and collagen domain containing)

Visfatin precursor: Visfatin-1 [pre-B cell colony enhancing factor-1 (PBEF1), nicotinamide phosphoribosyltransferase]

Resistin precursor: Resistin (cysteine-rich secreted protein FIZZ3, adipose tissue-specific secretory factor, cysteine-rich secreted protein A12-α-like 2, ADSF, Xcp4)

Resistin- δ2 precursor: Resistin-δ2

Resistin-like molecule α precursor: Resistin-like molecule α (found in inflammatory zone 1; FIZZ1, hypoxia-induced mitogenic factor, Xcp2)

TABLE 1.1 (CONTINUED)
Neuropeptide Gene Families: Classical Neuropeptides

Resistin-like β precursor: Resistin-like molecule β (cysteine-rich secreted protein FIZZ2, colon and small intestine-specific cysteine-rich protein, cysteine-rich secreted protein A12-α-like 1, colon carcinoma-related gene protein, Xcp3)

Resistin-like molecule γ precursor: Resistin-like molecule γ (cysteine-rich secreted protein FIZZ3, Xcp1)

Nucleobindin-2: Nesfatin-1

Beacon precursor: Beacon

25. Insulin Family

Prepro insulin: Insulin

IGF-1 precursor: IGF-1 (somatomedin C)

IGF-2 precursor (multiple precursors): IGF-2 (somatomedin A)

that C-terminal amidation is a hall mark for many neuropeptides. They, through a series of studies, were able to isolate, purely on this chemical principle, that is, without bioassay, several new neuropeptides, including peptide histidine isoleucine (PHI), peptide tyrosine tyrosine (PYY), NPY, pancreastatin, and galanin. In Italy, Erspamer and coworkers isolated a large number of bioactive peptides from frog skin and discovered that many of them had mammalian counterparts with similar structures and pharmacological activities (Ersparmer 1984).

Using the tools of molecular biology, Rosenfeld, Amara, and coworkers in their work on the calcitonin gene discovered another putative peptide in the precursor molecule (Amara et al. 1982; Rosenfeld et al. 1983). They deduced the amino acid sequence, synthesized the peptide, raised antibodies, and were able to demonstrate that this peptide, called calcitonin gene-related peptide (CGRP), was produced in a tissue-specific way, that is, in neurons as compared with the biosynthesis of calcitonin in the thyroid gland. CGRP has subsequently become a highly studied peptide.

Based on receptor deorphanization and reverse pharmacology (Civelli 2005), the neuropeptide nociceptin/orphanin FQ (N/OFQ) was identified as the transmitter for the orphan G-protein-coupled receptor (GPCR) and opioid receptor-like 1 (ORL-1) receptor (Meunier et al. 1995; Reinscheid et al. 1995). This approach has generated further interesting peptides. Among other fairly recently discovered neuropeptides, hypocretin/orexin and its receptors may be mentioned (de Lecea et al. 1998; Sakurai et al. 1998).

1.4 HOW NEUROPEPTIDES ARE MONITORED

Early works on peptides were mainly based on immunological principles. Thus, quantitative biochemical evaluations on peptide distribution and levels in various tissues were obtained using radioimmunoassay techniques originally described by Berson et al. (1956). At the histochemical level, the same antibodies can often be used to demonstrate the cellular and subcellular distribution of peptides using immunohistochemical techniques, originally developed by Coons et al. (1942) and then complemented by more sensitive techniques, for example, tyramide signal amplification (Adams 1992). Subsequently, molecular biological approaches were introduced. Thus, by monitoring mRNA levels for various peptides using Northern blot analysis and solution hybridization, as well as *in situ* hybridization, using both radioactively labeled and nonradioactive probes, levels and distribution/cellular localization of transcripts can be demonstrated. Advanced molecular biological/genetic approaches are now used to further visualize localization of neuropeptides and their receptors (and other molecules). They include the use of various reporter gene strategies, that is, genetically modified mice; for example, replacement of a seven transmembrane receptor by an active receptor-enhanced green fluorescent protein fusion (knock-in mice) (Scherrer et al. 2006) or introduction of marked bacterial artificial chromosomes (BAC mice) (Heintz 2001).

Release of peptides in the peripheral and central nervous system can be monitored using push–pull systems as well as microdialysis developed by Ungerstedt and coworkers (Zetterstrom et al. 1983). Another approach was pioneered by Duggan and his group—the microprobe technique (Duggan 1990). A micropipette coated with antibodies to, for example, substance P is inserted into the dorsal horn of the spinal cord, and the peripheral nerve of the corresponding segment is stimulated. Substance P released from primary afferents in the dorsal horn then binds to the antibody-coated micropipette. Subsequently, the micropipette is pulled out and dipped into radiolabeled substance P, which binds only to those areas of the pipette where no endogenous substance P is occupying the antibody-binding sites. In this way, quantitative information with a high spatial resolution can be obtained.

Another approach, the "FMRFamide tagging" method, based on molecular biological/genetic tools, that can measure peptide secretion on a millisecond timescale, was developed by Whim and Moss (2001). Here, the neuropeptide FMRFamide is added as an electrophysiological "Tag" to the DNA that codes for the peptide of interest, for example NPY. The FMRFamide-tagged neuropeptide prohormone is then expressed together with the FMRFamide receptor, a unique neuropeptide ionotropic receptor, which forms a sodium-permeable channel (Lingueglia et al. 1995). Triggering exocytosis from a single transfected cell results in cosecretion of both FMRFamide and NPY. The released FMRFamide then activates FMRFamide receptors, resulting in an influx of sodium ions that is detected as a rapid inward current, which reflects the cosecretion of NPY—the peptide of interest.

1.5   WHAT MAKES NEUROPEPTIDES UNIQUE AS MESSENGER MOLECULES?

Neuropeptides represent a type of chemical messengers, which, in several aspects, are different from acetylcholine (ACh), catecholamines, serotonin, and amino acids, which are sometimes collectively referred to as classic transmitters. The following is a summary of some of the main characteristics of neuropeptide signaling (Figure 1.1):

1. Neuropeptides are ribosomally synthesized as large precursor molecules (molecular weight ~25,000) in soma/dendrites and stored in large dense core vesicles (LDCVs, diameter ~1.000Å). The bioactive peptide is then

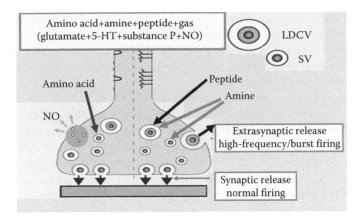

FIGURE 1.1 (See color insert) The one neuron multiple transmitters concept. This nerve ending stores four different types of messenger molecules, partly in different compartments: Peptides in large dense core vesicles (LDCVs, diameter ~1,000Å), amines in LDCVs and synaptic vesicles (SVs, diameter ~500Å), and glutamate in SVs. Nitric oxide (NO) is not stored but generated "upon demand" by NO synthase and diffuses through the membrane. Neuropeptides are released extrasynaptically in response to burst firing/high firing frequency, amino acids mainly into the synapse, and amines presumably in both ways. (From Hökfelt 1991; Lundberg 1996; Merighi 2002.)

excised by convertases. In contrast, classic transmitters are mainly stored in synaptic vesicles (SVs) (diameter 500Å), although amines also are present in LDCVs.

2. Neuropeptides can be released both from nerve endings and soma/dendrites, in fact similar to some classic transmitters. After release, they are mainly degraded by extracellular peptidases.

3. Each released molecule has to be replaced by new synthesis (=increased transcript levels) and is then intraaxonally transported to nerve endings, or to dendrites. Thus, there is no reuptake mechanism, neither at the cell membrane nor at the membrane of the storage vesicles.

4. Neuropeptides are not the main messengers but coexist with other molecules, classic transmitters (e.g., ACh, dopamine, serotonin, γ-aminobutyric acid (GABA), glutamate), nitric oxide, and others.

5. Neuropeptides are, when compared with classic transmitters, active at very low concentrations, in the nanomolar range.

1.6 PLASTICITY OF NEUROPEPTIDE EXPRESSION

It has been demonstrated in several systems that the expression of genes for peptides and peptide receptors is highly dynamic. Thus, there are marked changes under various experimental conditions, for example, in response to stimuli, such as nerve injury, electrical activity, and pharmacological manipulations. One reason for this is that, as mentioned earlier, every neuropeptide molecule released has to be replaced by ribosomal synthesis associated with a rapid increase in transcript levels. However, induction, even *de novo* synthesis of neuropeptides, may occur in response to nerve injury. A highly uneconomic system, one may think, perhaps due to an ancient origin of peptide signaling. Subsequently more efficient working mechanisms, such as synapses, have evolved, allowing rapid transmission in the mammalian nervous system.

Plasticity and expression of neuropeptides occur in many systems, for example, in primary sensory neurons (Hökfelt et al. 1994) and sympathetic neurons (Zigmond and Sun 1997). The former contain, under normal circumstances, high levels of CGRP, substance P, and somatostatin, and low levels of galanin. VIP and NPY can hardly be detected. After axotomy (transection of the sciatic nerve), there is appearance of VIP (mainly in small neurons) and NPY (mainly in large neurons), and upregulation of galanin (in small and large neurons) in the corresponding dorsal root ganglia (L4, L5). In contrast, substance P and CGRP are downregulated. Because the latter two peptides have an excitatory function in the dorsal horn, and NPY and galanin mainly are inhibitory, these regulations are thought to cause attenuation of dorsal horn excitability, perhaps to suppress pain. VIP (Waschek 1995), galanin (Hobson et al. 2008), and NPY (Hansel et al. 2001) can also have trophic effects and promote survival and regeneration. This reaction pattern is in agreement with the general view that the synthetic machinery of a neuron after injury is converted from transmitter synthesis to production of molecules of importance for survival and recovery. Interestingly, the expression of neuropeptide receptors also changes after nerve injury.

1.7   NEUROPEPTIDE RECEPTORS

For quite some time, the question of peptide receptors was open, in spite of extensive evidence for the presence of neuropeptide-binding sites based on experiments, both with isolated membranes and ligand-binding autoradiography (Kuhar et al. 1991). Could it be that the peptides act on specific sites on major transmitter receptor proteins, such as the benzodiazepines/diazepam-binding inhibitory peptide on the GABA receptor? This idea gained some credibility, when it was realized that neuropeptides are costored and coreleased with classic transmitters, such as monoamines and GABA (see aforementioned). Nakanishi and coworkers gave a clear answer to that question by cloning the first neuropeptide receptor, a substance K receptor (Masu et al. 1987). This receptor turned out to belong to the 7-transmembrane (7-TM) GPCR family. Subsequent work showed that virtually all other, several hundreds or so, neuropeptide receptors also are members of this family. However, there is at least one exception: the peptide Phe-Met-Arg-Phe-NH$_2$ (FMRFamide), which induces a fast excitatory depolarizing response via direct activation of an amiloride-sensitive sodium channel (Green et al. 1994; Lingueglia et al. 1995). Needless to say, the multiplicity of receptors offers unique and important openings for the development of specific agonists and antagonists, which are potential candidates for treatment of various diseases.

The cloning of the peptide receptors allowed mapping at the message level with *in situ* hybridization and, after production of antisera, defining the exact localization and trafficking of the receptor protein, using immunohistochemistry.

An early, impressive example was the dramatic internalization of the substance P (NK1) receptor in second-order spinal dorsal horn neurons after peripheral nerve stimulation (Mantyh et al. 1995).

1.8 NEUROPEPTIDE DRUGS AND PHARMACOLOGY

Our knowledge of the functional roles of classic transmitters has to a large extent been based on the large extent on the availability of a huge number of drugs influencing transmitter mechanisms. It has taken a long time to generate this type of tools in the peptide field, and still there is a long way to go. Peptide drugs can be classified into at least three categories, antagonists, agonists, and peptidase inhibitors, the latter preventing peptide breakdown and thus strengthening peptidergic transmission. Peptide conversion, an interesting mechanism, is a potential pathway for modulating G-protein signaling (Nyberg and Hallberg 2007). Also, early on, antisense methodology and, more recently, methods based on microRNA, siRNA, and viral approaches have been successfully used.

Following modified peptide antagonists (often D-amino acid substitutions), nonpeptide substances were developed for most neuropeptides, mostly within the framework of the pharmaceutical industry. By penetrating the BBB, they raise hope of treating various brain disorders. Perhaps, the most spectacular progress apparently was the report that an NK1 (substance P) receptor antagonist has clinical efficacy in major depression, virtually without side effects (Kramer et al. 1998). This could not be confirmed in the phase 3 trial, but a recent study gives new hope for NK1 antagonists in the treatment of depression (Ratti et al. 2011). Nevertheless, there are now a number of peptide antagonists close to the clinic, for example, an orexin antagonist in phase 3 trials as a novel type of sleeping pill (Brisbare-Roch et al. 2007) and CGRP antagonists as a new-generation antimigraine medicine (Diener et al. 2011).

1.9 WHY ARE NEUROPEPTIDE RECEPTORS ATTRACTIVE TARGETS FOR DRUG DEVELOPMENT?

Just the sheer number of neuropeptides and the correspondingly large number of 7-TM GPCRs make them interesting, also because more than half of all drugs prescribed today are acting via this type of receptor. Moreover, their wide distribution in discrete systems, many directly associated with diseases, such as neuropathic pain and depression, provides a basis for drug development. An important aspect is that neuropeptide systems are prone to species variations (Bowers 1994). Thus, targets based on animal experiments may not be valid for design of drugs for treatment of human diseases. And, the concept of cotransmission presents some advantages, as well as problems, when considering novel pharmacological treatment strategies for disease-involving peptides. Here are some aspects:

1. Neuropeptides are comparatively large molecules and have difficulties in passing through the BBB, reaching receptors in brain and spinal cord only to a limited extent. Thus, small, BBB-penetrating, nonpeptide molecules

have been developed. Even if they are efficacious in animal experiments, it has been difficult to perform clinical tests in several cases, as such compounds are associated with serious side effects in humans, not rarely liver toxicity.

2. Using antagonists, it may not be sufficient to block one receptor to obtain the expected effect, if several transmitters are released from the same nerve ending. For example, early animal work on substance P antagonists on pain pathways clearly indicated that an antagonist would be analgesic. However, when tested in clinical trials, no such effect was observed (Hill 2000). Perhaps release of costored, excitatory transmitters, for example, glutamate, CGRP, and others, from central sensory nerve endings conveyed the pain message even if NK1 receptors were blocked.

3. Intervention with antagonists may be preferred, because

   a. Antagonists will only affect deranged (upregulated) systems; peptide transmission is mostly silent under normal conditions, which should result in less side effects.

   b. Agonists will act on receptors in the entire body, resulting in more adverse reactions. A good example is morphine, which, in addition to its well-established and unsurpassed antinociception, can produce serious side effects.

1.10 NEUROPEPTIDE SYSTEMS AS TARGETS FOR TREATMENT OF NEUROPATHIC PAINS

Targeting neuropeptide receptors for disease treatment represents a promising venue, but can be full of pitfalls. As an example, here we focus on neuropathic pain, which still is of major concern, as no efficacious and safe pharmacological treatment is available (Costigan et al. 2009) (see Chapter 8 by Dr. Ueda in this book) (Figure 1.2). A research focus in this field has been on the manifold and often dramatic phenotypic neurochemical changes occurring in dorsal root gangila (DRGs) and spinal cord in various types of rodent neuropathic pain models (Costigan et al. 2002; Hökfelt et al. 1994; Xiao et al. 2002). One such molecule is the neuropeptide galanin (Tatemoto et al. 1983), which is strongly upregulated in lumbar L4 and L5 DRGs after transection of the sciatic nerve (Hökfelt et al. 1987; Villar et al. 1989). In a long series of studies, it has become clear that this upregulation serves at least two purposes, to modulate pain signaling and to enhance regeneration.

1.11 GALANIN AS A PAIN TRANSMITTER

It has been proposed that galanin after nerve injury, via the GalR1 receptor, serves as an endogenous analgesic, attenuating the excitatory tone at the spinal level (Xu et al. 2008). Galanin also has a pronociceptive effect, exerted via the GalR2 receptor, stimulation of phospholipase C (PLC) β3, increase of intracellular Ca^{2+}, and release of excitatory transmitters, for example, glutamate and CCK (Liu et al. 2001) (Figure 1.3). However, a GalR2 antagonist does not distinctly attenuate pain (Shi et al. 2012, unpublished data). One reason may be that in neuropathic pain,

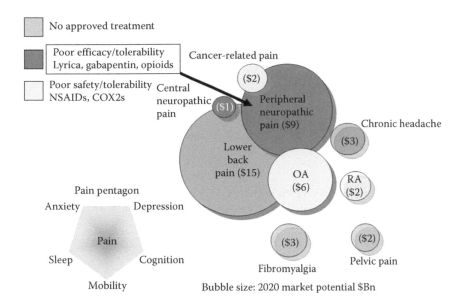

FIGURE 1.2 (See color insert) Schematic presentation of various pain states and their estimated values in the market. Dark grey indicates that available drugs have poor efficacy, as is the case for those presently used to treat peripheral neuropathic pain. The market for neuropathic pain drugs is large, here estimated to be 9 billion USD in 2020. NSAID, nonsteroidal anti-inflammatory drugs; COX, cyclooxygenase; OA, osteoarthritis; RA, rheumatoid arthritis. (Courtesy of Dr Andy, Dray, AstraZeneca, Montreal.)

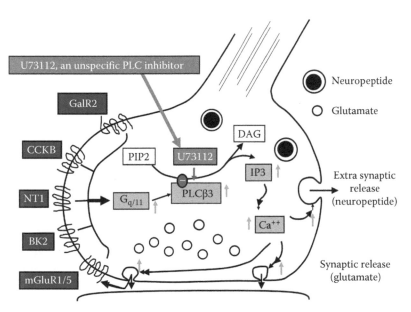

FIGURE 1.3 Schematic drawing proposing that one and the same nerve ending (or neuron) has five different receptors, all acting via $G_{q/11}$ G-protein, PLCβ3, IP3, and increased intracellular Ca^{2+}, resulting in release of glutamate and neuropeptides and eventually pain. DAG, diacylglycerol.

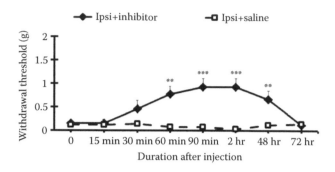

FIGURE 1.4 The unspecific PLCβ inhibitor U73112, administered as one dose (30 mg/kg), 14 days after performing unilateral SNI, causes a long-lasting, ~48 hours, increase in pain threshold. (From Shi, T.J., et al., *Proc. Natl. Acad. Sci. USA,* 105, 20004–20008, 2008. With permission.)

many other receptors, which converge on PLCβ3, are activated (Figure 1.3). Thus, blocking one of them may not be sufficient to achieve pain relief. When pain is induced after various types of peripheral nerve injuries, several dramatic changes occur in rat DRGs, including (i) increased number of CCK mRNA-positive DRG neuron profiles (up from a few to 30%) (Verge et al. 1993); (ii) increase in CCK subtype B (2) receptor mRNA-positive DRG neuron profiles (up by 20-fold) (Zhang et al. 1993); and (iii) increased bradykinin subtype 1 receptor (BK1) (up by 8-fold) and BK 2 (up by 1.5-fold) binding (Petcu et al. 2008). All these changes would appear to be pronociceptive. One approach is therefore to test more than one antagonist, for example, a GalR2 plus a BK1 antagonist. Such experiments are ongoing in our laboratory.

PLCβ, in particular the PLCβ3, has been shown to be expressed in rodent DRG neurons and associated with pain and itch (Han et al. 2006; Joseph et al. 2007; Shi et al. 2008). We have administered an unspecific PLCβ3 inhibitor, U73112, to mice exposed to spared nerve injury, a neuropathic pain model (Decosterd and Woolf 2000), and observed a dramatic antinociceptive effect, the pain threshold being elevated for ~48 hours (Shi et al. 2008) (Figure 1.4). To our knowledge, this is a uniquely long-lasting effect on neuropathic pain. As PLCβ3 is expressed in human dorsal root ganglia, both at the protein (Shi et al. 2008) and transcript (Barde and Hökfelt 2012, unpublished data) levels, inhibition of PLC may represent a novel and powerful treatment strategy for neuropathic pain.

1.12   GALANIN AS A TROPHIC MOLECULE

The first evidence that galanin also could have trophic functions was presented by Wynick and coworkers. Thus, it was noted that in galanin knockout mice, there was a clear reduction in the number of DRG neurons (down by 13%) (Holmes et al. 2000). Here, GalR2 seems to play a critical role, as genetic deletion of this receptor results in a similar reduction in the percentage of DRG neurons (Hobson et al. 2006; Shi et al. 2006). Wynick and coworkers also showed that galanin is involved in regeneration. Thus, following a crush injury, sciatic nerve regeneration is reduced by 35% in galanin (Holmes et al. 2000) and GalR2 (Hobson et al. 2006) knockout mice, deficits that can be rescued by administration of galanin and the GalR2 agonist galanin(2-11) (Mahoney et al. 2003). Galanin has survival effects also in the brain, for example, in galanin knockout mice a third of the cholinergic forebrain neurons are lost (O'Meara et al. 2000).

Several studies have demonstrated a neuroprotective role of galanin. For example, galanin knockout mice show a higher loss of pyramidal neurons in the hippocampus than wild-type mice after peripheral injection of kainic acid, and galanin or galanin(2-11) counteracts cell death induced by glutamate in hippocampal cultures (Elliott-Hunt et al. 2004; Pirondi et al. 2005).

1.13 CONCLUDING REMARKS

Serious research on neuropeptides has been in progress for some four decades and remarkable strides have been made. Today, we encounter a wealth of neuropeptides and corresponding receptors in animal experiments associated with both normal brain function and many pathologies. They include the topic of this book, neuroprotection and neuroregeneration. Clinical trials targeting neuropeptide mechanisms have been initiated and, even if not reaching prescription drugs, there are many promising candidates that could take the decisive step into the clinic in the near future.

ACKNOWLEDGMENTS

This study was supported by the Swedish Research Council, The Marianne and Marcus, and the Knut and Alice Wallenberg Foundations.

REFERENCES

Adams, J. C. 1992. Biotin amplification of biotin and horseradish peroxidase signals in histochemical stains. *J Histochem Cytochem* 40: 1457–1463.

Amara, S. G., Jonas, V., Rosenfeld, M. G., Ong, E. S., and Evans, R. M. 1982. Alternative RNA processing in calcitonin gene expression generates mRNAs encoding different polypeptide products. *Nature* 298: 240–244.

Berson, S. A., Yalow, R. S., Bauman, A., Rothschild, M. A., and Newerly, K. 1956. Insulin-I[131] metabolism in human subjects. Demonstration of insulin binding globulin in circulation of insulin-treated subjects. *J Clin Invest* 35: 170–190.

Bowers, C. W. 1994. Superfluous neurotransmitters? *Trends Neurosci* 17: 315–320.

Brenneman, D. E. 2007. Neuroprotection: A comparative view of vasoactive intestinal peptide and pituitary adenylate cyclase-activating polypeptide. *Peptides* 28: 1720–1726.

Brisbare-Roch, C., Dingemanse, J., Koberstein, R., et al. 2007. Promotion of sleep by targeting the orexin system in rats, dogs and humans. *Nat Med* 13: 150–155.

Burbach, J. P. 2010. Neuropeptides from concept to online database www.neuropeptides.nl. *Eur J Pharmacol* 626: 27–48.

Civelli, O. 2005. GPCR deorphanizations: The novel, the known and the unexpected transmitters. *Trends Pharmacol Sci* 26: 15–19.

Coons, A. H., Creech, H. J., Jones, R. N., and Berliner, E. 1942. The demonstration of pneumoccoccal antigen in tissues by the use of fluorescent antibody. *J Immunol* 45: 159–170.

Costigan, M., Befort, K., Karchewski, L., et al. 2002. Replicate high-density rat genome oligonucleotide microarrays reveal hundreds of regulated genes in the dorsal root ganglion after peripheral nerve injury. *BMC Neurosci* 3: 16.

Costigan, M., Scholz, J., and Woolf, C. J. 2009. Neuropathic pain: A maladaptive response of the nervous system to damage. *Annu Rev Neurosci* 32: 1–32.

de Lecea, L., Kilduff, T. S., Peyron, C., et al. 1998. The hypocretins: Hypothalamus-specific peptides with neuroexcitatory activity. *Proc Natl Acad Sci U S A* 95: 322–327.

Decosterd, I. and Woolf, C. J. 2000. Spared nerve injury: An animal model of persistent peripheral neuropathic pain. *Pain* 87: 149–158.

Diener, H. C., Barbanti, P., Dahlof, C., et al. 2011. BI 44370 TA, an oral CGRP antagonist for the treatment of acute migraine attacks: Results from a phase II study. *Cephalalgia* 31(3): 573–584.

Duggan, A. W. 1990. Detection of neuropeptide release in the central nervous system with antibody microprobes. *J Neurosci Methods* 34: 47–52.

Elliott-Hunt, C. R., Marsh, B., Bacon, A., et al. 2004. Galanin acts as a neuroprotective factor to the hippocampus. *Proc Natl Acad Sci U S A* 101: 5105–5110.

Ersparmer, V. 1984. Half a century of comparative research on biogenic amines and active peptides in amphibian skin and molluscan tissues. *Comp Biochem Physiol* 79: 1–7.

Gaumann, D. M., Grabow, T. S., Yaksh, T. L., Casey, S. J., and Rodriguez, M. 1990. Intrathecal somatostatin, somatostatin analogs, substance P analog and dynorphin A cause comparable neurotoxicity in rats. *Neuroscience* 39: 761–774.

Gozes, I., Fridkinb, M., Hill, J. M., and Brenneman, D. E. 1999. Pharmaceutical VIP: Prospects and problems. *Curr Med Chem* 6: 1019–1034.

Green, K. A., Falconer, S. W. P., and Cottreil, G. A. 1994. The neuropeptide Phe-Met-Arg-Phe-NH$_2$ (FMRF amide) directly gates two ion channels in an identified helix neurone. *Pflügers Arch* 428: 232–240.

Han, S. K., Mancino, V., and Simon, M. I. 2006. Phospholipase Cbeta 3 mediates the scratching response activated by the histamine H1 receptor on C-fiber nociceptive neurons. *Neuron* 52: 691–703.

Hansel, D. E., Eipper, B. A., and Ronnett, G. V. 2001. Regulation of olfactory neurogenesis by amidated neuropeptides. *J Neurosci Res* 66: 1–7.

Heintz, N. 2001. BAC to the future: The use of bac transgenic mice for neuroscience research. *Nat Rev Neurosci* 2: 861–870.

Hill, R. 2000. NK1 (substance P) receptor antagonists—Why are they not analgesic in humans? *Trends Pharmacol Sci* 21: 244–246.

Hobson, S. A., Bacon, A., Elliot-Hunt, C. R., et al. 2008. Galanin acts as a trophic factor to the central and peripheral nervous systems. *Cell Mol Life Sci* 65: 1806–1812.

Hobson, S. A., Holmes, F. E., Kerr, N. C., Pope, R. J., and Wynick, D. 2006. Mice deficient for galanin receptor 2 have decreased neurite outgrowth from adult sensory neurons and impaired pain-like behaviour. *J Neurochem* 99: 1000–1010.

Holmes, F. E., Mahoney, S., King, V. R., et al. 2000. Targeted disruption of the galanin gene reduces the number of sensory neurons and their regenerative capacity. *Proc Natl Acad Sci U S A* 97: 11563–11568.

Hökfelt, T. 1991. Neuropeptides in perspective: the last ten years. *Neuron* 7: 867–79.

Hökfelt, T., Stanic, D., Sanford, S. D., et al. 2008. NPY and its involvement in axon guidance, neurogenesis, and feeding. *Nutrition* 24: 860–868.

Hökfelt, T., Wiesenfeld-Hallin, Z., Villar, M., and Melander, T. 1987. Increase of galanin-like immunoreactivity in rat dorsal root ganglion cells after peripheral axotomy. *Neurosci Lett* 83: 217–220.

Hökfelt, T., Zhang, X., and Wiesenfeld-Hallin, Z. 1994. Messenger plasticity in primary sensory neurons following axotomy and its functional implications. *Trends Neurosci* 17: 22–30.

Joseph, E. K., Bogen, O., Alessandri-Haber, N., and Levine, J. D. 2007. PLC-beta 3 signals upstream of PKC epsilon in acute and chronic inflammatory hyperalgesia. *Pain* 132: 67–73.

Kramer, M. S., Cutler, N., Feighner, J., et al. 1998. Distinct mechanism for antidepressant activity by blockade of central substance P receptors. *Science* 281: 1640–1645.

Kuhar, M. J., Lloyd, D. G., Appel, N., and Loats, H. L. 1991. Imaging receptors by autoradiography: Computer-assisted approaches. *J Chem Neuroanat* 4: 319–327.

Lingueglia, E., Champigny, G., Lazdunski, M., and Barbry, P. 1995. Cloning of the amiloride-sensitive FMRFamide peptide-gated sodium channel. *Nature* 378: 730–733.

Liu, H. X., Brumovsky, P., Schmidt, R., et al. 2001. Receptor subtype-specific pronociceptive and analgesic actions of galanin in the spinal cord: Selective actions via GalR1 and GalR2 receptors. *Proc Natl Acad Sci U S A* 98: 9960–9964.

Lundberg, J. M. 1996. Pharmacology of cotransmission in the autonomic nervous system: integrative aspects on amines, neuropeptides, adenosine triphosphate, amino acids and nitric oxide. *Pharmacol Rev* 48: 113–178.

Mahoney, S. A., Hosking, R., Farrant, S., et al. 2003. The second galanin receptor GalR2 plays a key role in neurite outgrowth from adult sensory neurons. *J Neurosci* 23: 416–421.

Mantyh, P. W., DeMater, E., Malhotra, A., et al. 1995. Receptor endocytosis and dendrite reshaping in spinal neurons after somatosensory stimulation. *Science* 268: 1629–1632.

Masu, Y., Nakayama, K., Tamaki, H., et al. 1987. cDNA cloning of bovine substance-K receptor through oocyte expression system. *Nature* 329: 836–838.

Merighi, A. 2002. Costorage and coexistence of neuropeptides in the mammalian CNS. *Prog Neurobiol* 66: 161–190.

Meunier, J. C., Mollereau, C., Toll, L., et al. 1995. Isolation and structure of the endogenous agonist of opioid receptor-like ORL1 receptor. *Nature* 377: 532–535.

Mollenholt, P., Post, C., Rawal, N., et al. 1988. Antinociceptive and 'neurotoxic' actions of somatostatin in rat spinal cord after intrathecal administration. *Pain* 32: 95–105.

Nyberg, F. and Hallberg, M. 2007. Peptide conversion—A potential pathway modulating G-protein signaling. *Curr Drug Targets* 8: 147–154.

O'Meara, G., Coumis, U., Ma, S. Y., et al. 2000. Galanin regulates the postnatal survival of a subset of basal forebrain cholinergic neurons. *Proc Natl Acad Sci U S A* 97: 11569–11574.

Petcu, M., Dias, J. P., Ongali, B., et al. 2008. Role of kinin B1 and B2 receptors in a rat model of neuropathic pain. *Int Immunopharmacol* 8: 188–196.

Pirondi, S., Fernandez, M., Schmidt, R., et al. 2005. The galanin-R2 agonist AR-M1896 reduces glutamate toxicity in primary neural hippocampal cells. *J Neurochem* 95: 821–833.

Ratti, E., Bellew, K., Bettica, P., et al. 2011. Results from randomized, double-blind, placebo-controlled studies of the novel NK1 receptor antagonist casopitant in patients with major depressive disorder. *J Clin Psychopharmacol* 31: 727–733.

Reinscheid, R. K., Nothacker, H. P., Bourson, A., et al. 1995. Orphanin FQ: A neuropeptide that activates an opioidlike G protein-coupled receptor. *Science* 270: 792–794.

Rosenfeld, M. G., Mermod, J. J., Amara, S. G., et al. 1983. Production of a novel neuropeptide encoded by the calcitonin gene via tissue-specific RNA processing. *Nature* 304: 129–135.

Sakurai, T., Amemiya, A., Ishii, M., et al. 1998. Orexins and orexin receptors: A family of hypothalamic neuropeptides and G protein-coupled receptors that regulate feeding behavior. *Cell* 92: 573–585.

Scherrer, G., Tryoen-Toth, P., Filliol, D., et al. 2006. Knockin mice expressing fluorescent delta-opioid receptors uncover G protein-coupled receptor dynamics in vivo. *Proc Natl Acad Sci U S A* 103: 9691–9696.

Schwartz, J. P. 1992. Neurotransmitters as neurotrophic factors: A new set of functions. *Int Rev Neurobiol* 34: 1–23.

Shi, T. J., Hua, X. Y., Lu, X., et al. 2006. Sensory neuronal phenotype in galanin receptor 2 knockout mice: Focus on dorsal root ganglion neurone development and pain behaviour. *Eur J Neurosci* 23: 627–636.

Shi, T. J., Liu, S. X., Hammarberg, H., et al. 2008. Phospholipase C{beta}3 in mouse and human dorsal root ganglia and spinal cord is a possible target for treatment of neuropathic pain. *Proc Natl Acad Sci U S A* 105: 20004–20008.

Skilling, S. R., Sun, X., Kurtz, H. J., and Larson, A. A. 1992. Selective potentiation of NMDA-induced activity and release of excitatory amino acids by dynorphin: Possible roles in paralysis and neurotoxicity. *Brain Res* 575: 272–278.

Stewart, P. and Isaac, L. 1989. Localization of dynorphin-induced neurotoxicity in rat spinal cord. *Life Sci* 44: 1505–1514.

Strand, F. L. 1999. New vistas for melanocortins. Finally, an explanation for their pleiotropic functions. *Ann N Y Acad Sci* 897: 1–16.

Tatemoto, K. and Mutt, V. 1980. Isolation of two novel candidate hormones using a chemical method for finding naturally occurring polypeptides. *Nature* 285: 417–418.

Tatemoto, K., Rökaeus, A., Jörnvall, H., McDonald, T. J., and Mutt, V. 1983. Galanin—A novel biologically active peptide from porcine intestine. *FEBS Lett* 164: 124–128.

Verge, V. M., Wiesenfeld-Hallin, Z., and Hökfelt, T. 1993. Cholecystokinin in mammalian primary sensory neurons and spinal cord: In situ hybridization studies in rat and monkey. *Eur J Neurosci* 5: 240–250.

Villar, M. J., Cortes, R., Theodorsson, E., et al. 1989. Neuropeptide expression in rat dorsal root ganglion cells and spinal cord after peripheral nerve injury with special reference to galanin. *Neuroscience* 33: 587–604.

Waschek, J. A. 1995. Vasoactive intestinal peptide: An important trophic factor and developmental regulator? *Dev Neurosci* 17: 1–7.

Whim, M. D. and Moss, G. W. 2001. A novel technique that measures peptide secretion on a millisecond timescale reveals rapid changes in release. *Neuron* 30: 37–50.

Xapelli, S., Agasse, F., Ferreira, R., Silva, A. P., and Malva, J. O. 2006. Neuropeptide Y as an endogenous antiepileptic, neuroprotective and pro-neurogenic peptide. *Recent Pat CNS Drug Discov* 1: 315–324.

Xiao, H. S., Huang, Q. H., Zhang, F. X., et al. 2002. Identification of gene expression profile of dorsal root ganglion in the rat peripheral axotomy model of neuropathic pain. *Proc Natl Acad Sci U S A* 99: 8360–8365.

Xu, X. J., Hökfelt, T., and Wiesenfeld-Hallin, Z. 2008. Galanin and spinal pain mechanisms: Where do we stand in 2008? *Cell Mol Life Sci* 65: 1813–1819.

Zetterström, T., Sharp, T., Marsden, C. A., and Ungerstedt, U. 1983. In vivo measurement of dopamine and its metabolites by intracerebral dialysis: changes after d-amphetamine. *J Neurochem* 41: 1769–1773.

Zhang, X., Dagerlind, A., Elde, R. P., et al. 1993. Marked increase in cholecystokinin B receptor messenger RNA levels in rat dorsal root ganglia after peripheral axotomy. *Neuroscience* 57: 227–233.

Zigmond, R. E. and Sun, Y. 1997. Regulation of neuropeptide expression in sympathetic neurons. Paracrine and retrograde influences. *Ann N Y Acad Sci* 814: 181–197.

2 Neuropeptide Biosynthesis

Margery C. Beinfeld

CONTENTS

2.1 INTRODUCTION

Neuropeptide biosynthesis takes place in the regulated secretory pathway of neurons. Prepro-neuropeptides are cotranslationally inserted into the endoplasmic reticulum (ER), where the signal sequence that targeted them to the ER is removed by the signal peptidase, producing a proneuropeptide. Proneuropeptides are further modified in the rough endoplasmic reticulum by protein folding, formation of disulfide bonds, and glycosylation of asparagines. The proneuropeptides move to the Golgi apparatus and are posttranslationally modified by addition of carbohydrate groups to serine and threonine residues and sulfates to tyrosine residues. In the trans-Golgi network (TGN), they are sorted into large dense core synaptic vesicles. In the Golgi and in these vesicles, they are cleaved in a specific order by endoproteases, exoproteases, and, finally, in some cases, further modified by carboxyl-terminal amidation, *N*-octylation, amino-terminal acetylation, and conversion of amino-terminal glutamines to pyroglutamate. The "finished" products are stored in synaptic vesicles and transported to the synaptic cleft until they are released by an appropriate stimulus. Once released, they are degraded by extracellular proteases.

2.2 STRUCTURE OF PRONEUROPEPTIDES

Proneuropeptides vary in size; most are small proteins. In most cases, there is a single gene, which produces a single mRNA and a single prepro-neuropeptide. In a few cases, it was evident that gene duplication occurs during evolution and multiple genes exist, encoding related peptides like gastrin and cholecystokinin (CCK) and the melanin-concentrating hormone peptides.

There are multiple examples (tachykinins, cocaine- and amphetamine-regulated transcript, adrenomedullin, and calcitonin/calcitonin gene-related peptide; CGRP) of a single gene that produces multiply spliced mRNA, resulting in the production of a different proneuropeptide and different corresponding active products. Some proneuropeptides, proopiomelanocortin (POMC), CCK, and neurotensin/neuromedin are processed differently in different tissues, presumably by a different set of processing enzymes and perhaps to serve different physiological functions.

In the majority of cases, the biologically active neuropeptides are flanked on their amino- and carboxyl-termini by single or double basic residues (lysine and arginine). Rarely, triple or quadruple basic residues are found. These basic residues are recognized by the endoproteolytic enzymes, which are responsible for their cleavage during processing. Not all of the basic residues in proneuropeptides are cleaved. Presumably, it is the three-dimensional structure of the proneuropeptide that informs the enzyme where to cleave. When the proneuropeptide is cleaved, it is cleaved in a strict temporal order. This also implies that the structure determines where the cleavages occur; subsequent cleavages are favored by a change in structure after the first site is cleaved. Specific sequences also act as recognition sequences for subsequent glycosylation, octylation, and tyrosine sulfation. The major steps in neuropeptide biosynthesis are shown in Figure 2.1.

The biologically active product(s) can occur at the extreme carboxyl-neuropeptide Y (NPY) or amino-terminals, vasopressin, of the proneuropeptide or in the middle. There are some proneuropeptides, which have multiple identical or related biological products (proenkephalin- and prothyrotrophin-releasing hormone), and some that

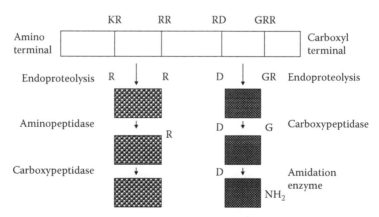

FIGURE 2.1 Schematic representation of the cleavage of a proneuropeptide to produce two products—a nonamidated peptide (left) and an amidated peptide (right). The enzymes involved in each process are indicated.

have a single product (NPY). It is conceivable that other portions of the proneuro-peptides that are not associated with a biological activity have as yet undiscovered biological activity.

It has been hypothesized that the structure of the proneuropeptide is essential for its recognition as secretory material destined for regulated dense core vesicles.

2.3 REGULATED SECRETORY PATHWAY AND PRONEUROPEPTIDE SORTING MECHANISMS

Neuropeptide biosynthesis takes place in the regulated secretory pathway. Neuropeptide mRNAs contact ribosomes adjacent to the ER and protein transla-tion begins from the amino-terminal. All proteins that pass through this pathway have a signal sequence that is recognized by the proteins of the signal recognition machinery and insert the growing peptide chain into the lumen of the ER. The sig-nal sequence is removed by the signalase enzyme, which is located in the lumen of the ER. The proneuropeptide is then folded and any disulfide bonds that need to be formed are made. This folding is facilitated by chaperone proteins and the reducing environment of the ER.

It is now known that any secretory proteins that are misfolded are removed from the ER by the same signal recognition machinery operating in reverse. They are ubiquitinated in the cytosol and degraded by proteosomes. Glycosylation of amino groups of asparagines also occurs in the ER.

The proneuropeptides move from the ER to the various Golgi compartments in transport vesicles that involve the action of many proteins. A number of posttrans-lational modifications of proneuropeptides are made in the Golgi, many of which occur in only a few neuropeptides. These modifications include oligosaccharide trimming and maturation, and O-linked glycosylation. Endoproteolysis may begin in the Golgi.

In the TGN, proteins that will be secreted in the regulated pathway, including pro-neuropeptides, are sorted into large dense core vesicles. The process by which this occurs is not completely clear, although some recent experiments do shed light on possible mechanisms (Zhang 2010). Based on studies on PC12 cells and other mod-els, it appears likely that in neuronal cells, immature secretory vesicles are formed initially by budding off in the TGN. These vesicles fuse and constitutive vesicles bud off, removing proteins and receptors targeted to the cell surface. The regulated secretory material, including proneuropeptides, is retained; this is called *sorting by retention*. As the secretory material mature, the contents are concentrated, and inter-nal pH falls and calcium levels increase.

One model involves recognition of a sorting motif in a prohormone by a membrane receptor that could direct the prohormone to the vesicle. In the sorting of POMC and proenkephalin, amino-terminal disulfide bonds in the proregion may bind to carboxy-peptidase E (CPE), which assists in delivering them to vesicles. The α-helical region in the amino-terminal region of prosomatostatin may act as a sorting signal. This may be a common feature of sorting signals. The dibasic cleavage site of pro-NPY may serve as a sorting signal. The neurophysin portion of provasopressin and prooxytocin are involved in sorting but mutations in these regions block sorting.

Electrical polarity of the proneuropeptide may be important for proneuropeptide sorting. Both pro-CGRP and protachykinin have positively charged proregions and negatively charged middle and carboxyl-terminal regions. The electrostatic force between these regions may facilitate aggregation to form complexes that segregate from the constitutive material. The increasing acidity and calcium concentration of the TGN may further promote aggregation. This pattern of electrical charge is found on other proneuropeptides as well.

It has been suggested that neuropeptide receptors may be sorted to neuropeptide vesicles by their interaction with these aggregated proneuropeptides such as tachykinin.

Once these vesicles are formed, the internal milieu is rendered more acidic by the sodium–potassium ATPases on the membrane surface. The calcium concentration also increases. At the same time, the vesicle contents are greatly concentrated to the point at which they may be said to be almost crystalline. It is in these large dense core vesicles that most of the posttranslational processing occurs.

2.4 POSTTRANSLATIONAL MODIFICATIONS: ENDOPROTEOLYSIS

When the amino acid sequence of the proneuropeptides was determined by cDNA cloning, starting in the 1980s, it became apparent that most of the known biologically active peptides were flanked by single or double basic amino acids that would need to be cleaved to yield the active peptides. Rarely, triple or quadruple basic amino acid residues were found.

This, of course, suggested that the enzyme(s) responsible were trypsin-like proteases. Knowledge of the likely milieu where these cleavages take place suggested that the enzymes would have an acidic pH optimum and might require calcium. A number of candidate enzymes were identified, but the big breakthrough came with the isolation and cloning of the yeast enzyme responsible for the processing of α-mating factor, Kex-2. α-Mating factor precursor contains multiple copies of the mating factor, flanked by dibasic cleavage sites.

The availability of the cDNA sequence of Kex-2 made it possible to clone furin, the first of a family of mammalian enzymes called the *prohormone convertases* (PCs). This family now has nine members, of which four (PC1, PC2, PC5, and PC7) are most likely to be involved in proneuropeptide processing as they are expressed in the brain and colocalized with specific neuropeptides. It is possible that furin is also involved in proneuropeptide processing. It is known to play a major role in processing growth factors and bacterial and viral virulence factors.

All these PCs have structural homology, in particular in their catalytic domain. The prodomain ensures that they are routed into the ER, and this domain actually remains with the enzyme (and inhibits its activity) until it reaches the TGN or the secretory vesicles where the lowered pH and increased calcium concentration promote an autocatalytic cleavage of the inhibitory prodomain, yielding an active enzyme. PC2 is not autocatalytically activated but requires other PCs to cleave its prodomain. It is the only member of the family that requires the accessory protein 7B2 for its activation.

Numerous studies using tumor cells in culture, recombinant enzymes and precursors, and knockout mice have documented the role of these enzymes in proneuropeptide processing. These studies have been complicated by the fact that furin and PC5 knockout mice are embryonic lethals, whereas most PACE4 knockouts are viable but have multiple developmental defects (Creemers and Khatib 2008).

Recently, evidence has emerged that an additional enzyme, which is not in the PC family, is involved in proneuropeptide processing. This enzyme, cathepsin L, is a lysosomal cysteine protease, which is present in secretory vesicles in adrenal chromaffin cells and in neuropeptide containing vesicles in the brain. It was identified as the enzyme responsible for proenkephalin processing in chromaffin cells. Recent studies with cathepsin L knockout mice have indicated that it is involved in the processing of enkephalin, dynorphin, β-endorphin, POMC, CCK, and NPY in the brain (Hook et al. 2009). Cathepsin L tends to cleave the dibasic residues in the middle or on the carboxyl-terminal, leaving basic residues attached to the amino-terminal of the resulting peptide. This necessitates the action of an aminopeptidase enzyme; the best candidate is aminopeptidase B (Piesse et al. 2002).

Some of these studies have revealed that the processing of individual proneuropeptides is complex in the sense that when the loss of one individual PC causes the level of the processed peptide to decrease, this decrease is rarely complete, and it is usually restricted to some regions of the brain and not others. It appears that multiple enzymes are involved in the processing of some proneuropeptides, a redundancy, which ensures that some processed peptide is made even if one enzyme is gone. For example, the loss of PC1, PC2, PC7, or cathepsin L results in decreases (but not elimination) in CCK levels in some, but not all, brain regions.

2.5 POSTTRANSLATIONAL MODIFICATIONS: EXOPROTEOLYSIS

Exoproteolytic cleavage of proneuropeptides can occur at the carboxyl- and amino-termini of the dibasic residues, or between the dibasic residues. Peptide intermediates that are produced may require removal of basic residues from carboxyl- or amino-termini by carboxypeptidase (CP) and aminopeptidase enzymes, respectively.

The CPs responsible for removal of the carboxyl-terminal basic residues are better understood than the aminopeptidase enzymes. CPE was discovered as the defective enzyme in the fat/fat mouse, a mouse strain discovered at Jackson Labs (Naggert et al. 1995). Examination of the processing of a large number of neuropeptides in this strain showed that loss of CPE resulted in defective processing, accumulation of peptides with their carboxyl-terminals extended by glycine and basic amino acids, and a decrease in the content of correctly amidated peptides. Interestingly, even in the absence of CPE, levels of amidated peptides were reduced, but they were not zero. This indicated that another CP was involved. The most likely candidates, which replace CPE, are carboxypeptidase N, M, and D.

In some cases, where cathepsin L acts, it cleaves at a different site than the PCs and basic amino acids are left on the amino-terminal of cleaved peptides. These amino acids are generally removed to render the peptides biologically active. The most likely candidate for this cleavage is aminopeptidase B (Piesse et al. 2002).

2.6   POSTTRANSLATIONAL MODIFICATIONS: AMIDATION

Carboxyl-terminal amidation occurs in about half of the neuropeptides. In most cases where it has been examined, amidation is required for receptor binding and biological activity. Exceptions may be ghrelin and vasoactive intestinal polypeptide.

Glycine-extended peptides coexist with their amidated counterparts. This is presumably because of an incomplete conversion of the glycine peptides to the amidated counterparts. It is also possible that in some biological systems, glycine-extended peptides have a biological activity that is distinct from the amidated peptides; typically, the glycine-extended peptides may be trophic for some target tissues. This area is still controversial, as evidence of which receptor these peptides bind to has not yet been established.

Neuropeptides that are destined to be amidated are flanked on their carboxyl-terminal by a glycine residue, followed by single or double basic residues. This neuropeptide is cleaved from the proneuropeptide by the action of endoproteases; extra basic residues on the carboxyl-terminal are removed by CPs and the glycine-extended peptide is ready for amidation.

The amidation is achieved by peptidylalpha-amidating monooxygenase (PAM), a copper and ascorbate requiring enzyme (Eipper et al. 1993). This bifunctional enzyme converts the glycine to an amide in a two-step process. The first step is hydroxylation performed by peptidylglycine α-hydroxylating monooxygenase and the second is a lyase reaction with removal of glyoxylate performed by peptidylglycine α-amidating lyase.

Mice in which PAM has been inactivated die *in utero*; PAM heterozygotes display a number of behavioral abnormalities, which are consistent with a partial loss of amidated peptides (Czyzyk et al. 2005). Dietary copper can partially ameliorate the effects of reduced PAM expression.

2.7   OTHER POSTTRANSLATIONAL MODIFICATIONS: TYROSINE SULFATION, *N*-OCTYLATION, PYROGLUTAMATE FORMATION, SERINE PHOSPHORYLATION, AND *N*-ACETYLATION

In the TGN, tyrosine residues are sulfated by tyrosyl-protein-sulfotransferase (TPST). Tyrosine sulfation is a common posttranslational modification of secretory proteins; CCK is the only neuropeptide that is sulfated. The vast majority of CCK isolated has this modification. It is essential for maximal binding activity at the CCK 1 receptor, less important for binding to the CCK 2 receptor, which is identical with the gastrin receptor.

There are two different enzymes that are responsible for this modification, TPST1 and TPST2. They use phosphoadenosine phosphosulfate as the sulfate donor. The presence of multiple acidic residues adjacent to the tyrosines that are sulfated serves as a recognition signal for the enzyme. Knockout mice in which TPST1 or TPST2 are inactivated have a number of defects, whereas mice in which both are inactivated survive briefly after birth but die apparently of pulmonary hypertension (Ouyang et al. 2002).

Ghrelin, a hypothalamic peptide, has a very unusual modification. The third amino acid from its amino-terminal is a serine, which is modified by the addition of an N-octanoyl group. This modification is essential for its known biological activity. The enzyme responsible for this modification, ghrelin O-acyltransferase, has been identified by two different groups working independently (Gutierrez et al. 2008; Yang et al. 2008). This acyltransferase is the only one, of a family of 16 related enzymes, which is capable of modifying ghrelin. It uses a broad range of acyl substrates, which are conjugated to concanavalin A. The enzyme has a basic pH optimum, so it is highly likely that acylation occurs before endoproteolytic cleavage. Significant amounts of ghrelin without the acyl modification are also found in tissues expressing ghrelin, including the hypothalamus. It is possible that this form of the peptide binds to different receptors than ghrelin and has distinct biological activity, although this has not been demonstrated.

In peptides with an amino-terminal glutamate, in particular thyrotrophin-releasing hormone, gonadotrophin-releasing hormone, luteinizing hormone-releasing hormone, apelin, and human hypocretin 1, glutamate can be converted to a pyroglutamate by a glutaminyl cyclase. Some proportion of this conversion may be spontaneous and nonenzymatic.

Some peptides, including gastrin and CCK, have a serine residue in the carboxyl-terminal flanking peptide that is phosphorylated by a casein kinase-like enzyme. It is known that this phosphorylated serine is not required for the processing of pro-CCK.

The amino-terminal serine of α-melanocyte-stimulating hormone (α-MSH) can be mono- or diacteylated by N-acetyltransferase. Acetylation of α-MSH increases the biological activity of this peptide. In the α-MSH cell body regions of the hypothalamus, the majority of this peptide is nonacetylated, whereas mono and diacetylated peptides predominate in the areas where these neurons project, implying that this modification occurs late in vesicle during transport or just before or during release.

The amino-terminal tyrosine on β-endorphin can also be acetylated. This renders the peptides less active in analgesia assays.

2.8 NEUROPEPTIDE TRANSPORT, RELEASE, AND INACTIVATION

Neuropeptides in secretory vesicles of neurons can be transported fairly long distances from the cell bodies to the axon terminals where they are stored and released. The rate of biosynthesis and transport can apparently be influenced by demands for increased release.

Neuropeptide release is regulated by depolarization, which is accompanied by internal calcium release from the ER. Neuropeptide release has been studied in brain slice preparations, isolated synaptosomes, and in intact awake rodents by microdialysis. Release of neuropeptide is tightly regulated; a maximal stimulus releases only a small percentage (1–3%) of the total neuropeptide content of a brain slice.

Neuropeptides are frequently colocalized in individual neurons with classical neurotransmitters like catecholamines. The stimulus required to release the neuropeptides is considerably greater than the stimulus required for the release of catecholamines, and this is attributed to the fact that catecholamines are present in small synaptic vesicles that have a higher release probability than neuropeptides, which are stored in large dense core vesicles.

After the neuropeptides are released into the synaptic cleft, they are thought to be degraded by proteases. It is not completely understood which proteases are responsible for the degradation of individual peptides.

Enkephalin-degrading enzymes may include aminopeptidase N (EC 3.4.11.12), neutral endopeptidase, and dipeptidyl peptidase 3. They may also work on other neuropeptides.

Dipeptidyl peptidase 4 (EC 3.4.14.2) is also able to degrade a number of neuropeptides, including NPY and substance P. Januvia, an inhibitor of this enzyme, is now used to treat type-2 diabetes, because it increases the stability of endogenous glucagon-like peptide I in circulation. A soluble enzyme, EC 3.4.24.15, and a membrane-bound enzyme, EC 3.4.24.16., are thought to be responsible for somatostatin degradation. Various aminopeptidases, such as aminopeptidase A and M, may also be responsible for somatostatin degradation.

REFERENCES

Creemers, J. W. and Khatib, A.-M. 2008. Knock-out models of proprotein convertases: Unique functions or redundancy? *Front Biosci* 13: 4960–4971.

Czyzyk, T. A., Ning, Y., Hsu, M. S., Peng, B., Mains, R. E., Eipper, B. A., and Pintar, J. E. 2005. Deletion of peptide amidation enzymatic activity leads to edema and embryonic lethality in the mouse. *Dev Biol* 287: 301–313.

Eipper, B. A., Milgram, S. L., Husten, E. J., Yun, H.-Y., and Mains, R. E. 1993. Peptidylglycine alpha-amidating monooxygenase: A multifunctional protein with catalytic, processing and routing domains. *Protein Sci* 2: 489–497.

Gutierrez, J. A., Solenberg, P. J., Perkins, D. R., Willency, J. A., Knierman, M. D., Jin, Z., Witcher, D. R., Luo, S., Onyia, J. E., and Hale, J. E. 2008. Ghrelin octanoylation mediated by an orphan lipid transferase. *Proc Natl Acad Sci U S A* 105: 6320–6325.

Hook, V., Funkelstein, L., Toneff, T., Mosier, C., and Hwang, S. R. 2009. Human pituitary contains dual cathepsin L and prohormone convertase processing pathway components involved in converting POMC into the peptide hormones ACTH, alpha-MSH, and beta-endorphin. *Endocrine* 35: 429–437.

Naggert, J. K., Fricker, L. D., Varmalov, O., Nishina, P. M., Rouille, Y., Steiner, D. F., Carroll, R. J., Paigen, B. J., and Leiter, E. H. 1995. Hyperproinsulinaemia in obese fat/fat mice associated with a carboxypeptidase E mutation which reduces enzyme activity. *Nat Genet* 10: 135–142.

Ouyang, Y. B., Crawley, J. T., Aston, C. E., and Moore, K. L. 2002. Reduced body weight and increased postimplantation fetal death in tyrosylprotein sulfotransferase-1-deficient mice. *J Biol Chem* 277: 23781–23787.

Piesse, C., Tymms, M., Garrafa, E., Gouzy, C., Lacasa, M., Cadel, S., Cohen, P., and Foulon, T. 2002. Human aminopeptidase B (rnpep) on chromosome 1q32.2: Complementary DNA, genomic structure and expression. *Gene* 292: 129–140.

Yang, J., Brown, M. S., Liang, G., Grishin, N. V., and Goldstein, J. L. 2008. Identification of the acyltransferase that octanoylates ghrelin, an appetite-stimulating peptide hormone. *Cell* 132: 387–396.

Zhang, X., Bao, L., and Ma, G.-Q. 2010. Sorting of neuropeptides and neuropeptide receptors into secretory pathways. *Prog Neurobiol* 90: 276–283.

3 Neuropeptide Degradation Related to the Expression of the Physiological Action of Neuropeptides

Chikai Sakurada, Hirokazu Mizoguchi,
Takaaki Komatsu, Shinobu Sakurada,
and Tsukasa Sakurada

CONTENTS

3.1 INTRODUCTION

The neuropeptides play a major role in the transmission or modulation of signals in the central and peripheral nervous system (Kastin et al. 1996). They are involved in several neurologic functions, including those related to memory, pain, reward, stress, and food intake. Neuropeptides are released from nerve terminals of peptidergic neurons, and they induce their effects by binding to pre- or postsynaptically located receptors, which are embedded in the nerve cell membrane.

A typical pathway for neuropeptide biosynthesis starts at the ribosomes in the nerve cell bodies, where the messenger RNA is translated into peptide. Generally, the neuropeptides are synthesized as large, biologically inert protein precursors (prepropeptides). The neuropeptide precursors are subsequently cleaved by a series of proteolytic steps in a sequence-specific and tissue-specific fashion to generate smaller bioactive peptides (Canaff, Bennett, and Hendy 1999). During this process, additional modifications, such as amidation and truncation of the released peptide may also occur. The neuroactive peptides that are released from nerve terminals bind and activate their receptors.

The physiological action of neuropeptides as a neurotransmitter or neuromodulator in the central nervous system may be related to degradation of neuropeptides by peptidase(s). The physiological effect of neuropetides is probably terminated by a membrane-bound protease capable of degrading neuropeptide in the synaptic region, by analogy with membrane-bound acetyl cholinesterase functioning in acetylcholine degradation in the synapse. For instance, Met- and Leu-enkephalins are metabolized by membrane-bound peptidases, endopeptidase-24.11 and aminopeptidase. In addition, inhibitors of endopeptidase-24.11 are analgesic and substitute in opiate abstinence (Roques, Noble, and Dauge 1993). Definition of inactivation pathway allows the design of enzyme inhibitor that may be of pharmacological interest.

Degradation of neuropeptides may also generate products that are active but less potent or, more interestingly, molecules with another activity. For example, substance P (SP) is enzymatically cleaved into several biologically active fragments.

This chapter discusses neuropeptide degradation related to the expression of physiological action of neuropeptides. In particular, it focuses on degradation of endomorphin-2, an endogenous opioid peptide, and SP as a pain transmitter or modulator.

3.2 ROLE OF ENDOMORPHIN IN MODULATING PAIN AT THE SPINAL AND SUPRASPINAL LEVELS

Endomorphin-1 (Tyr-Pro-Trp-PheNH$_2$) and endomorphin-2 (Tyr-Pro-Phe-PheNH$_2$) have been isolated from bovine and human brain as the endogenous ligands for the μ-opioid receptor (Hackler et al. 1997; Zadina et al. 1997). Of all known mammalian opioids, these peptides exhibit the highest affinity and selectivity for the μ-opioid receptor (Zadina et al. 1997). Immunohistological studies have shown that immunoreactivity of endomorphins was observed in the caudal nucleus of the spinal trigeminal tract, the parabrachial nucleus, the nucleus *tractus solitarius*, the periaqueductal gray, ambiguous nucleus, locus coeruleus, midline thalamic nuclei, and the amygdala (Schreff, Schulz, and Wilborny 1998; Martin-Schild et al. 1999; Pierce and Wessendorf 2000). Many of these regions where high densities of μ-opioid receptors occur are known to be involved in the processing and modulation of nociceptive transmission (Mansour et al. 1995). In behavioral studies, intracerebroventricularly (i.c.v.) or intrathecally (i.t.) administered endomorphins significantly increase nociceptive thresholds (Stone et al.

1997; Zadina et al. 1997; Goldberg et al. 1998; Sakurada et al. 1999, 2002), which are blocked by μ-receptor antagonists, such as naloxone and β-funaltrexamine (Zadina et al. 1997). Neither supraspinal endomorphin-1 nor endomorphin-2 elicited antinociception in μ-receptor knockout mice (Mizoguchi et al. 1999) and in μ_1-receptor-deficient mice (Goldberg et al. 1998). Thus, these reports suggest that endomorphins may play a physiological role in modulating pain at the spinal and supraspinal levels.

3.3   INVOLVEMENT OF DIPEPTIDYL PEPTIDASE IV AND AMINOPEPTIDASE IN THE DEGRADATION OF ENDOMORPHIN-2 BY SYNAPTIC MEMBRANES OF MOUSE BRAIN

Neuropeptides are degraded extracellulary in the central nervous system probably by a limited number of enzymes with relatively broad specificities, although one cannot exclude an intracellular cleavage of the peptide after internalization. Most of these extracellular neuropeptide-degrading enzymes are membrane-bound peptidases (ectoenzyme), that is, integral membrane proteins that have active sites facing the extracellular space. Endomorphin-2 functioning as a neurotransmitter or neuromodulator could be degraded at the plasma membrane of neuronal cells to abolish their physiological function.

Major products of endomorphin-2 produced through the action of mouse synaptic membranes were free tyrosine, free phenylalanine, PheNH$_2$, and Tyr-Pro (Figure 3.1).

The effect of various peptidase inhibitors on the degradation of endomorphin-2 by synaptic membranes of mouse brain is summarized in Table 3.1. The effect of peptidase inhibitors on the initial cleavage rate of endomorphin-2 was analyzed by measuring the effects of decrease of the high-performance liquid chromatography (HPLC) peak for endomorphin-2.

A serine protease inhibitor, diisopropyl fluorophosphate (DFP), and specific inhibitors for dipeptidyl peptidase IV, diprotin A and B, caused approximately 50% inhibition of endomorphin-2 degradation by synaptic membranes. Other inhibitors, including an angiotensin-converting enzyme inhibitor (enalapril), an aminopeptidase inhibitor (bestatin), an endopeptiase-24.11 inhibitor (phosphoramidon), and a prolyl endopetidase inhibitor (Z-321), had little inhibitory effects on endomorphin-2 degradation. The generation of major metabolites, free tyrosine, PheNH$_2$, free phenylalanine, and Tyr-Pro was strongly inhibited by the addition of diprotin A and B, whereas enalapril exerted little inhibitory effect on the generation of almost all the major metabolites (Table 3.2).

Incubation of endomorphin-2 with synaptic membranes in the presence of bestatin gave a large amount of Phe-PheNH$_2$ and Pro-Phe-PheNH$_2$ as the incubation time increased, whereas the appearance of free phenylalanine and PheNH$_2$ was inhibited by the addition of bestatin (Figure 3.2).

Evaluation of the effect of several peptidase inhibitors on the degradation of endomorphin-2 allowed us to propose that two membrane-bound peptidases,

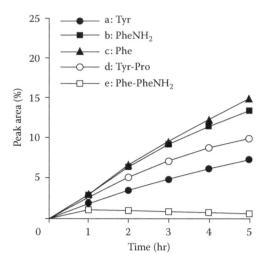

FIGURE 3.1 Time course of endomorphin-2 degradation products produced through the action of synaptic membranes of mouse brain. The reaction mixture containing 25 mM Tris-HCl (pH 7.5), 155 mM NaCl, 50 μM endomorphin-2, and 10 μg protein of the membrane preparation was incubated at 37°C for 0, 1, 2, 3, 4, and 5 hours. After heating at 100°C for 10 minutes, 10-μL aliquots of the mixture were analyzed by HPLC on a reversed phase column (4.6 × 150 mm) of Symmetry C18, which had been equilibrated with 1.0% acetonitrile in 0.05% TFA. Elution was carried out at room temperature with a 60-minute linear gradient of 1.0–65% acetonitrile in 0.05% TFA at a flow rate of 1.0 mL/minute. The absorbance at 210 nm was monitored. Cleavage products were separated by HPLC and their peak areas traced on the chart were measured. The vertical coordinate indicates the relative value of the peak area calculated on the basis of that of the substrate, endomorphin-2 at 0 hours. The results are the mean of triplicate determinations. The sequence of endomorphin-2 as a substrate is Tyr-Pro-Phe-PheNH$_2$.

TABLE 3.1

Effect of Peptidase Inhibitors on the Degradation of Endomorphin-2 by Synaptic Membranes of Mouse Brain[a]

Inhibitors	Concentration (mM)	Inhibition (%)
Diprotin A	1.0	51
Diprotin B	1.0	53
DFP	1.0	47
Enalapril	0.1	4
Bestatin	0.1	10
Phosphoramidon	0.1	13
Z-321	0.1	23

[a] The activity was measured on the basis of the disappearance of endomorphin-2, as detected by HPLC.

TABLE 3.2

**Effect of Peptidase Inhibitors on the Formation of
Endomorphin-2 Metabolites Following Incubations with
Synaptic Membranes of Mouse Brain[a]**

	Inhibition (%)			
Metabolites	Diprotin A (1.0 mM)	Diprotin B (1.0 mM)	Enalapril (1.0 mM)	Bestatin (0.1 mM)
Tyr	60	46	0	6
PheNH$_2$	65	52	0	91
Phe	62	53	0	91
Tyr-Pro	80	68	0	4

[a] The extent of degradation was determined by measuring the decrease of
major peaks derived from endomorphin-2 degraded.

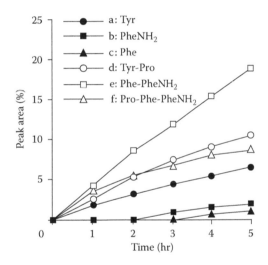

FIGURE 3.2 Time course of endomorphin-2 degradation products produced through the
action of synaptic membranes of mouse brain in the presence of 0.1 mM bestatin. The reaction mixture was incubated and treated with the same procedure as described in Figure 3.1.
Cleavage products were separated by HPLC as described in Figure 3.1 and their peak areas
traced on the chart were measured. The vertical coordinate indicates the relative value of
the peak area calculated on the basis of that of the substrate, endomorphin-2 at 0 hours. The
results are the mean of triplicate determinations.

dipeptidyl peptidase IV and aminopeptidase, would be mainly responsible for
the degradation of endomorphin-2 at the synapse. We have shown several lines
of evidence to suggest that the degradation of endomorphine-2 by the synaptic
membranes of mouse brain may be initially triggered by the action of dipeptidyl peptidase IV. Both degradation of endomorphin-2 and accumulation of major

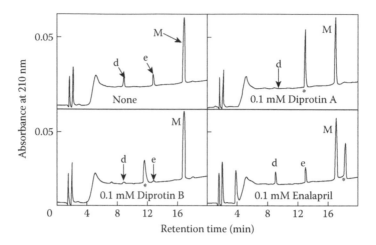

FIGURE 3.3 Effect of peptidase inhibitors on the degradation of endomorphin-2 by purified dipeptidyl peptidase IV. The reaction mixture (pH = 7.5) containing 1.0 ng of purified
dipeptidyl peptidase IV from porcine kidney was incubated at 37°C for 3 hours in the absence
of an inhibitor, or in the presence of 0.1 mM diprotin A, 0.1 mM diprotin B, and 0.1 mM
enalapril. After heating at 100°C for 10 minutes, 10 mL aliquots of the mixture were analyzed
by HPLC as described in Figure 3.1. * indicates the peak derived from the inhibitor added. d,
Tyr-Pro; e, Phe-PheNH$_2$; M, endomorphin-2 (Tyr-Pro-Phe-PheNH$_2$).

metabolites were suppressed by the presence of inhibitors of dipeptidyl peptidase
IV, such as diprotin A and B. It is important to note that DFP inhibited the generation of endomorphin-2 (Table 3.1), since dipeptidyl peptidase IV belongs to the
serine class of peptidases and is inhibited by serine enzyme inhibitor such as DFP
that covalently modifies their active site (Mentlein 1999). Dipeptidyl peptidase
IV removes dipeptides from the N-terminal of peptides, which contain proline
as the penultimate amino acid (Mentlein 1999). Neuropeptide Y (Mentlein et al.
1993), peptide YY (Mentlein et al. 1993), and β-casomorphin (Nausch, Mentlein,
and Heymann 1990), all of which share a common N-terminal dipeptide sequence
(Tyr-Pro-) with endomorphin-2, are good substrates for dipetidyl peptidase IV.
Furthermore, the purified dipeptidyl peptidase IV hydrolyzed endomorphin-2 at
the cleavage site, Pro^2-Phe^3 bond. The generation of Tyr-Pro and Phe-PheNH$_2$ was
strongly inhibited by the addition of diprotin B, an inhibitor of dipeptidyl peptidase IV (Figure 3.3). In addition, diprotin A completely inhibited the occurrence
of peak d (Tyr-Pro) except for peak e (Phe-PheNH$_2$) whose size could not be estimated because peak e was overlaid by a large peak due to the inhibitor added.

These results indicate that the first enzyme, dipeptidyl peptidase IV, cleaves endo-
morphin-2 at the Pro^2-Phe^3 bond to generate Tyr-Pro and Phe-PheNH$_2$, from which
free phenylalanine and PheNH$_2$ are produced by aminopeptidase (Figure 3.4).

The generation of Phe-PheNH$_2$ was strongly enhanced by bestatin, an aminopeptidase inhibitor, whereas it was decreased in free phenylalanine and PheNH$_2$,
although bestatin had little effect on the degradation of endomorphin-2. These
results may imply the involvement of aminopeptidase in secondary cleavage of

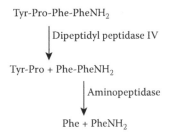

Tyr-Pro-Phe-PheNH$_2$

Dipeptidyl peptidase IV

Tyr-Pro + Phe-PheNH$_2$

Aminopeptidase

Phe + PheNH$_2$

FIGURE 3.4   Scheme for metabolism of endomorphin-2 by synaptic membranes of mouse brain.

the initially formed cleavage products by the action of dipeptidyl peptidase IV. It is noteworthy that bestatin also gave a large amount of Pro-Phe-PheNH$_2$ whose peak could not be detected in the absence of bestatin. This result suggests that the conversion of Pro-Phe-PheNH$_2$ into free phenylalanine, Phe-PheNH$_2$, and PheNH$_2$ may be catalyzed by aminopeptidase and implies the existence of peptidase related to remove tyrosine, N-terminal amino acid residue of endomorphin-2. It is extremely important to detect a peptidase related to cleave tyrosine, an N-terminal amino acid residue of endomorphin-2, as known mammalian opioid peptides, such as enkephalins, possess tyrosine as N-terminal amino acid residue, which is essential for opioid activity (Horn and Rodgers 1977; Schwartz, Malfroy, and De La Baume 1981). Peter, Toth, and Tomboly (1999) have reported that aminopeptidase M obtained from rat kidney microsomes cleaves Pro^2-Phe^3 bond of intact endomorphin-2 and the C-terminal fragment, Phe-PheNH$_2$ hydrolyzed further, giving free phenylalanine and PheNH$_2$. When synaptic membranes are used, bestatin leads to an increase in Pro-Phe-PheNH$_2$ with no change in free tyrosine. It is difficult to understand how free tyrosine becomes a major metabolite. It is therefore inferred that bestatin-insenstive peptidase may be involved in cleaving Tyr^1-Pro^2 bond of intact endomorphin-2, or cleaving Tyr^1-Pro^2 bond of initially formed Tyr-Pro by the action of dipeptidyl peptidase IV. However, it is speculated that the former is a minor pathway, whereas the latter is a major pathway as both degradation of endomorphin-2 and accumulation of free tyrosine were inhibited by the presence of diprotin A and B.

3.4   DIPEPTIDYL PEPTIDASE IV MAY BE A KEY ENZYME RESPONSIBLE FOR TERMINATING ENDOMORPHIN-2-INDUCED ANTINOCICEPTION AT SUPRASPINAL LEVEL

The i.c.v. administration of endomorphin-2 produced a dose-dependent antinociception in the paw withdrawal test (Sakurada et al. 2003). The peak effect of endomorphin-2-induced antinociception appeared at 5 minutes after i.c.v. injection and decayed at 15–20 minutes postinjection (Figure 3.5a). The ED$_{50}$ value for endomorphin-2 was 13.0 nmol (8.14–20.74) at 5 minutes after i.c.v. administration. Endomorphin-2-induced antinociception was remarkably enhanced in the paw

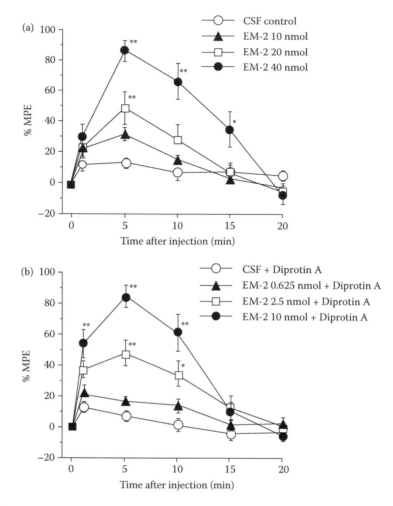

FIGURE 3.5 (a) Time course of antinociceptive effect of i.c.v.-administered endomorphin-2 in the mouse paw withdrawal test. Antinociception was expressed as percent of maximum possible effect (%MPE) = (postdrug responsive latency − predrug responsive latency) / (10 − predrug responsive latency) × 100. Each value represents the mean ± SEM for 10 mice. $^{**}p < .01$, $^{*}p < .05$, when compared to the respective value in the CSF-control group. (b) Effect of diprotin A on endomorphin-2 (EM-2)-induced antinociception in the mouse paw withdrawal test. Mice were coadministered CSF or various doses of endomorphin-2 in combination with 4.0 nmol of diprotin A. Each value represents the mean ± SEM for 10 mice. $^{**}p < .01$, $^{*}p < .05$, when compared to diprotin A alone.

withdrawal test when diprotin A, a dipeptidyl peptidase IV inhibitor, was simultaneously injected with endomorphin-2 (Figure 3.5b).

The ED_{50} value for endomorphin-2 in combination with diprotin A was 2.6 nmol (1.51–4.46) at 5 minutes after i.c.v. coadministration. Combined injection of cerebrospinal fluid (CSF) and diprotin A had no antinociceptive effect. Judging from the ED_{50} value of endomorphin-2, endomorphin-2 in combination with diprotin

A was fivefold more potent than endomorphin-2 alone in producing antinociceptive effects. Shane, Wilk, and Bodnar (1999) have reported that i.c.v.-administered Ala-pyrrolidonyl-2-nitrile, a specific inhibitor of dipeptidyl peptidase IV, potentiated endomorphin-2-induced antinociception in the rat tail flick test. Furthermore, Ronai et al. (1999) have shown that i.c.v.-administered diprotin A produced a dose-dependent, short-lasting, and naloxone-reversible analgesia in the rat tail flick test. These behavioral data support the involvement of dipeptidyl peptidase IV as an enzyme that is important in the supraspinal catabolism of endomorphin-2. In conclusion, a physiological role for the degradation of endomorophin-2 by dipeptidyl peptidase IV at the supraspinal level may be responsible for the inactivation of endomorphin-2.

These findings also suggest that selective dipeptidyl peptidase IV inhibitors or dipeptidyl peptidase-IV-resistant endomorphin-2 analogs (Janecka et al. 2006) have the potential for clinical usefulness as analgesics.

3.5   SUBSTANCE P AS A PAIN TRANSMITTER OR MODULATOR

SP is an undecapeptide (Arg-Pro-Lys-Pro-Gln-Gln-Phe-Phe-Gly-Leu-MetNH$_2$) that is widely distributed in the central nervous system and peripheral tissues (Maggio 1988). A member of the tachykinins, SP is the endogenous ligand for the tachykinin NK$_1$ receptor, whereas neurokinin A and B are endogenous ligands for NK$_2$ and NK$_3$ receptors, respectively (Maggio et al. 1993; Regoli and Nantel 1991). Considerable evidence has implicated SP as a major neurotransmitter or neuromodulator of pain in the mammalian central nervous system. SP is located in primary afferent C-fiber and released into the spinal cord after noxious stimulation (Duggan et al. 1987; Schaible et al. 1990; Radhakrishnan and Henry 1991). Spinal NK$_1$ receptor-expressing cells play a critical role in injury-induced hyperalgesia or pain (Mantyh et al. 1997; Khasabov et al. 2002; Suzuki et al. 2002; Vera-Portocarrero et al. 2007). The physiologically significant action of SP is probably terminated by a membrane-bound protease capable of degrading SP in the synaptic region. Several membrane-bound proteases in brain have been reported to be involved in the degradation of SP. These enzymes include endopeptidase-24.11 (Matsus et al. 1983; Matsus, Kenny, and Turner 1984; Skidgel et al. 1984), SP-degrading enzyme from human brain (Lee et al. 1980), SP-degrading endopeptidase from rat brain (Endo, Yokosawa, and Ishii 1988), postproline dipeptidyl aminopeptidase from guinea pig brain (O'Connor and O'Connor 1986), and angiotensin-converting enzyme (Yokosawa et al. 1983). However, the physiological function of these neuropeptidases from a membrane fraction remains obscure.

For activation of NK$_1$ receptor, it seems that the SP C-terminal sequence is essential. However, several effects of SP in the central nervous system are not mediated through its SP C-terminal sequence. For instance, SP N-terminal fragments, such as SP(1-7) and SP(1-8), have been shown to modulate or oppose several behaviors elicited by SP (Hall and Stewart 1983). These reports suggest that SP N-terminal fragments are not inactive substances degraded by enzymatic processes. It is of importance to explore the spinal metabolic pathway of SP to elucidate the physiological functions of SP and SP N-terminal fragments as a pain transmitter or modulator.

3.6 DEGRADATION OF SUBSTANCE P BY SYNAPTIC MEMBRANES OF MOUSE SPINAL CORD IS TRIGGERED BY ENDOPEPTIDASE-24.11: AN *IN VITRO* AND *IN VIVO* STUDY

SP was degraded by synaptic membranes of mouse spinal cord (Sakurada, Watanabe, and Sakurada 2004). Cleavage products were separated by reversed phase HPLC and identified by amino acid composition analyses as shown in Figure 3.6.

SP *N*-terminal fragments, SP(1-6), SP(1-7), and SP(1-9), were found to be major products, in addition to free phenylalanine, SP(8-9) and SP(10-11), when incubated with synaptic membranes of mouse spinal cord.

It is, therefore, likely that SP *C*-terminal fragments released through the initial cleavage of SP may be readily susceptible to the action of neuropeptidases, as recoveries of SP *C*-terminal fragments were low compared with those of SP *N*-terminal fragments. These results are in accordance with our previous data showing that SP(1-7) is concentrated in the dorsal horn; the ratio between SP and SP(1-7) ranged from approximately 5:1 to 10:1 in the rat spinal cord (Sakurada, Le Greves, and Terenius 1985).

Next, we examined the effect of various peptidase inhibitors on the degradation of SP by synaptic membranes of the mouse spinal cord.

First, the inhibitory effect on the initial cleavage rate of SP was analyzed by measuring the effects on the decrease of HPLC peak for SP. The results are shown in Table 3.3.

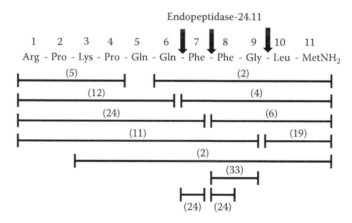

FIGURE 3.6 Summary of cleavage products of SP by synaptic membranes of mouse spinal cord. The degradation of SP by the synaptic membranes of the mouse spinal cord was carried out at 37°C in a mixture containing 25 mM Tris-HCl (pH 7.5), 155 mM NaCl, 50 μM SP, and 50 μg protein of the membrane preparation. After heating at 100°C for 10 minutes, 10-μL aliquots of the mixture were analyzed by HPLC on a reversed phase column (4.6 × 150 mm²) of symmetry C18, which had been equilibrated with 1.0% acetonitrile in 0.05% TFA. Elution was carried out at room temperature with a 60-minute linear gradient of 1.0–65% acetonitrile in 0.05% TFA at a flow rate of 1.0 mL/minute. The absorbance at 210 nm was monitored. Numbers in parentheses represent yield of cleavage products determined on the basis of degraded SP. The extent of SP degradation was 75%.

TABLE 3.3

Effect of Peptidase Inhibitors on the Degradation of SP by Synaptic Membranes of Mouse Spinal Cord[a]

Inhibitors	Concentration (mM)	Inhibition (%)
O-Phenthroline	1.0	97
Phosphoramidon	0.1	83
Thiorphan	0.1	66
PCMBS	1.0	20
PMSF	1.0	11
Captopril	0.1	0
Bestatin	0.1	0
Chymostatin	0.1	20
Z-321	0.1	0

[a] The activity was measured on the basis of the disappearance of SP, as detected by HPLC.

PCMBS, *p-chloromercribenzenesulfonic acid;* PMSF, *phenyl methyl sulfonyl fluoride.*

A metal chelator, O-phenanthroline, and specific inhibitors for endopeptidase-24.11, phosphoramidon and thiorphan, inhibited SP degradation by synaptic membranes. Other inhibitors, including an angiotensin-converting enzyme inhibitor (captopril), a prolylendopeptidase inhibitor (Z-321), an aminopeptidase inhibitor (bestatin), and inhibitors for serine and cysteine proteases (*p*-chloromercuribenzensulfonic acid, phenylmethyl sulfonyl fluoride, and chymostatin) had little inhibitory effects on SP degradation. Second, the inhibitory effect on the generation of cleavage products of SP separated by HPLC was analyzed (Table 3.4).

The generation of all major metabolites, including major peaks of SP *N*-terminal fragments, was strongly inhibited by the addition of thiorphan and phosphoramidon, whereas other inhibitors, such as captopril and Z-321, exerted little inhibitory effect on the generation of major SP metabolites.

Finally, we examined the effect of peptidase inhibitors on the characteristic behavioral response (scratching, biting, and licking) elicited by i.t. injection of SP in mice (Sakurada et al. 1990). Phosphoramidon (0.002–2.0 nmol), an endopeptidase-24.11 inhibitor, injected along with SP, remarkably enhanced and prolonged SP-induced behavioral response in a dose-dependent manner. SP-induced behavioral response was not enhanced by bestatin and captopril.

We have presented several lines of evidence to suggest that the degradation of SP by mouse spinal cord synaptic membranes may be initially triggered by the action of endopeptidase-24.11. Both the degradation of SP and the accumulation of major metabolites were suppressed by the presence of endopeptidase-24.11 inhibitors,

TABLE 3.4

Effect of Peptidase Inhibitors on the Formation of SP
Metabolites Following Incubations with Synaptic
Membranes of Mouse Spinal Cord[a]

	Inhibition (%)			
Metabolites	Phosphoramidon (0.1 mM)	Thiorphan (0.1 mM)	Captopril (0.1 mM)	Z-321 (0.1 mM)
SP(1-6)	78	70	0	0
SP(1-7)	85	84	8	5
SP(1-9)	91	86	0	0
SP(8-9)	93	95	24	4
SP(10-11)	84	91	0	0
Phe	84	62	20	5

[a] The extent of degradation was determined by measuring the decrease of major peaks derived from endomorphin-2 degraded.

such as phosphoramidon and thiorphan. It has also been reported that purified endopeptidase-24.11 from glioma C_6 cells cleaved the Gln^6-Phe^7, Phe^7-Phe^8 and Gly^9-Leu^{10} bonds (Endo, Yokosawa, and Ishii 1989), as was the case for cleavage by the synaptic membranes of mouse spinal cord.

Neuronal localization of endopetidase-24.11 has been demonstrated by an immunohistochemical technique. The highest density of endopeptidase-24.11 immunoreactivity was observed in the superficial layers of the spinal nucleus, located in the trigeminal nerve of the dorsal horn in the spinal cord, which also displays the highest immunoreactivity for both enkephalins and SP (Pollard et al. 1989). Taking into account these biochemical and behavioral data, it is likely that endopeptidase-24.11 present in spinal synaptic membranes plays a major role in the initial stage of SP metabolism at the spinal cord level in mice.

3.7   BIOLOGICAL ACTIVITY OF SUBSTANCE P N-TERMINAL FRAGMENT

SP is enzymatically cleaved into several C-terminal and N-terminal fragments, which have biological activity. SP(1-7), the predominant N-terminal metabolite of SP, has been found to be active in several different biological functions, including antinociception (Goettl and Larson 1996; Sakurada et al. 2004), inflammation (Wiktelius, Khalil, and Nyberg 2006), memory (Tomaz, Silva, and Nogueira 1997), and opioid withdrawal reactions (Kreeger and Larson 1993; Zhou et al. 2003).

SP C-terminal fragments, SP(4-11), SP(5-11), and SP(6-11), injected i.t. into mice, elicit scratching, biting, and licking responses similar to that observed with SP, whereas SP N-terminal fragment, SP(1-7) did not elicit a behavioral response. However, when low doses of SP(1-7)(1.0–4.0 pmol) were injected along with SP or

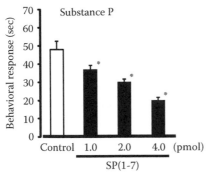
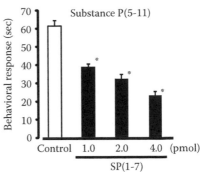

FIGURE 3.7   The effect of SP(1-7) on intrathecally (i.t.-injected SP and SP(5-11)-induced behavioral response. The duration of scratching, biting, and licking induced by SP and SP(5-11) was determined over a 5-minute period starting immediately after injection. Mice received i.t. injection of SP(1-7) in doses ranging from 1.0 to 4.0 pmol in combination with i.t. injection of SP (100 pmol) or SP(5-11) (100 pmol). Each value represents the mean ± SEM for 10 mice. *$p < .01$ when compared to SP or SP(5-11) alone. (From Sakurada, T., et al., *Neurosci. Lett.,* 95, 281–285, 1988. With permission.)

SP(5-11) (100 pmol), the characteristic behaviors induced by SP or SP(5-11) were significantly reduced (Figure 3.7) (Sakurada et al. 1988).

These results indicate that SP(1-7) formed endogenously could modulate action of SP or SP(5-11) in the spinal cord.

We have recently shown the effect of SP(1-7) on spontaneous pain-related behavior evoked by high-dose i.t. morphine in rats (Sakurada et al. 2007) and on morphine-induced antinocicepting in mice (Komatsu et al. 2009). Injection of high-dose morphine (500 nmol) into the spinal lumbar i.t. space of rats elicits an excitatory behavioral syndrome indicative of severe vocalization and agitation. We examined the effect of i.t. SP(1-7) on both the nociceptive response and the extracellular concentrations of glutamate and nitric oxide (NO) metabolites (nitrite/nitrate) evoked by high-dose i.t. morphine (Sakurada et al. 2007). The induced behavioral responses were attenuated dose-dependently by i.t. pretreatment with SP(1-7) (100–400 pmol). The inhibitory effect of SP(1-7) was reversed significantly by pretreatment with [D-Pro², D-Phe⁷] SP(1-7), a D-isomer and antagonist of SP(1-7). *In vivo* microdialysis analysis showed a significant elevation of extracellular glutamate and NO metabolites in the spinal cord after i.t. injection of high-dose morphine. Pretreatment with SP(1-7) produced a significant reduction of the elevated concentrations of glutamate and NO metabolites evoked by high-dose i.t. morphine. The reduced levels of glutamate and NO metabolites were significantly reversed by the SP(1-7) antagonist. These results suggest that i.t. SP(1-7) may attenuate the excitatory behavior of high-dose i.t. morphine by inhibiting the presynaptic release of glutamate and reducing NO production in the dorsal spinal cord.

Next, we examined the mechanism of SP to modulate the antinociceptive action of i.t. morphine in paw-licking and biting responses evoked by subcutaneous injection of capsaicin into the plantar surface of the hindpaw in mice (Komatsu et al.

2009). The i.t. injection of morphine inhibited capsaicin-induced nociception in a dose-dependent manner. SP (25 and 50 pmol), injected i.t., did not alter capsaicin-induced nociception, whereas SP at a higher dose of 100 pmol significantly reduced the capsaicin response. The N-terminal fragment of SP(1-7) was more effective than SP on capsaicin-induced nociception. Combination treatment with SP (50 pmol) and morphine at a subthreshold dose enhanced the antinociceptive effect of morphine. The enhanced effect of the combination of SP with morphine was reduced significantly by coadministration of phosphoramidon, an inhibitor of endopeptidase-24.11. Administration of the SP(1-7) antagonist, or antisera against SP(1-7), reversed the enhanced antinociceptive effect by coadministration of SP and morphine. Taken together, these data suggest that morphine-induced antinociception may be enhanced through SP(1-7) formed by the enzymatic degradation of i.t.-injected SP in the spinal cord.

These behavioral data show that SP(1-7) is not simply an inactive substance of the enzymatic process. Although the mechanism of actions of SP(1-7) is still unclear, it is of interest to note that the antagonistic effect of SP(1-7) on SP-induced behavioral responses in mice was inhibited by naloxone, an opioid receptor antagonist, and [D-Pro^2, D-Phe^7]SP(1-7), a D-isomer and antagonist of SP(1-7), but not by selective antagonists of μ-, δ-, and k-opioid receptors (Mousseau and Larson 1992). Furthermore, the effect of SP(1-7) may not be mediated by action at the NK_1 receptor, as SP(1-7) has no binding for NK_1 receptors in brain and spinal cord (Hanley et al. 1980). One could speculate that the physiological action of SP(1-7) may not be mediated by the action at a typical opioid receptor, but by the action at the SP(1-7)-binding site or its own receptor. A specific binding site for SP(1-7) has been shown in the spinal cord of mice and rats (Igwe et al. 1990; Botros et al. 2006). This binding site was very specific for SP(1-7), whereas SP and other SP fragments exhibited negligible affinity for this binding site. Some μ-opioid receptor agonists, including DAMGO, Tyr-MIF-1, and endomorphin-2, exhibited comparatively high affinity for the SP(1-7) site. It is noteworthy that endomorphin-2 exhibited high potency for binding to the SP(1-7) site, in contrast to endomorphin-1 (Botros et al. 2006, 2008).

In conclusion, a physiological role for the degradation of SP by endopeptidase-24.11 is not only responsible for inactivation of the parent peptide SP but may also lead to the formation of SP(1-7) with another activity.

REFERENCES

Botros, M., Hallberg, M., Johansson, T., et al. 2006. Endomorphin-1 and endomorphin-2 differentially interact with specific binding sites for substance P (SP) aminoterminal SP_{1-7} in the rat spinal cord. *Peptides* 27: 753–759.

Botros, M., Johansson, T., Zhou, Q., et al. 2008. Endomorphins interacts with the substance P (SP) aminoterminal SP_{1-7} binding in the ventral tegmental area of the rat brain. *Peptides* 29: 1820–1824.

Canaff, L., Bennett, H. P., and Hendy, G. N. 1999. Peptide hormone precursor processing: Getting sorted? *Mol Cell Endocorinol* 156: 1–6.

Duggan, A. W., Morton, C. R., Zhao, Z. Q., et al. 1987. Noxious heating of the skin releases immunoreactive substance P in the substantina gelatinosa of the cat: A study with antibody microprobes. *Brain Res* 403: 345–349.

Endo, S., Yokosawa, H., and Ishii, S. 1988. Purification and characterization of a substance P-degrading endopeptidase from rat brain. *J Biochem* (Tokyo) 104: 999–1006.

Endo, S., Yokosawa, H., and Ishii, S. 1989. Involvement of endopeptidase-24.11 in degradation of substance P by glioma cells. *Neuropeptides* 14: 31–37.

Goettl, V. M. and Larson, A. A. 1996. Nitric oxide mediates long-term hyperalgesic and antinociceptive effects of the N-terminus of substance P in the formalin assay in mice. *Pain* 67: 435–441.

Goldberg, I. E., Rossi, G. C., Letchworth, S. R., et al. 1998. Pharmacological characterization of endomorohin-1 and endomorphin-2 in mouse brain. *J Pharmacol Exp Ther* 286: 1007–1013.

Hackler, L., Zadina, J. E., Ge, L. J., et al. 1997. Isolation of relatively large amounts of endomorphin-1 and endomorphin-2 from human brain cortex. *Peptides* 18: 1635–1639.

Hall, M. E. and Stewart, J. M. 1983. Substance P and behavior: Opposite effects of N-terminal and C-terminal fragments. *Peptides* 4: 763–768.

Hanley, M. R., Sandberg, B. E., Lee, C. M., et al. 1980. Specific binding of ^{3}H-substance P to rat brain membranes. *Nature* 286: 810–812.

Horn, A. S. and Rodgers, J. R. 1977. The enkephalins and opiates: Structure–activity relations. *J Pharm Pharmacol* 29: 257–265.

Igwe, O. J., Kim, D. C., Seybold, V. S., et al. 1990. Specific binding of substance P aminoterminal heptapeptide [SP(1-7)] to mouse brain and spinal cord membranes. *J Neurosci* 10: 3653–3663.

Janecka, A., Kruszynski, R., Fichna, J., et al. 2006. Enzymatic degradation studies of endomorphin-2 and its analogs containing N-methylated amino acids. *Peptides* 27: 131–135.

Kastin, A. J., Zadina, J. E., Olson R. D., et al. 1996. The history of neuropeptide research. *Ann NY Acad Sci* 780: 1–18.

Khasabov, S. G., Rogers, S. D., Ghilardi, J. R., et al. 2002. Spinal neurons that possess the substance P receptor are required for the development of central sensitization. *J Neurosci* 22: 9086–9098.

Komatsu, T., Sasaki, M., Sanai, K., et al. 2009. Intrathecal substance P augments morphine-induced antinociception: Possible relevance in the production of substance P N-terminal fragments. *Peptides* 30: 1689–1696.

Kreeger, J. S. and Larson, A. A. 1993. Substance P (1-7), a substance P metabolite, inhibits withdrawal jumping in morphine-dependent mice. *Eur J Pharmacol* 238: 111–115.

Lee, C. M., Sandberg, B. E. B., Hanley, M. R., and Iversen, L. L. 1980. Purification and characterization of a membrane-bound substance P-degrading enzyme from human brain. *Eur J Biochem* 114: 315–327.

Maggio, J. E. 1988. Tachykinins. *Annu Rev Neurosci* 11:13–28.

Maggio, J. E., Patacchini, R., Rovero, R., and Giachetti, A. 1993. Tachykinin receptors and tachykinin receptor antagonists. *J Autonom Pharmacol* 13:23–93.

Mansour, A., Fox, C. A., Burke, S., et al. 1995. Immunohistochemical localization of the cloned μ opioid receptor in the rat CNS. *J Chem Neuroanat* 8: 283–305.

Mantyh, P. W., Rogers, S. D., Honore, P., et al. 1997. Inhibition of hyperalgesia by ablation of lamina I spinal neurons expressing the substance P receptors. *Science* 278: 275–279.

Martin-Schild, S., Gerall, A. A., Kastin, A. J., et al. 1999. Differential distribution of endomorphin 1- and endomorphin 2-like immunoreactivities in the CNS of the rodent. *J Comp Neurol* 405: 450–471.

Matsus, R., Fulcher, I. S., Kenny, A. J., et al. 1983. Substance P and [Leu] enkephalin are hydrolyzed by an enzyme in pig caudate synaptic membranes that is identical with the endopeptidase of kidney microvilli. *Proc Natl Acad Sci U S A* 80: 3111–3115.

Matsus, R., Kenny, A. J., and Turner, A. J. 1984. The metabolism of neuropeptides. The hydrolysis of peptides, including enkephalins, tachykinins and their analogues, by endopeptidase-24.11. *Biochem J* 223: 433–440.

Mentlein, M., Dahms, P., Grandt, D., et al. 1993. Proteolytic processing of neuropeptide Y and peptide YY by dipeptidyl peptidase IV. *Regul Pept* 49: 133–144.

Mentlein, R. 1999. Dipeptidyl peptidase IV (CD26)-role in the inactivation of regulatory peptides. *Regul Pept* 85: 9–24.

Mizoguchi, H., Narita, M., Oji, D. E., et al. 1999. The μ-opioid receptor gene-dose dependent reductions in G-protein activation in the pons/medulla and antinociception induced by endomorphins in m-opioid receptor knockout mice. *Neuroscience* 94: 203–207.

Mousseau, D. D. and Larson, A. A. 1992. Identification of a novel receptor mediating substance P-induced behavior in the mouse. *Eur J Pharmacol* 271: 197–201.

Nausch, I., Mentlein, R., and Heymann, B. 1990. The degradation of bioactive peptides and proteins by dipeptidyl peptidase IV from human placenta. *Biol Chem Hoppe-Seyler* 371: 1113–1118.

O'Connor, B. and O'Connor, G. 1986. Post-proline dipeptidyl-aminopeptidase from synaptosomal membranes of guinea-pig brain. *Eur J Biochem* 154: 329–335.

Peter, A., Toth, G., and Tomboly, G. 1999. Liquid chromatographic study of the enzymatic degradation of endomorphins, with identification by electrospray ionization mass spectrometry. *J Chromatogr* 846: 39–48.

Pierce, T. L. and Wessendorf, M. W. 2000. Immunocytochemical mapping of endomorphin-2-immunoreactivity in rat brain. *J Chem Neuroanat* 18: 181–207.

Pollard, H., Bouthenet, M. L., Moreau, J. et al. 1989. Detailed immunoautoradiographic mapping of enkephalinase (EC3.4.24.11) in rat central nervous system: Comparison with enkephalins and substance P. *Neuroscience* 30: 339–376.

Radhakrishnan, V. and Henry, J. L. 1991. Novel substance P antagonist, CP-96,345, blocks responses of cat spinal dorsal horn neurons to noxious cutaneous stimulation and to substance P. *Neurosci Lett* 132: 39–43.

Ronai, A. Z., Timar, J., Mako, E., et al. 1999. Diprotin A, an inhibitor of dipeptidyl aminopeptidase IV (EC 3.4.14.5) produces naloxone-reversible analgesia in rats. *Life Sci* 64: 145–152.

Roques, B. P., Noble, F., and Dauge, V. 1993. Neutral endopeptidase 24.11: Structure, inhibition, and experimental and clinical pharmacology. *Pharmacol Rev* 45: 87–146.

Schreff, M., Schulz, S., and Wilborny, D. 1998. Immunofluorescent identification of endomorphin-2-containing nerve fibers and terminals in the rat brain and spinal cord. *NeuroReport* 9: 1031–1034.

Sakurada, C., Sakurada, S., Hayashi, T., et al. 2003. Degradation of endomorphin-2 at the supraspinal level in mice is initiated by dipeptidyl peptidase IV: An *in vitro* and *in vivo* study. *Biochem Pharmacol* 66: 653–661.

Sakurada, C., Watanabe, C., and Sakurada, T. 2004. Occurrence of substance P(1-7) in the metabolism of substance P and its antinociceptive activity at the mouse spinal cord level. *Methods Find Exp Clin Pharmacol* 26: 171–176.

Sakurada, S., Hayashi, T., Yuhki, M., et al. 2002. Differential antagonism of endomorphin-1 and endomorphin-2 supraspinal antinociception by naloxonazine and 3-methylnaltrexone. *Peptides* 23: 895–901.

Sakurada, S., Zadina, J. E., Kastin, A. J. et al. 1999. Differential involvement of μ opioid receptor subtypes in endomorphin-1- and -2-induced antinociception. *Eur J Pharmacol* 372: 25–30.

Sakurada, T., Komatsu, T., Kuwahara, H., et al. 2007. Intrathecal substance P(1-7) prevents morphine-evoked spontaneous pain behavior via spinal NMDA-NO cascade. *Biochem Pharmacol* 74: 758–767.

Sakurada, T., Kuwahara, H., Takahashi, K., et al. 1988. Substance P(1-7) antagonizes substance P-induced aversive behavior in mice. *Neurosci Lett* 95: 281–285.

Sakurada, T., Le Greves, P., and Terenius, L. 1985. Measurement of substance P metabolites in rat CNS. *J Neurochem* 44: 718–722.

Sakurada, T., Tan-No, K., Yamada, T., et al. 1990. Phosphoramidon potentiates mammalian tachykinin induced biting, licking and scratching behaviour in mice. *Pharmacol Biochem Behav* 37: 779–783.

Schaible, H. G., Jarrott, B., Hope, P. J., et al. 1990. Release of immunoreactive substance P in the spinal cord during development of acute arthritis in the knee joint of the cat: A study with antibody microprobes. *Brain Res* 529: 214–223.

Schwartz, J. C., Malfroy, B., and De La Baume, S. 1981. Biological inactivation of enkephalins and the role of enkephalin-dipeptidyl-carboxypeptidase (enkephalinase) as neuropeptidase. *Life Sci* 29: 1715–1740.

Shane, R., Wilk, S., and Bodnar, R. J. 1999. Modulation of endomorphin-2-induced analgesia by dipeptidyl peptidase IV. *Brain Res* 815: 278–286.

Skidgel, R. A., Engelbrecht, S., Johnson, A. R., et al. 1984. Hydrolysis of substance P and neurotensin by converting enzyme and neutral endopeptidase. *Peptides* 5: 769–776.

Stone, L. S., Fairbanks, C. A., Laughlin, T. M., et al. 1997. Spinal analgesic actions of the new endogenous opioid peptides endomorphin-1 and -2. *NeuroReport* 8: 3131–3135.

Suzuki, R., Morcuende, S., Webber, M., et al. 2002. Superficial NK$_1$-expressing neurons control spinal excitability through activation of descending pathways. *Nat Neurosci* 5: 1319–1326.

Tomaz, C., Silva, A. C., and Nogueira, P. J. 1997. Long-lasting mnemotropic effect of substance P and its N-terminal fragment (SP1-7) on avoidance learning. *Braz J Med Biol Res* 30: 231–233.

Vera-Portocarrero, L. P., Zhang, E. T, King, T., et al. 2007. Spinal NK-1 receptor expressing neurons mediate opioid-induced hyperalgesia and antinociceptive tolerance via activation of descending pathways. *Pain* 129: 35–45.

Wiktelius, D., Khalil, Z., and Nyberg, F. 2006. Modulation of peripheral inflammation by the substance P, N-terminal metabolite substance P 1-7. *Peptides* 27: 1490–1497.

Yokosawa, H., Endo, S., Ogura, Y., et al. 1983. A new feature of angiotensin converting enzyme: Hydrolysis of substance P. *Biochem Biophys Res Commun* 116: 735–742.

Zadina, J. E., Hackler, L., Ge, L. J., et al. 1997. A potent and selective endogenous agonist for the μ-opiate receptor. *Nature* 386: 499–502.

Zhou, Q., Frandberg, P. A., Kindlundh, A. M., et al. 2003. Substance P (1-7) affects the expression of dopamine D2 receptor mRNA in male rat brain during morphine withdrawal. *Peptides* 24: 147–153.

4 Neuropeptides, Isolation, Identification, and Quantitation

Jolanta Kotlinska and Jerzy Silberring

CONTENTS

4.1 INTRODUCTION

Neuropeptides are considered as the largest class of neuromessengers in the central nervous system (CNS) and are classically defined by their synthesis in the neuronal cell body, storage, and transportation in dense-core vesicles and release by high-frequency neuronal activity. Neuropeptides show a more protracted transmission than classic neurotransmitters, are highly plastic and responsive to circumstance, modulate classic neurotransmitter systems, and exert trophic effects (Hökfelt et al. 2000). Neuropeptides are ubiquitously distributed and most often associated with one or more classic neurotransmitters, sustaining the notion of their modulatory functions. They control our mood, energy levels, pain and pleasure reception, body weight, and ability to solve problems; they also form memories and regulate our immune system. Together, all these properties render neuropeptides as a suitable mechanism for mediating adaptation and responding to homeostatic challenges (Alldredge 2010).

Neuropeptides act via specific G-protein-coupled receptors. Thus, their action is slower in nature (seconds to minutes) than many "classical" neurotransmitters, due to the longer time of the second messenger system activation, such as cAMP, cGMP, or IP3. It is now well established that most of the physiologically active neuropeptides are derived from large inactive precursor proteins (propeptides). For example, the opioid peptides are first synthesized as part of larger inactive prohormone precursors and are subsequently cleaved at basic residues to release biologically active peptides. However, sometimes, unidentified peptides are present in sequence of the precursor molecules. Such peptides are called cryptic peptides.

One of the pioneers who significantly contributed to the research on neuropeptides was Victor Mutt (1923–1998) (Gozes 1999). His work was focused on gut/brain peptides, leading to a discovery of more than 50 naturally occurring biologically active sequences, including secretin, cholecystokinin, vasoactive intestinal polypeptide, neuropeptide Y (NPY), peptide tyrosine tyrosine, intestinal peptide ZF-1, and so on (Jörnvall et al. 2008).

Discovery of opioid receptors and, later, their endogenous ligands, endorphins, in 1975, boosted the research to find novel mechanisms of pain transmission, drug dependence, and memory formation. Recent discoveries, including imaging techniques and mass spectrometry, add even more evidence on the multiple roles of neuropeptides, depending on the balance between various fragments, intracellular localization, and local concentration. It is, therefore, not a coincidence that the market for peptidergic drugs is constantly increasing, going far beyond the commonly known classical example—insulin.

Thorough identification of molecules, among them neuropeptides, includes a complete amino acid sequence and all modifications. Mass spectrometry has revolutionized our definition of endogenous neuropeptides from "peptide-like immunoreactivity" to a true peptide sequence (Renlund et al. 1993; Näslund et al. 1994). Such a change of view led to the clarification of what was previously measured, thus eliminating false-positive data by initiating a broader outlook on many endogenous pathways.

This chapter describes, in brief, strategies and methodologies for discovery and characterization of neuropeptides, including their pharmacological aspects.

4.2   MULTIPLE FUNCTIONS OF NEUROPEPTIDES AND ENZYMATIC CONVERSION

Neuropeptide precursors are inactive and contain multiple copies of various biologically active sequences. These sequences are usually flanked by a pair of basic amino acid residues (combinations of Arg, Lys); however, single residues indicating cleavage sites are also present as in the case of prodynorphin (for review, see Mansour et al. 1995). Trypsin-like cleavage and excision of a sequence is followed by further truncation of the C-terminal Arg or Lys, and in some cases, by amidation, if C-terminal Gly is present. According to a commonly accepted theory, N-terminal fragment of a given peptide is responsible for its specificity toward receptor (address), and the remaining sequence determines the function involved in activation (message domain) and potency enhancing (address domain). An example might be nociceptin recognition by its OP4 receptor (Guerrini et al. 1997; Lapalu et al. 1997).

As stated earlier, neuropeptides undergo various posttranslational events, including proteolytic cleavage, C-terminal amidation, and so on. Thus, it can be stated that the level and bioactivity of a given sequence depends on a number of processes, including conversions and modifications, leading to the release of mature neuropeptides. Moreover, conversion of one bioactive sequence may lead to the release of another, shorter peptide, possessing distinct activity. A number of such findings were observed, among them dynorphins, substance P, nociceptin/orphanin FQ, and cocaine- and amphetamine-regulated transcript (CART). Such enzymatic truncation leads to a clear change of bioactivity, often toward distinct receptors (e.g., dynorphins and enkephalins), or influences other pathways in an opposite way to the native peptide. Another example might be nociceptin/orphanin FQ and nocistatin. Both sequences derive from the same precursor but have opposite functions: nociceptin possesses algesic properties, whereas nocistatin possesses analgetic properties. Thus, one precursor may serve as a "self-regulator" of various senses, depending on the balance between appropriate fragments.

4.3   ENZYME INHIBITORS

Proteolytic enzymes are regulated in many ways. These include their biosynthesis as inactive zymogens, further processed to an active enzyme at the destination site, pH dependence (e.g., lyzosomal enzymes), and endogenous inhibitors, to name a few possibilities. The latter is of particular importance, as synthetic analogs may serve as therapeutic agents. For instance, RB-101 is a mixed inhibitor of enkephalin-degrading enzymes, which causes a decrease in spontaneous morphine abstinence and possesses analgesic properties by inhibition of enkephalins degradation (Roques et al. 1996; Ruiz et al. 1996). In a series of studies, Tan-No and coworkers (1996, 2009, 2010) have described the application of various inhibitors of serine and cysteine peptidases to prolong the action of neuropeptides.

Endogenous inhibitors may indicate the complete metabolic/catabolic pathways leading to the release, maturation, and degradation of neuropeptides. Such molecules may also serve as a lead for the design of novel therapeutics. Recently, Suder et al. (2006) discovered an endogenous inhibitor of dynorphin-converting enzyme (DCE) in cerebrospinal fluid (CSF). It appeared to be bikunin, which coexists in CSF with the enzyme. It is interesting to note that even a fraction of DCE activity, in a free, active form, is able to process dynorphins (Suder et al. 2006), despite the presence of its inhibitor in the same fluid.

Another discovery includes identification of an inhibitor of prohormone convertase-1. Two micromoles proSAAS selectively inhibited PC1 but not furin, PACE4, PC5A, or PC7 (Qian et al. 2000).

4.4 CRYPTIC PEPTIDES—FROM HYPOTHETICAL SEQUENCE TO BIOLOGICAL ACTIVITY

For many years, there was a consensus that the most commonly known neuropeptide precursors contain an already known, finite number of bioactive sequences. Recent discoveries and sequencing of complete genomes shed new light on many other potential excision sites (i.e., Arg, Lys flanking) and, thus, new, "hidden" peptides were described. Such sequences are referred to as cryptic peptides or crypteins (Autelitano et al. 2006; Ng and Ilag 2006; Pimenta and Lebrun 2007).

The simplest approach to find such cryptein is to search the full sequence of neuropeptide precursors in the flanking basic amino acid residues. The scheme in Figure 4.1 shows such an approach.

Recently, we demonstrated that the rat neuropeptide FF_A ($NPFF_A$) precursor contains another bioactive sequence, NAWGPWSKEQLSPQA, referred to as NPNA (Neuropeptide NA) spanning between positions 85 and 99 (Figure 4.1). Synthetic NPFF precursor (85–99) (10 and 20 nmol, intracerebroventricularly—i.c.v.) blocked the expression of conditioned place preference induced by morphine (5 mg/kg, s.c.).

This peptide alone (10 and 20 nmol, i.c.v.) had no influence on the baseline latency of a nociceptive reaction but reversed the antinociceptive activity of morphine (5 mg/kg, s.c.) in the tail-immersion test in rats (Dylag et al. 2008). Synthetic peptide also influenced the expression of mRNA coding for $G\alpha$(i1), (i2), and (i3) subunits (Suder et al. 2008).

The discovered sequence is not C-terminally amidated in contrast to other NPFF-released sequences; neither does it bind to NPFF receptors (Dr. Jean-Marie Zajac 2010, personal communication). Despite this observation, it still possesses antiopioid activity, perhaps by acting on the aforementioned G subunits.

The most important step is validation of the obtained sequence, despite the fact that it shows biological activity in pharmacological tests. Affinity (or other) purification/extraction need to be performed to verify whether the endogenous counterpart exists. This might be the most difficult part of the research, as there are many combinations of elongated sequences that might not be recognized by antibodies with similar affinity. Further (or parallel) validation should also include

Human	MDSRQAAALLVLLLLIDG–GCAEGPGGQQE–DQLSAEEDSEPLPPQDA------QTSGSL	52
Bovine	MDARQAAALLLVLLLVTDWSHAEGPGGRDGGDQIFMEEDSGAHPAQDA------QTPRSL	54
Rat	MDSK–WAAVLLLLLLLLRNWGHAEEAGSWGE–DQVFAEEDKGPHPSQYAHTPDRIQTPGSL	58
Mouse	MDSK–WAALLLLLLLLLLNWGHTEEAGSWGE–DQVFAGEDKGPHPPQYAHIPDRIQTPGSL	58

	NPFF		NPAF	NPSF	
Human	LHYLLQAMERPGRSQA·FLFQPQR·FGRNTQGSWRNEWLSPR·AGEGLNSQFW·SLAAPQRF·GKK	112
Bovine	LRSLLQAMQRPGRSPA·FLFQPQR·FGRNTRGSWSNKRLSPR·AGEGLSSPFW·SLAAPQRF·GKK	114
Rat	MRVLLQAMERPRRNPA·FLFQPQR·FGRNAWGPWSKEQLSPQARE-----FW·SLAAPQRF·GKK	113
Mouse	FRVLLQAMDTPRRSPA·FLFQPQR·FGRSAWGSWSKEQLNPQARQ-----FW·SLAAPQRF·GKK	113

FIGURE 4.1 Search for cryptic peptides. As an example, NPNA derived in the $NPFF_A$ precursor is shown. The cryptic peptide NPNA is underlined. (From Dylag, T., et al., *Peptide*, 29(3), 473–8, 2008. With permission.)

specificity to the specific receptor or clear influence of other pathways. An elegant example of such studies includes discovery and function of corticostatin (de Lecea et al. 1996). The authors found new, complementary DNA clone in the brain, the sequence of which suggested the existence of mRNA encoding preprocortistatin. The active peptide has been synthesized and compared with biological potency of somatostatin.

There are also other approaches, which lead to the discovery of novel sequences, utilizing the knowledge on neuropeptide conversion. Experiments using cell cultures, brain homogenates, or specific brain structures show that neuropeptides are cleaved at various sites, not necessarily at basic residues. Such observation and time-dependent studies can reveal natural, consecutive proteolytic pathways that can further be studied using synthetic analogs, molecular biology, pharmacology, and biochemistry.

An aforementioned case, which concerns the discovery of nociceptin/orphanin FQ, is a classical example of the so-called reverse pharmacology (Meunier et al. 1995; Reinscheid et al. 1995). This strategy uses receptors as a "bait" to find new ligands. It is cost-effective and is applied in the search for novel drugs (Kotarsky and Nilsson 2004).

4.5 STRATEGIES FOR ISOLATION AND PURIFICATION FOR MASS SPECTROMETRY ANALYSIS (INCLUDING AFFINITY CHROMATOGRAPHY)

It is trivial, though important, to state that purification and identification of endogenous neuropeptides is a challenging task. This is because of their low (picomolar to attomolar) levels in the brain, rapid degradation during extraction procedures, and the handling of complex biological mixtures, often containing high amounts of lipids (nervous tissue). As the final result and eventual success strongly depends on many phases, including tissue isolation, extraction, and prevention of proteolysis, these key steps are discussed in this chapter.

4.5.1 TISSUE SELECTION

It is always beneficial to search for the tissue or structure with high content of the given peptide. This might be difficult, as mRNA expression is not necessarily related to the peptide content, unless at least the peptide-like immunoreactivity and distribution is known. A convenient model includes cell cultures that can be multiplied to the sufficient amount of peptide, are less complex, and can be handled under strictly controlled conditions. It is also important that biological material is not contaminated by blood or surrounding tissues. There are several modern and precise techniques, such as laser tweezers and laser capture microdissection, to avoid at least some problems associated with tissue isolation. A good indicator of improper collection of material is detection of hemoglobin, indicating that the sample is contaminated by blood proteins.

4.5.2 Procedures for Peptide Extraction

This step should be performed as fast as possible. Otherwise, peptides undergo rapid degradation. To prevent their loss, the tissue is frequently homogenized in hot 1 M acetic acid, followed by boiling at 95°C for 10 minutes. Addition of inhibitory cocktail might be beneficial in some cases. The most efficient method involves application of microwave irradiation before the tissue is isolated. This procedure inactivates proteolytic enzymes before isolation and preserves higher amounts of neuropeptides for further studies (Nylander et al. 1997).

4.5.3 Isolation of Peptides from Complex Biological Mixtures

There is no single consensus strategy with regard to isolating endogenous molecules. Size exclusion chromatography or ultrafiltration on the cutoff centrifugal devices might be a good start. Another efficient approach is application of strong cation exchanger and volatile eluents (Nyberg and Terenius 1985).

This volatility is crucial for further purification steps, as it does not add any salts to the sample, which may interfere with consecutive purification steps or analysis by means of chromatographic procedures, identification by mass spectrometry, or even quantitation by radioimmunoassay and enzyme-linked immunosorbent assay (ELISA), which are also sensitive to certain buffers and enzyme inhibitors, giving false-positive results. Each purification step must be monitored for the substance of interest and such an assay should be highly specific.

The best and most selective isolation procedure, not only limited to neuropeptide isolation, is affinity chromatography. While searching for a known or predicted sequence, as in the case of cryptic peptides, it is relatively easy to prepare antibodies toward such fragments. Antibodies can then be coupled with the insoluble support, forming a column for purification, or mixed directly with the sample, followed by separation of the complex on a microcentrifugal concentrating device with a low cutoff limit (e.g., 10 kDa). After separation and concentration of the antibody-bound peptide (high molecular mass above 100 kDa) from the sample, the complex undergoes dissociation under mild acidic conditions. The released peptide is centrifuged through the membrane and collected for further studies. This elegant approach has been developed by Przybylski and coworkers for identification of epitopes (Suckau et al. 1990).

Affinity purification has also been successfully applied for purification of protein complexes, followed by their analysis by mass spectrometry. Peptides were used as affinity ligands in a "reverse" strategy, where isolation of the large protein complex from the postsynaptic density fraction of brain synapses were described (Husi et al. 2000). A very good overview on such methods was presented by Bauer and Kuster (2003).

More recent approaches also involve application of activated magnetic beads, where specific antibodies or other molecules can be attached (Mirre et al. 2005; Villanueva et al. 2006).

Antibodies against a given sequence will also detect all elongated (and sometimes truncated) forms of neuropeptide, provided the epitope is not hidden.

This is beneficial because discovery of various fragments may immediately suggest processing pathway(s) and specificity of potential peptidases involved in these processes. Extraction protocol itself may produce artifacts, leading to similar, but false, conclusions.

4.5.4 PEPTIDE IDENTIFICATION

Immunological methods (RIA, ELISA) are still the commonly applied techniques for identification and quantitation of endogenous material. It should be stressed here that such methods may only detect peptide-like immunoreactive material and not a given, specific sequence, including its complete characterization. Often quantitation is performed directly, without even a single purification step. This may lead to serious problems where the neuropeptide is not present in the sample, though it is still "detected" by antibodies (Silberring et al. 1998). On the other hand, immunological methods are easy to perform, rapid, and can analyze hundreds of samples in one assay. Therefore, they are ideal for quantitation under circumstances where the selectivity of the assay is well defined and validated.

The ultimate technique to fully identify a molecule is mass spectrometry, often in combination with other separation methods (liquid chromatography, gas chromatography, and capillary electrophoresis), supplemented by NMR, Raman, X-ray crystallography, and so on. These topics are beyond the scope of this chapter, therefore, we will only focus on mass spectrometry and an interested reader may refer to other chapters in this book.

The most important feature of mass spectrometry is its unambiguity in determination of sequence/structure of molecules. Often, samples need not be purified to absolute homogeneity. In addition, high sensitivity at femtomolar (or lower) level makes mass spectrometry the method of choice for the analysis of neuropeptides. A complete identification of neuropeptide (similarly to other molecules) includes its sequencing, posttranslational modifications, and all other changes within primary structure (e.g., mutations, modifications of the *N*-, *C*-terminuses).

The typical protocol for identification of a molecule by mass spectrometry is shown in Figure 4.2. It should be taken into account that this scheme picturizes only important steps that may vary significantly while handling different molecules. Specific protocol depends on type of sample (body fluid, tissue), nature of a molecule (size, shape, hydrophobicity, and modifications), previous experience, and laboratory instrumentation.

There are also experiments based on release of neuropeptides upon stimulation of "living" tissues. Such material differs from that present in tissues or body fluids. Usually, it is a diluted sample and contains a mixture of salts and, sometimes, pH indicators, as well as fetal calf serum (cell cultures). Despite the fact that fetal calf serum in commercial products is heat inactivated, it still has strong potential to degrade peptides. Therefore, it is advisable to use serum-free media. Moreover, pH indicators are known to irreversibly adsorb to the HPLC columns, destroying them permanently.

Mass spectrometry can also be used as a molecular scanner. Such imaging techniques were developed (Amstalden et al. 2010; Schwartz and Caprioli 2010),

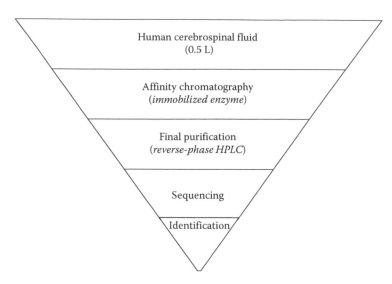

FIGURE 4.2 An example of a protocol peptide/protein identification by mass spectrometry. The endogenous inhibitor of dynorphin-converting enzyme present in human cerebrospinal fluid has been purified by affinity chromatography on the anhydrotrypsin column. (From Suder, P., et al., *FEBS J.*, 273(22), 5113–20, 2006. With permission.)

allowing identification of spatial distribution of various molecules simultaneously. Many techniques are presently available, including MALDI TOF, TOF-SIMS, DESI, and so on. These methods are also structure specific, which means they provide complete identification of given molecules, including amino acid sequence, modifications, and in most cases at least semiquantitation.

4.6 PEPTIDE ARRAYS, COMBINATORIAL LIBRARIES, AND COMPLEMENTARY SEQUENCES

Peptides and peptide analogs, such as substrates, inhibitors, and ligands, can be synthesized on the surfaces in the form of various arrays. This allows for rapid and simultaneous identification of enzyme activities, screening for selective proteinase inhibitors, selection of specific receptor ligands, and so on. Such surfaces are compatible with various mass spectrometry techniques for detection of cleavage sites, noncovalent complexes, and so on. The mass spectrometry (MS) methods involve MALDI or DESI to scan surfaces to identify fragments, including sequencing (MS^n). A more advanced, but also demanding more experience, is the application of various combinatorial libraries. Such libraries comprise many components (up to several millions), synthesized together, and, thus, saves time. Mass spectrometry is an ideal tool to select important sequences in such mixtures (Gurard-Levin et al. 2009). Combinatorial peptide ligand libraries (ProteoMiner application) are also utilized in an alternative process to capture proteomic samples, where they act as an equalizer, allowing for extraction and identification of diluted compounds, besides albumin and other highly abundant molecules (Righetti et al. 2010).

4.7 PHARMACOLOGICAL APPROACHES IN TESTING NEUROPEPTIDES AND THEIR ANALOGS

One of the major goals of proteomic research is to discover and ultimately validate new protein biomarkers of disease. Therefore, various "omics" techniques are utilized in diagnostic studies. Till now, extensive protein identifications have been performed in the framework of numerous proteome projects to study human pathogens (e.g., comparative proteome analysis of *Helicobacter pylori*) (Park et al. 2006) to identify disease-associated proteins (e.g., cancer or heart proteomics) (Jungblut et al. 1999) and to analyze proteins involved in signal transduction cascade (Soskic et al. 1999). However, the overall goal of peptidomic studies is to identify the biological activity of each peptide, and this knowledge could be used for drug discovery. As peptides are involved in many physiological processes, practically any disorder, from addiction to anxiety, would be a potential area that could benefit from drugs based on neuropeptide receptors. Therefore, proteomics and peptidomics need pharmacological experiments to extend knowledge about the role of peptides in physiological and pathological processes.

Pharmacological experiments (characterization) of a new peptide can be categorized into two types of studies: *in vivo* and *in vitro*. Animal testing (preclinical) and clinical trials are two forms of *in vivo* research. First, experiments are performed on animals (mice, rats, guinea pigs, rabbits, monkeys, and others) to ensure that the new compound/peptide is safe in animals and to determine the effectiveness and safety of such a drug/peptide in humans. Many examples showed that compounds effective for animals were noneffective when applied to humans, thus special care should be taken to eliminate such possibilities. Several clinical phases were devoted to test the biological potency of the introduced drugs, including possible side effects and toxicity. Furthermore, experiments on animals facilitate learning about the mechanism of action of drugs. All experiments on higher organisms (animals, humans) require an agreement of the local ethics committee. In the past, genetically modified animals were introduced to *in vivo* experiments. Such modified animals are used in pharmacology for identification of drug/peptide targets, distinction of target isoforms, differentiation of signaling pathways, and generation of disease model (Rudolph and Möhler 1999). Generally, such experiments may determine the function of a drug/protein in living organisms. However, some authors suggest that receptor deletion in the knockout animals may produce adaptive changes in the CNS during brain development that may be responsible for the lack of differences in the pharmacological response between knockout and wild-type animals (see Carroll 2008). Such compensatory mechanisms are evident even in the spectacular myoglobin knockout mice model (Gödecke et al. 1999).

In vivo testing is often preferred over the *in vitro* approaches (e.g., experiments on cultured nerve cells, isolated blood vessels, isolated smooth muscles, and receptor binding in brain homogenate) because the model is significantly simplified and is better suited for observation of the overall effects of a drug/peptide on a living subject. Although *in vitro* cell culture models help to discover the cellular mechanism of drug/peptide action and its permeability, metabolism, and toxicity allows to reduce animal usage in pharmacological experiments, it is important to verify such results as simple model does not necessarily reflect real changes in living organisms.

More recently, modern molecular biology brought about a strong impact on the study of the functional significance of neuropeptides. The histochemical approach has benefited, in particular, from various *in situ* hybridization procedures (Young 1990), including cloning neuropeptide precursors and peptide receptors. The discovery of numerous receptors has been crucial for assigning functional significance to neuropeptides, as was the discovery of peptide antagonists, which pass the blood–brain barrier. An alternative approach is the use of antisense probes, which are complementary to a sequence of mRNA and can be used to attenuate translation process (see Wahlestedt 1994). Newer and a more effective technique includes the siRNA method (Higuchi, Kawakami, and Hashida 2010).

The explosion of the research activity in the field of neuropeptides has led to the identification of numerous naturally occurring endogenous peptides that act as neurotransmitters, neuromodulators, or trophic factors to mediate nervous system functions (Levine 2007). For experimental or diagnostic research, neuropeptides are isolated from different biological samples (e.g., CSF, spinal cord, and different structures of the brain), purified, sequenced, chemically synthesized (including various analogs), and then used for biological studies. There are two approaches to pharmacological investigation of peptides/neuropeptides. The first approach is based on evaluation of biological function(s) of the newly discovered peptide or its metabolites. In this type of tests, classification of the peptide to one of the neuropeptide families or receptor-binding studies of peptide is important to reveal its biological function. Of course, the classification may suggest development of the appropriate behavioral tests for the new sequence to confirm (or not) its functional connection with the neuropeptide family. In behavioral procedure, selective antagonists toward receptors may be applied to verify the receptor engagement in the behavioral effect of a new peptide.

To identify biological activity of cryptic peptides, the newly discovered sequence is synthesized and validated in animals and other (e.g., biochemical) experiments. In such situations, scientists try to detect functional homology of new cryptic peptides with the parent peptide. Therefore, behavioral tests characteristic for the peptide family (often using reference compound, which is a known member of the family) are applied. Blockade of effects of the new peptide by family receptor antagonist confirms that it belongs to this peptide family. The cryptic peptides were identified in $NPFF_A$ precursor in mice and rats (Dylag et al. 2008; Suder et al. 2008).

A functional role for peptides may also be verified by determining the physiological states that regulate the levels of each peptide and then using this knowledge to predict the function of the peptide. Although one would still need to directly test the peptide(s) for the anticipated function, this approach would reduce the number of tests. For example, peptides present in hypothalamus that are regulated by food deprivation are potentially involved in the control of body weight/energy balance. Similarly, peptides upregulated or downregulated by the chronic administration of drugs of abuse may be important mediators of the reward mechanisms and may contribute to neurochemical changes that lead to addiction. Although this approach is not free from false positives, it can at least reduce the number of candidate functions for a particular peptide. For instance, the hypocretin/orexin was found in one study to be upregulated by food deprivation, although subsequent studies suggested that this peptide was involved with arousal states and not food intake. However, it makes sense as an animal would spend more time awake and forage for food during food deprivation; so the regulation of hypocretin/orexin by food deprivation is consistent with functions in both arousal and feeding (Fricker 2005).

The second approach to pharmacological studies on peptides/neuropeptides is synthesis of new peptides designed on the base of natural components (e.g., nondegradable analogs of natural peptides or PEG-ylated sequences). Structure–function investigations of the newly synthesized analogs of peptides/neuropeptides require application of a reference compound (original, endogenous peptide) and its antagonist for pharmacological studies, to compare the obtained effects. Therefore, protein/peptide identification in biological samples is an important task in drug discovery research, which has been described earlier in this chapter.

Most protein/peptide drugs exhibit short biological half-lives (because of their instability) and poor bioavailability through the blood–brain barrier (Pardridge 1999; Bickel et al. 2001). Although several approaches have been described for peptide drug delivery to the brain, such as local invasive delivery by direct injection or infusion, induction of enhanced permeability (e.g., PEG-ylation), and various physiological-targeting strategies (Begley 2004; Gaillard et al. 2005; Pardridge 2003), peptides are usually given intracerebrally (i.c.v. or into specific brain structure) or intrathecally in animal studies. Injection of peptides to brain structure requires the implantation of a guide cannula. Surgery, under anesthesia, can be performed using a map of the brain called stereotaxic atlas (Paxinos and Watson 2005) and the stereotaxic instrument. Stereotaxic atlas provides detailed coordinates for brain structures. This procedure may be performed in animals by operators skilled in such technique. In mice, peptide injection to one of the lateral ventricles may be performed as a "free hand," according to the procedure of Haley and McCormick (1957). In any case, the place of introduction of a needle should be verified postmortem, and all injections that do not fall into the predicted brain area should be rejected from final data.

4.7.1 Animal Behavioral Models in Peptide Studies

Peptides possess a wide range of physiological and pharmacological effects, and they are subjected to the same pharmacological tests as synthetic compounds. Behavioral screening tests are used in pharmacology for a rapid estimation of safety (assessment of toxicological threshold), to create a profile of acute effects of the compounds, and to investigate the dose–response effects. Screening tests are also applied when a drug's/peptide's pharmacokinetics are not known, and further observations are required over a time to determine whether an organism's response to the drug/peptide changes during chronic exposure can lead to physical dependence. Screening procedure comprises qualitative and quantitative measures of behavior, such as abnormal body posture, piloerection, body weight, body temperature, ataxia, aerial righting response, open-field activity, tremor, salivation, touch response, known as the functional observational battery (Moser 2000), and locomotor activity.

In behavioral experiments, more awareness is taken during preparation of solutions and other experimental details, whereas considerably less attention is given to problems concerning the acclimation of animals to the experimental testing conditions, handling, and so on, despite the fact that these variables can also be of critical significance. The decision about the selection of behavioral tests for investigation of a new peptide is easier when the structure of the peptide or the receptor-binding data is known. Many bioactive peptides derived from the same precursor protein share common behavioral effects. For example, opioid peptides derived from proenkephalin A indicate that antinociceptive effect (Akil et al. 1984) or two C-terminal CART-derived peptides, similar to the CART peptide, are involved in the regulation of ingestive behaviors (Bannon et al. 2001). However, a modification of the peptide sequence may partially change its behavioral efficacy or develop new properties (Bannon et al. 2001; Condamine et al. 2010).

A variety of behavioral tests have been developed to evaluate (detect) *in vivo* effects of peptide drugs using experimental animals. For example, anxiolytic-like (or anxiogenic) effects of the peptide may be estimated in the elevated plus-maze test (Rizzi et al. 2008), light-dark test (Vergura et al. 2008), or the open-field test (Sørensen et al. 2004) in mice and rats. These screening experiments are based on the natural aversion of animals to open spaces or brightly illuminated areas. This aversion generates an anxiety-like behavior. The extent to which the aversion is overcome by peptide injection, without any change in locomotor activity, is a measure of validation of the anxiety-like effect of peptide (an increase of time spent in open or bright area). In turn, experimental evaluation of pain behavior includes tests that measure pain threshold to high-intensity stimuli (acute pain tests) and changes in spontaneous or evoked behavioral responses in animals with peripheral injury or inflammation (persistent pain models). Acute thermal pain is modeled by the hot-plate and tail-flick tests, whereas persistent pain can be induced by the formalin test (Bannon et al. 2007). Analogs of deltorphins (Kotlinska et al. 2009) and other opioid peptides (Bodnar 2009) are effective in suppression of the acute pain evoked by noxious thermal stimuli (heat pain), although, recently, the role of opioid peptides in peripheral inflammatory and neuropathic pain has been described (Obara et al. 2009; Busch-Dienstfertig and Stein 2010).

Furthermore, antiepileptic effects of peptides are evaluated in the pentylenetetrazole test (Loacker et al. 2007) and antidepressant effects are screened in the forced swimming test (also called the Porsolt test) (Torregrossa et al. 2006; Berrocoso et al. 2009; Engin and Treit 2009).

4.7.2 VALIDATION OF ANIMAL MODELS USED IN STUDIES OF PEPTIDES

The number of methods employed to evaluate various behavioral processes and to examine the behavioral effects of drugs/peptides has increased dramatically in recent years. Animal models, except for behavioral tasks, were developed and validated to study human disorders, such as anxiety, schizophrenia, depression, and epilepsy. Generally, two criteria appear to be necessary and sufficient for validation of an animal model: reliability and predictive validity. Currently, the criteria for identifying animal models of depression rely on either of the two principles: actions of known antidepressants and responses to stress. The most widely used animal models of depression are learned helplessness, chronic mild stress, and social defeat paradigms (Yan et al. 2010). Behavioral and clinical investigations indicate an involvement of neuropeptides, such as corticotrophin-releasing factor, arginine vasopressin, galanin, substance P, neuropeptides S and Y, melanin-concentrating hormone, and oxytocin in the stress-related disorders, including anxiety and depression (Alldredge 2010).

Animal models in the field of addiction are considered to be among the best available models of neuropsychiatric diseases. These models underwent a number of refinements that allow for a deeper understanding of the circuitries involved in initiating drug seeking and relapse (Kalivas et al. 2006). Currently, models for the positive reinforcing properties of drug of abuse, such as drug self-administration, brain stimulation reward, or place preference, provided a major improvement in studies focused on the neurobiology of addiction. Many studies indicated that peptides play an important role in drug addiction (Kotlinska et al. 2002, 2008; Sakoori and Murphy 2004; Economidou et al. 2008; Ciccocioppo et al. 2009; Simmons and Self 2009), and peptides or drugs with peptide-like activity could be promising candidates in drug abuse therapy. Kindling is an experimental model for epilepsy, in which repeated stimuli (electrical brain stimulation) induce longer electrographic seizures and eventually cause behavioral convulsions. This model of chronic spontaneous limbic seizures has a number of similarities to human limbic epilepsy (Bertram 2007). It has been shown that, in this model, NPY and several other peptides are strongly upregulated, especially in hippocampal formation (Gall et al. 1990; Schwarzer et al. 1995).

4.8 FINAL CONCLUSIONS AND FUTURE PROSPECTS

Novel peptide therapeutics are increasingly making their way into clinical applications. Indeed, certain naturally derived peptides have been successful drugs for many years (e.g., oxytocin, vasopressin, insulin and gramicidins, or more recent conotoxins) (Wieland 1995). With the advent of large biological and synthetic peptide libraries and high-throughput screening, many promising candidates could soon be added to the list of peptides under development. This progress introduced new strategies for the administration of peptide drugs and improvements of the clearance half-lives *in vivo*. Despite the potential obstacles that still remain, peptide therapeutics are poised to play a significant role in the treatment of diseases, ranging from Alzheimer's disease to cancer (Lien and Lowman 2003).

The field of behavioral pharmacology is far from static. These developments, coupled with the results obtained from clinical studies, will soon result in newer techniques that provide better sensitivity and selectivity for studying behavioral effects of drugs. Therefore, with the extensive discovery of new peptides, pharmacological characterization, and clinical investigations that is underway, the future of new peptidergic therapeutics and multiplexed diagnostic tools (peptide arrays) seems very promising.

REFERENCES

Akil H, Watson SJ, Young E, Lewis ME, Khachaturian H, and Walker JM. Endogenous opioids: biology and function. *Annu Rev Neurosci.* 1984;7:223–55.

Alldredge B. Pathogenic involvement of neuropeptides in anxiety and depression. *Neuropeptides.* 2010;44(3):215–24.

Amstalden van Hove ER, Smith DF, Heeren RM. A concise review of mass spectrometry imaging. *J Chromatogr A.* 2010;1217(25):3946–54.

Autelitano DJ, Rajic A, Smith AI, Berndt MC, Ilag LL, and Vadas M. The cryptome: a subset of the proteome, comprising cryptic peptides with distinct bioactivities. *Drug Discov Today.* 2006;11(7–8):306–14.

Bannon AW and Malmberg A B. Models of nociception: hot-plate, tail-flick, and formalin tests in rodents. *Curr Protoc Neurosci.* 2007;41:8.9.1–8.9.16.

Bannon AW, Seda J, Carmouche M, Francis JM, Jarosinski MA, and Douglass J. Multiple behavioral effects of cocaine- and amphetamine-regulated transcript (CART) peptides in mice: CART 42-89 and CART 49-89 differ in potency and activity. *J Pharmacol Exp Ther.* 2001;299(3):1021–6.

Bauer A. and Kuster B. Affinity purification-mass spectrometry. *Eur J Biochem.* 2003;270: 570–8.

Begley DJ. Delivery of therapeutic agents to the central nervous system: the problems and the possibilities. *Pharmacol Ther.* 2004;104(1):29–45.

Berrocoso E, Sánchez-Blázquez P, Garzón J, and Mico JA. Opiates as antidepressants. *Curr Pharm Des.* 2009;15(14):1612–22.

Bertram E. The relevance of kindling for human epilepsy. *Epilepsia.* 2007;48(Suppl 2):65–74.

Bickel U, Yoshikawa T, and Pardridge WM. Delivery of peptides and proteins through the blood-brain barrier. *Adv Drug Deliv Rev.* 2001;46(1–3):247–79.

Bodnar RJ. Endogenous opiates and behavior: 2008. *Peptides.* 2009;30(12):2432–79.

Busch-Dienstfertig M and Stein C. Opioid receptors and opioid peptide-producing leukocytes in inflammatory pain—basic and therapeutic aspects. *Brain Behav Immun.* 2010;24(5):683–94.

Carroll FI. Antagonists at metabotropic glutamate receptor subtype 5: structure activity relationships and therapeutic potential for addiction. *Ann N Y Acad Sci.* 2008;1141:221–32.

Ciccocioppo R, Gehlert DR, Ryabinin A, Kaur S, Cippitelli A, Thorsell A, Lê AD, Hipskind PA, Hamdouchi C, Lu J, et al. Stress-related neuropeptides and alcoholism: CRH, NPY, and beyond. *Alcohol.* 2009;43(7):491–8.

Condamine E, Courchay K, Rego JC, Leprince J, Mayer C, Davoust D, Costentin J, and Vaudry H. Structural and pharmacological characteristics of chimeric peptides derived from peptide E and beta-endorphin reveal the crucial role of the C-terminal YGGFL and YKKGE motifs in their analgesic properties. *Peptides.* 2010;31(5):962–72.

de Lecea L, Criado JR, Prospero-Garcia O, Gautvik KM, Schweitzer P, Danielson PE, Dunlop CL, Siggins GR, Henriksen SJ, and Sutcliffe JG. A cortical neuropeptide with neuronal depressant and sleep-modulating properties. *Nature.* 1996;381(6579):242–5.

Dylag T, Pachuta A, Raoof H, Kotlinska J, and Silberring J. A novel cryptic peptide derived from the rat neuropeptide FF precursor reverses antinociception and conditioned place preference induced by morphine. *Peptide.* 2008;29(3):473–8.

Economidou D, Hansson AC, Weiss F, Terasmaa A, Sommer WH, Cippitelli A, Fedeli A, Martin-Fardon R, Massi M, Ciccocioppo R, Heilig M. Dysregulation of nociceptin/orphanin FQ activity in the amygdala is linked to excessive alcohol drinking in the rat. *Biol Psychiatry.* 2008;64(3):211–8.

Engin E and Treit D. Anxiolytic and antidepressant actions of somatostatin: the role of sst2 and sst3 receptors. *Psychopharmacology (Berl).* 2009;206(2):281–9.

Fricker LD. Neuropeptide-processing enzymes: applications for drug discovery. *AAPS J.* 2005;7(2):449–55.

Gaillard PJ, Visser CC, and de Boer AG. Targeted delivery across the blood-brain barrier. *Expert Opin Drug Deliv.* 2005;2(2):299–309.

Gall C, Lauterborn J, Isackson P, and White J. Seizures, neuropeptide regulation, and mRNA expression in the hippocampus. *Prog Brain Res.* 1990;83:371–90.

Gödecke A, Flögel U, Zanger K, Ding Z, Hirchenhain J, Decking UK, and Schrader J. Disruption of myoglobin in mice induces multiple compensatory mechanisms. *Proc Natl Acad Sci U S A.* 1999;96(18):10495–500.

Gozes I. In memory of Victor Mutt. *J Mol Neurosci.* 1999;11:105–8.

Guerrini R, Calo G, Rizzi A, Bianchi C, Lazarus LH, Salvadori S, Temussi PA, and Regoli D. Address and message sequences for the nociceptin receptor: a structure-activity study of nociceptin-(1-13)-peptide amide. *J Med Chem.* 1997;40(12):1789–93.

Gurard-Levin ZA, Kim J, and Mrksich M. Combining mass spectrometry and peptide arrays to profile the specificities of histone deacetylases. *ChemBioChem.* 2009;10(13):2159–61.

Haley TJ and McCormick WG. Pharmacological effects produced by intracerebral injection of drugs in the conscious mouse. *Br J Pharmacol Chemother.* 1957;12(1):12–5.

Higuchi Y, Kawakami S, and Hashida M. Strategies for in vivo delivery of siRNAs: recent progress. *BioDrugs.* 2010;24(3):195–205.

Hökfelt T, Broberger C, Xu ZQ, Sergeyev V, Ubink R, and Diez M. Neuropeptides—an overview. *Neuropharmacol.* 2000;39(8):1337–56.

Husi H, Ward MA, Choudhary JS, Blackstock WP, and Grant SG. Proteomic analysis of NMDA receptor-adhesion protein signaling complexes. *Nat Neurosci.* 2000;3(7):661–9.

Jörnvall H, Birgitta A, and Michael Z. Viktor Mutt: a giant in the field of bioactive peptides. *Comp Biochem.* 2008;46:397–416.

Jungblut PR, Zimny-Arndt U, Zeindl-Eberhart E, Stulik J, Koupilova K, Pleissner KP, Otto A, Müller EC, Sokolowska-Köhler W, Grabher G, et al. Proteomics in human disease: cancer, heart and infectious diseases. *Electrophoresis.* 1999;20(10):2100–10.

Kalivas PW, Jamie P, and Lori K. Animal models and brain circuits in drug addiction. *Mol Interv.* 2006;6:339–44.

Kotarsky K and Nilsson NE. Reverse pharmacology and the de-orphanization of 7TM receptors. *Drug Discov Today: Technol.* 2004;1(2):99–104.

Kotlinska J, Bochenski M, Lagowska-Lenard M, Gibula-Bruzda E, Witkowska E, and Izdebski J. Enkephalin derivative, cyclo[Nepsilon,Nbeta-carbonyl-D-Lys2, Dap5] enkephalinamide (cUENK6), induces a highly potent antinociception in rats. *Neuropeptides.* 2009;43(3):221–8.

Kotlinska J, Pachuta A, and Silberring J. Neuropeptide FF (NPFF) reduces the expression of cocaine-induced conditioned place preference and cocaine-induced sensitization in animals. *Peptides.* 2008;29:933–9.

Kotlinska J, Wichmann J, Legowska A, Rolka K, and Silberring J. Orphanin FQ/nociceptin but not Ro 65-6570 inhibits the expression of cocaine-induced conditioned place preference. *Behav Pharmacol.* 2002;13:229–35.

Lapalu S, Moisand C, Mazarguil H, Cambois G, Mollereau C, and Meunier J-C. Comparison of the structure-activity relationships of nociceptin and dynorphin A using chimeric peptides. *FEBS Lett.* 1997;417:333–6.

Levine BA (Ed). *Neuropeptide Research Trends.* New York: Nova Science, 2007:1–293.

Lien S and Lowman HB. Therapeutic peptides. *Trends Biotechnol.* 2003;21(12):556–62.

Loacker S, Sayyah M, Wittmann W, Herzog H, and Schwarzer C. Endogenous dynorphin in epileptogenesis and epilepsy: anticonvulsant net effect via kappa opioid receptors. *Brain.* 2007;130:1017–28.

Mansour A, Hoversten MT, Taylor LP, Watson SJ, and Akil H. The cloned mu, delta and kappa receptors and their endogenous ligands: evidence for two opioid peptide recognition cores. *Brain Res*. 1995;700(1–2):89–98.

Meunier JC, Mollereau C, Toll L, Suaudeau C, Moisand C, Alvinerie P, Butour JL, Guillemot JC, Ferrara P, Monsarrat B, et al. Isolation and structure of the endogenous agonist of opioid receptor-like ORL1 receptor. *Nature*. 1995;377:532–5.

de Noo ME, Tollenaar RAEM, Özalp A, Kuppen PJK, Bladergroen MR, Eilers PH C, and Deelder AM. Reliability of human serum protein profiles generated with C8 magnetic beads assisted Maldi-TOF mass spectrometry. *Anal Chem*. 2005;77(22):7232–41.

Moser VC. Observational batteries in neurotoxicity testing. *Int J Toxicol*. 2000;19:407–11.

Näslund J, Schierhorn A, Hellman U, Lannfelt L, Roses A, Tjernberg L, Silberring J, Gandy S, Winblad B, Greengard P, et al. Relative abundance of Alzheimer A beta amyloid peptide variants in Alzheimer disease and normal aging. *Proc Natl Acad Sci U S A*. 1994;91:8378–82.

Ng JH and Ilag LL. Cryptic protein fragments as an emerging source of peptide drugs. *IDrugs*. 2006;9(5):343–6.

Nyberg F and Terenius L. Identification of high molecular weight enkephalin precursor forms in human cerebrospinal fluid. *Neuropeptides*. 1985;5(4–6):537–40.

Nylander I, Stenfors C, Tan-No K, Mathé AA, and Terenius L. A comparison between microwave irradiation and decapitation: basal levels of dynorphin and enkephalin and the effect of chronic morphine treatment on dynorphin peptides. *Neuropeptides*. 1997;31(4):357–65.

Obara I, Parkitna JR, Korostynski M, Makuch W, Kaminska D, Przewlocka B, and Przewlocki R. Local peripheral opioid effects and expression of opioid genes in the spinal cord and dorsal root ganglia in neuropathic and inflammatory pain. *Pain*. 2009;141(3):283–91.

Pardridge WM. Vector-mediated drug delivery to the brain. *Adv Drug Deliv Rev*. 1999;36(2–3):299–321.

Pardridge WM. Blood-brain barrier drug targeting: the future of brain drug development. *Mol Interv*. 2003;3(2):90–105, 51.

Park SA, Lee HW, Hong MH, Choi YW, Choe YH, Ahn BY, Cho YJ, Kim DS, and Lee NG. Comparative proteomic analysis of Helicobacter pylori strains associated with iron deficiency anemia. *Proteomics*. 2006;6(4):1319–28.

Paxinos G and Watson C. *The Rat Brain in Stereotaxic Coordinates* (5th ed). San Diego, CA: Elsevier Academic Press, 2005:1–367.

Pimenta DC and Lebrun I. Cryptides: buried secrets in proteins. *Peptides*. 2007;28(12):2403–10.

Qian Y, Devi LA, Mzhavia N, Munzer S, Seidah NG, and Fricker LD. The C-terminal region of proSAAS is a potent inhibitor of prohormone convertase 1. *J Biol Chem*. 2000 4;275(31):23596–601.

Reinscheid RK, Nothacker HP, Bourson A, Ardati A, Henningsen RA, Bunzow JR, Grandy DK, Langen H, Monsma FJ Jr, and Civelli O. Orphanin FQ: a neuropeptide that activates an opioid-like G protein-coupled receptor. *Science*. 1995;270:792–4.

Renlund S, Erlandsson I, Hellman U, Silberring J, Lindström L, and Nyberg F. Micropurification of immunoreactive beta-casomorphins present in milk from women with post-partum psychosis. *Peptides*. 1993;14:1125–32.

Righetti PG, Fasoli E, Aldini G, Regazzoni L, Kravchuk AV, Citterio A. Les Maîtres de l'Orge: the Proteome Content of Your Beer Mug. *J Proteome Res*. 2010;9(10):5262–9.

Rizzi A, Vergura R, Marzola G, Ruzza C, Guerrini R, Salvadori S, Regoli D, and Calo G. Neuropeptide S is a stimulatory anxiolytic agent: a behavioural study in mice. *Br J Pharmacol*. 2008;154(2):471–9.

Roques BP and Noble F. Association of enkephalin catabolism inhibitors and CCK-B antagonists: a potential use in the management of pain and opioid addiction. *Neurochem Res*. 1996;21(11):1397–410.

Rudolph U and Möhler H. Genetically modified animals in pharmacological research: future trends. *Eur J Pharmacol.* 1999;375:327–37.

Ruiz F, Fournié-Zaluski MC, Roques BP, and Maldonado R. Similar decrease in spontaneous morphine abstinence by methadone and RB 101, an inhibitor of enkephalin catabolism. *Br J Pharmacol.* 1996;119(1):174–82.

Sakoori K and Murphy NP. Central administration of nociceptin/orphanin FQ blocks the acquisition of conditioned place preference to morphine and cocaine, but not conditioned place aversion to naloxone in mice. *Psychopharmacology (Berl).* 2004;172:129–36.

Schwartz SA and Caprioli RM. Imaging mass spectrometry: viewing the future. *Methods Mol Biol.* 2010;656:3–19.

Schwarzer C, Williamson JM, Lothman EW, Vezzani A, and Sperk G. Somatostatin, neuropeptide Y, neurokinin B and cholecystokinin immunoreactivity in two chronic models of temporal lobe epilepsy. *Neuroscience.* 1995;69(3):831–45.

Silberring J, Li YM, Terenius L, and Nylander I. Characterization of immunoreactive dynorphin B and beta-endorphin in human plasma. *Peptides.* 1998;19(8):1329–37.

Simmons D and Self DW. Role of mu- and delta-opioid receptors in the nucleus accumbens in cocaine-seeking behavior. *Neuropsychopharmacology.* 2009;34:1946–57.

Soskic V, Görlach M, Poznanovic S, Boehmer FD, and Godovac-Zimmermann J. Functional proteomics analysis of signal transduction pathways of the platelet-derived growth factor beta receptor. *Biochemistry.* 1999;38(6):1757–64.

Suckau D, Köhl J, Karwath G, Schneider K, Casaretto M, Bitter-Suermann D, and Przybylski M. Molecular epitope identification by limited proteolysis of an immobilized antigenantibody complex and mass spectrometric peptide mapping. *Proc Natl Acad Sci U S A.* 1990;87(24):9848–52.

Suder P, Bierczynska-Krzysik A, Kraj A, Brostedt P, Mak P, Stawikowski M, Rolka K, Nyberg F, Fries E, and Silberring J. Identification of bikunin as an endogenous inhibitor of dynorphin convertase in human cerebrospinal fluid, *FEBS J.* 2006;273, 5113–20.

Suder P, Nawrat D, Bielawski A, Zelek-Molik A, Raoof H, Dylag T, Kotlinska J, Nalepa I, and Silberring J. Cryptic peptide derived from the rat neuropeptide FF precursor affects G-proteins linked to opioid receptors in the rat brain. *Peptides.* 2008;29(11):1988–93.

Sørensen G, Lindberg C, Wörtwein G, Bolwig TG, and Woldbye DP. Differential roles for neuropeptide Y Y1 and Y5 receptors in anxiety and sedation. *J Neurosci Res.* 2004;77(5):723–9.

Tan-No K, Sato T, Shimoda M, Nakagawasai O, Niijima F, Kawamura S, Furuta S, Sato T, Satoh S, Silberring J, Terenius L, and Tadano T. Suppressive effects by cysteine protease inhibitors on naloxone-precipitated withdrawal jumping in morphine-dependent mice. *Neuropeptides.* 2010;44(3):279–83.

Tan-No K, Taira A, Sakurada T, Inoue M, Sakurada S, Tadano T, Sato T, Sakurada C, Nylander I, Silberring J, Terenius L, and Kisara K. Inhibition of dynorphin-converting enzymes prolongs the antinociceptive effect of intrathecally administered dynorphin in the mouse formalin test. *Eur J Pharmacol.* 1996;314(1–2):61–7.

Tan-No K, Takahashi H, Nakagawasai O, Niijima F, Sakurada S, Bakalkin G, Terenius L, and Tadano T. Nociceptive behavior induced by the endogenous opioid peptides dynorphins in uninjured mice: evidence with intrathecal N-ethylmaleimide inhibiting dynorphin degradation. *Int Rev Neurobiol.* 2009;85:191–205.

Torregrossa MM, Jutkiewicz EM, Mosberg HI, Balboni G, Watson SJ, and Woods JH. Peptidic delta opioid receptor agonists produce antidepressant-like effects in the forced swim test and regulate BDNF mRNA expression in rats. *Brain Res.* 2006;1069(1):172–81.

Vergura R, Balboni G, Spagnolo B, Gavioli E, Lambert DG, McDonald J, Trapella C, Lazarus LH, Regoli D, Guerrini R, Salvadori S, and Caló G. Anxiolytic- and antidepressant-like activities of H-Dmt-Tic-NH-CH(CH2-COOH)-Bid (UFP-512), a novel selective delta opioid receptor agonist. *Peptides.* 2008;29(1):93–103.

Villanueva J, Shaffer DR, Philip J, Chaparro CA, Erdjument-Bromage H, Olshen AB, Fleisher M, Lilja H, Brogi E, Boyd J, et al. Differential exoprotease activities confer tumor-specific serum peptidome patterns. *J Clin Invest*. 2006;116(1):271–84.

Wieland T. The history of peptide chemistry. In: Gutte B (Ed), *Peptides: Synthesis, Structures and Applications*. San Diego, CA: Academic Press, 1995:1–38.

Yan HC, Cao X, Das M, Zhu XH, and Gao TM. Behavioral animal models of depression. *Neurosci Bull*. 2010;26(4):327–37.

Young WSI. *In situ* hybridization histochemistry. In: Björklund A, Hökfelt T, Wouterlood FG, and Van den Pol AN (Eds), *Handbook of Chemical Neuroanatomy*. Amsterdam, The Netherlands: Elsevier, Vol. 8, 1990:481–512.

Wahlestedt C. Antisense oligonucleotide strategies in neuropharmacology. *Trends Pharmacol Sci*. 1994;15(2):42–6.

5 Cellular and Molecular Processes of Neuronal Damage and Mechanisms in Neuroprotection and Neuroregeneration— Focus on Brain Ischemia

Gregor Zündorf, Daniel Förster, and Georg Reiser

CONTENTS

5.1   INTRODUCTION

Cellular signaling requires harnessing the protein functions in response to the changing environment. In neurons, electrical pulses or receptor-mediated stimuli generate signals with distinct spatial dimensions, temporal extension, amplitude, subcellular localization, and, in some cases, oscillations. The subsequent readout of signals employs downstream signaling proteins, which transmit the message to cellular effectors for regulating neuronal plasticity underlying learning and memory and neuronal survival.

Neurodegenerative diseases are among the most important causes of disability worldwide. Although the understanding of the biochemical events is continuously growing, there is a shocking lack of effective therapies, which are more and more in demand. Despite a wide spectrum of appearance of these diseases, there are striking similarities in their molecular pathogenesis. Thus, all these pathological states are associated with perturbed Ca^{2+} homeostasis triggering cell death by activating catabolic enzymes. Furthermore, another common feature is that mitochondria are intimately involved in neuronal cell death, as they are severely affected by deprivation of substrate or energy. Finally, the immune system is involved in neurological disorders. As microglia determines the progression and final outcome of the neuropathological process, these disorders should also be considered as gliopathologies. As a further common feature, intrinsic neuroprotective and neuroregenerative pathways are found that antagonize, interrupt, or delay the sequence of injurious biochemical and molecular events.

The progression of detrimental effects after stroke is well described and is obviously similar to that in other neurodegenerative diseases. Hence, the pathogenic route after ischemia with ensuing energy deprivation is used in this review to illustrate the evolution of the destructive cascades over time and to describe the impact of intrinsic protective cascades on the final outcome. Based on these pathways, novel therapeutic approaches to enhance survival and regeneration are presented. In this context, several neuropeptides are described to be intimately involved in the prevention of neuronal apoptosis.

5.2   PATHOPHYSIOLOGY

The pathological processes after ischemia can be separated into acute effects occurring within hours, subacute effects within hours to days, and the chronic phase, which lasts from days to months. An overview about the degenerative acute and subacute processes is given in Figure 5.1.

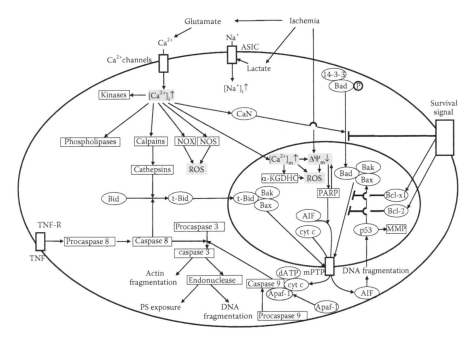

FIGURE 5.1 Ischemia-induced degenerative acute and subacute processes. Decrease of oxygen and nutrients induces oxidative stress and glutamate release. Cellular depolarization is boosted by opening of ionotropic Ca^{2+} channels. In addition, an increased lactate level activates ASICs that contribute to further disturbance of the cellular ion homeostasis. Increased $[Ca^{2+}]_i$ activates several cytosolic and mitochondrial enzymes in the intrinsic pathway of apoptosis. Calpains activate cathepsins that mediate the proteolysis of Bid, allowing t-Bid to translocate to the mitochondria. CaN-mediated dephosphorylation of Bad causes release of Bad from protein 14-3-3 and allows its translocation to the mitochondria. t-Bid and Bad support proapoptotic proteins Bak and Bax to induce mPTP formation and cytochrome c release. Cytochrome c forms the apoptosome complex by binding to dATP, Apaf-1, and caspase 9. This complex activates caspase 3, which induces actin fragmentation, DNA fragmentation and the exposure of PS. PARP-1 overstimulation induces translocation of AIF to the nucleus and, thus, activates caspase-independent cell death. DNA degradation induces p53 to activate Bax and MMP. TNF activation of its receptor (TNF-R) induces the extrinsic pathway of apoptosis by activation of caspases and cleavage of Bid. Survival factors prevent mitochondrial pore formation via inhibition of apoptotic proteins and stimulation of antiapoptotic proteins Bcl-xl and Bcl-2. AIF, apoptosis-inducing factor; α-KGDHC, α-ketoglutarate dehydrogenase complex; Apaf-1, apoptotic protease-activating factor 1; ASIC, acid-sensing ion channels; $[Ca^{2+}]_i$, intracellular cytosolic free Ca^{2+} concentration; $[Ca^{2+}]_m$, mitochondrial free Ca^{2+} concentration; CaN, calcineurin; cyt c, cytochrome c; mPTP, mitochondrial permeability transition pore; $\Delta\Psi_m$, mitochondrial potential; MMP, matrix metalloproteinase; NOS, NO-synthase; NOX, NAD(P)H oxidase; PARP, poly ADP-ribose polymerase; PS, phosphatidylserine; ROS, reactive oxygen species; TNF, tumor necrosis factor.

5.2.1 ACUTE EFFECTS

A few minutes after brain ischemia, decrease of oxygen and nutrients induces depolarization leading to interwoven mechanisms, primarily glutamate release (excitotoxicity) and oxidative stress (Lo, Moskowitz, and Jacobs 2005). Excess glutamate activates ionotropic receptors for N-methyl-D-aspartate (NMDA) and thus triggers deregulation of cellular ion homeostasis, mainly $[Ca^{2+}]_i$, in a vicious cycle, as reviewed earlier (Zündorf and Reiser 2011). Brain tissue is not sufficiently equipped with antioxidant defense. Therefore, reactive oxygen species (ROS) released by cells of both neural and myeloid descent threaten tissue viability in the vicinity of the ischemic core. Propagating waves of depolarization induce peri-infarct depolarization and cause an expansion of the core-infarcted tissue into the adjacent penumbral regions (Hertz 2008).

Increase in ROS does not necessarily represent the primary cause, but can be a consequence of disturbances in Ca^{2+} homeostasis, mitochondrial malfunction, or activation of specific ROS-producing enzymes, namely, NADPH oxidases (NOX), α-ketoglutarate dehydrogenase (Zündorf et al. 2009), NO synthase (NOS), and others (Feissner et al. 2009). Excessive rise of intramitochondrial Ca^{2+} is a trigger for mitochondrial permeability transition pore (mPTP) opening and thus induces enhanced cytochrome c dissociation from the inner mitochondrial membrane. As cytochrome c mediates electron transport from complex III to complex IV, lowered cytochrome c levels lead to ROS formation in the Krebs cycle (Lemasters et al. 2009). mPTP opening also causes loss of antioxidant defense in mitochondria by depletion of glutathione and NADPH (Murphy 2009). Oxidative stress contributes to degeneration by afflicting damage to lipids, DNA, and proteins and by activating detrimental signaling pathways. Furthermore, in cerebral ischemia, an increased lactate level activates acid-sensing ion channels (ASICs). These channels are permeable to Na^{2+} and thereby contribute to further disturbance of the ion homeostasis (Pignataro, Simon, and Xiong 2007). During stroke, activation of polyADP-ribose polymerases (PARP) leads to formation of ADP-ribose polymers and, thus, further inhibits NAD^+(NADH)-dependent ATP production (Duan, Gross, and Sheu 2007).

5.2.2 SUBACUTE EFFECTS

5.2.2.1 Apoptosis

Stroke produces an ischemic core with necrotic cells and a surrounding area called the penumbra with preserved ion homeostasis and membrane potential but reduced electrical activity. Within this zone, cell death occurs predominantly by apoptosis (Azarashvili, Stricker, and Reiser 2010; Doyle, Simon, and Stenzel-Poore 2008; Mehta, Manhas, and Raghubir 2007; Moroni 2008; Okouchi et al. 2007; Pradelli, Beneteau, and Ricci 2010).

Increased $[Ca^{2+}]_i$ activates phospholipases, protein kinases, endonucleases, neuronal NOS, and calpains in the intrinsic pathways of apoptosis. These pathways are presented in Figure 5.1. Calpains activate cathepsins that mediate the proteolysis of the protein Bid, allowing truncated Bid (t-Bid) to translocate to the mitochondria. Furthermore, Ca^{2+} induces dephosphorylation of Bad by calcineurin, which releases Bad from the protein 14-3-3 and allows its translocation to the mitochondria. Bid and Bad support the proapoptotic proteins Bak and Bax to induce formation of pores in the outer mitochondrial membrane with cytochrome c release. Cytochrome c forms the apoptosome complex by binding to deoxyadenosine triphosphate (dATP), apoptotic protease-activating factor 1 (Apaf-1), and procaspase 9. This complex activates caspase 3, which causes actin fragmentation, and endonuclease activation-inducing DNA fragmentation and the exposure of phosphatidylserine on the outer leaflet of the plasma membrane. Caspase 3 can also be activated by caspase 8. Mitochondrial membrane permeabilization as well as PARP-1 overstimulation induces translocation of apoptosis-inducing factor (AIF) to the nucleus and, thus, activates caspase-independent programmed cell death. DNA degradation activates p53, which further increases Bax and matrix metalloproteinase (MMP) expression.

The extrinsic pathway of apoptosis involves death receptor activation by tumor necrosis factor (TNF), leading to activation of caspase 8 and 3. Caspase 8 also mediates the limited proteolysis of Bid and allows t-Bid to translocate to the mitochondria. Caspase 3 cleaves several substrates and, thus, initiates chromatin condensation and cell blebbing. Various survival factors prevent mitochondrial pore formation via activation of the antiapoptotic proteins Bcl-xl and Bcl-2.

5.2.2.2 Inflammation

One of the important pathophysiological mechanisms unleashed after stroke is immune activation (Amor et al. 2010; Jin, Yang, and Li 2010). Microglia and blood inflammatory cells play complex and multiphasic roles. They display both beneficial and adverse effects, as outlined in Figure 5.2. During the acute phase, activation of microglia, the resident macrophages of the central nervous system, is seen (del Zoppo et al. 2007). Microglial cell proliferation peaks at 2–3 days after onset of stroke. There is a release of the inflammatory mediators, interleukin 1β (IL-1β) and TNFα and tissue damage through expression of MMP and production of ROS via NOX. ROS triggers the expression of a number of proinflammatory genes, including cytokines and adhesion molecules, by oxidant-sensitive transcription factors. These play an important role in leukocyte–endothelium interactions and secondary brain damage. On the other hand, microglia produces neurotrophins and growth factors supporting neuronal survival and regeneration. Microglia cells can change their morphology to a phagocyte-like phenotype, indistinguishable from macrophages, and macropahges are able to acquire a ramified form, which is not distinguishable from microglia.

The blood–brain barrier (BBB) is damaged by activated microglia-evoking edema and increased infiltration of leukocytes into the infarct area (Pun, Lu, and Moochhala 2009). Recruitment of blood-derived immune cells takes place with a delay of hours to days by expression of adhesion molecules on leukocytes and on microvascular endothelial cells. Neutrophils are among the first leukocytes to infiltrate the ischemic brain, peaking at day one to three, and they are present until day 15 after stroke (Yilmaz and Granger 2010). Negative effects of neutrophils are induced by ROS production via NOX, production of proinflammatory cytokines, and

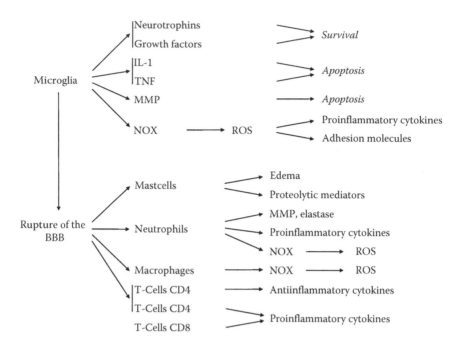

FIGURE 5.2 Involvement of microglia and blood cells in ischemic stroke. Inflammatory cells play both beneficial and adverse effects in neurodegenerative diseases. Activated microglia produces neurotrophins and growth factors supporting neuronal survival and regeneration. On the other hand, microglia cells express MMP and releases inflammatory mediators, IL-1β and TNFα. ROS from NOX trigger oxidant-sensitive transcription factors to express a number of proinflammatory genes and adhesion molecules on leukocytes and on vascular endothelial cells. The BBB is damaged by activated microglia evoking infiltration of blood cells into the brain. Infiltrated neutrophils release elastase and MMP-9, proinflammatory cytokines and produce ROS from NOX. Phagocytosis of neutrophils by microglia and thereby elimination of excitotoxins might be beneficial for neuronal survival (not shown). Furthermore, macrophages and T-cells accumulate in the ischemic brain. CD4+ TH2 T-cells provide neuroprotection through antiinflammatory cytokines. On the other hand, inflammation is mediated substantially by CD4+ TH1- and CD8+ T-cells. Mast cells release proteolytic and fibrinolytic mediators, thereby supporting neutrophil infiltration and edema formation. BBB, blood–brain barrier; IL-1, interleukin 1; MMP, matrix metalloproteinase; NOX, NAD(P)H oxidase; ROS, reactive oxygen species; TNF, tumor necrosis factor.

release of elastase and MMP-9 (del Zoppo et al. 2007). Phagocytosis of neutrophils by microglia and thereby elimination of excitotoxins might be beneficial for neuronal survival (McColl, Allan, and Rothwell 2009). Blood-derived macrophages infiltrate the stroke tissue abundantly from day three to seven after stroke.

T-cells accumulate in the ischemic brain within the first 24 hours after focal cerebral ischemia. Regulatory T-cells are both cerebroprotective and degenerative immunomodulators. Thus, CD4+ TH2 T-cells provide neuroprotection through antiinflammatory cytokines. On the other hand, inflammation in the postacute phase of ischemic stroke is mediated substantially by CD4+ TH1 and CD8+ T-cells (Arumugam, Granger, and Mattson 2005).

Mast cells are located at the diencephalic parenchyma, thalamus, and cerebral cortex boundary between extravascular and intravascular space. They respond to blood- and tissue-derived signals with release of proteolytic and fibrinolytic mediators thereby increasing the permeability of the BBB, supporting neutrophil infiltration and edema formation (Strbian et al. 2009).

5.2.3 Chronic Phase

After stroke, a glial scar consisting of reactive astrocytes, microglia, and fibroblasts forms a physical barrier around the infarcted tissue. Noncellular components of the scar are proteoglycans, tenascin, fibronectin, laminin, and collagen. Endogenous repair mechanisms are induced and, thus, contribute to the recovery of neurological function (Candelario-Jalil 2009; Kernie and Parent 2010; Zhang and Chopp 2009). The intrinsic regenerative processes are presented in Figure 5.3. Neurogenesis involves proliferation of neural stem cells and progenitor cells, differentiation of neural progenitor cells, and migration of neuroblasts to the ischemic boundary, where neuroblasts mature into resident neurons and integrate into the parenchymal tissue. In the ischemic boundary, neurogenesis and angiogenesis involving endothelial cell proliferation, migration, tube formation, branching, anastomosis, and stimulation of blood flow are highly interdependent processes.

Cerebral endothelial cells secrete factors that regulate the biological activity of neural progenitor cells. Thus, the laminin receptor, α6β1 integrin, expressed by neural stem and progenitor cells interacts with laminin-containing vessels in the subventricular zone. Blockage of this interaction increases the proliferation of neural stem cells and progenitor cells. After stroke, endothelial cells secrete vascular endothelial growth factor (VEGF) to increase neurogenesis and guide neuroblast migration via its receptor, VEGFR2. Furthermore, upregulation of angiopoietin 1 and its receptor Tie2 promotes cerebral vessel sprouting. Activated endothelial cells secrete stromal-derived factor 1α (SDF-1α), the hypoxia-inducible factor (HIF), the monocyte chemoattractant protein (MCP), and MMPs, which are all associated with neurogenesis. MMPs degrade the extracellular matrix and thus enable neuroblasts to penetrate into the ischemic boundary.

Under normal conditions, myelin-associated inhibitors limit the axonal outgrowth. Three of these proteins, neurite outgrowth inhibitor (Nogo-A), myelin-associated glycoprotein, and oligodendrocyte-myelin glycoprotein (OMgp), share two common neuronal receptors, namely, NgR1 together with the paired immunoglobulin-like receptor B (PirB), thereby activating the RhoA pathway. Blocking of either receptor during neuronal damage improves the regeneration of fiber tracts (Llorens, Gil, and Del Rio 2010; Quarles 2007).

The synaptic activity increases in the border zone around the ischemic lesion. Morphological adaptations in ischemia are related to an increase in the production of neurotrophins that stimulate the growth and dendrite branching and the synaptogenesis

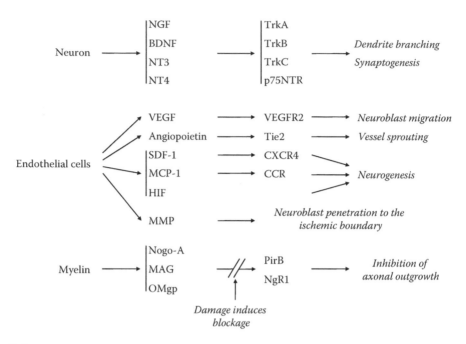

FIGURE 5.3 Intrinsic regenerative processes after various brain insults. Around the lesion area, neurotrophins NGF, BDNF, NT3, and NT4 activate Trk and p75NT receptors, thus sustain survival, and stimulate dendrite branching, synaptogenesis, and differentiation by activation of proregenerative transcriptions factors. Endothelial cells secrete VEGF to increase neurogenesis and guide neuroblast migration via its receptor VEGFR2. Upregulation of angiopoietin 1 and its receptor Tie2 promotes cerebral vessel sprouting. Activated endothelial cells secrete SDF-1α, HIF, MCP, and MMPs, which are all associated with neurogenesis. MMPs degrade the extracellular matrix and thus enable neuroblasts to penetrate into the ischemic boundary. Under normal conditions, myelin-associated proteins, Nogo-A, MAG, and OMgp, sharing the neuronal receptors NgR1 and PirB, inhibit axonal outgrowth. Blockade of the receptors during neuronal damage improves the regeneration of fiber tracts. BDNF, brain-derived neurotrophic factor; CCR, CC motif receptor; CXCR, CXC motif receptor; HIF, hypoxia-inducible factor; MAG, myelin-associated glycoprotein; MCP monocyte-chemoattractant protein; MMP, matrix metalloproteinase; NGF, nerve growth factor; NgR, Nogo receptor; NT, neurotrophin; OMgp, oligodendrocyte-myelin glycoprotein; p75NTR, p75 neurotrophin receptor; PirB, paired immunoglobulin-like receptor B; SDF, stromal cell-derived factor; Trk, tropomyosin-related kinase; VEGF, vascular endothelial growth factor.

in the lesion and perilesion area. Neurotrophins were identified to sustain morphology, differentiation, and survival of sensory and sympathetic neurons by activation of proregenerative transcription factors. One or more of the four neurotrophins, namely, nerve growth factor (NGF), brain-derived neurotrophic factor (BDNF), and neurotrophins 3 and 4 (NT3 and NT4), activate tropomyosin-related kinase (Trk) receptors, a family of three receptor tyrosine kinases. In addition, each neurotrophin activates the p75 neurotrophin receptor (p75NTR), a member of the TNF receptor superfamily (Huang and Reichardt 2003; Reichardt 2006). Trk receptor signaling activates phospholipase C-γ, PI3K, and signaling pathways controlled through these proteins, such as the MAP kinases. Activation of p75NTR results in activation of the nuclear factor-κB (NF-κB) and Jun kinase.

5.3 THERAPEUTIC APPROACHES

Endogenous neuroprotective mechanisms need to target those cascades that are critically involved in mediating damage, to prevent their induction or propagation. Restorative therapies support brain repair by encouraging endogenous capacity through pharmacological rehabilitation, but also address the exogenous administration of cellular material, the so-called cell therapy.

5.3.1 PRECONDITIONING

Preexposure of the brain to a short ischemic event (preconditioning) triggers a cerebroprotective state limiting the damage exerted by a subsequent ischemic insult (Dirnagl, Simon, and Hallenbeck 2003; Gidday 2006). Immunomodulation induces cerebral ischemic tolerance involving the proinflammatory Toll-like receptor, transcription factor NF-κB and its activator signals, such as proinflammatory cytokines TNF, IL-1β, IL-6, IL-8, and also the neurotrophic factors. Furthermore, the energy metabolism is adapted and even regenerative mechanisms on various levels are enhanced (Dirnagl and Meisel 2008). This may be a therapeutic option in individuals with a high risk for ischemic complications (Dirnagl, Becker, and Meisel 2009). Brief preconditioning ischemia applied to one organ or tissue can also protect distant organs, pointing to the fact that humoral mechanisms are involved in mediating the tissue-protective effect.

5.3.2 TREATMENT OF ACUTE EFFECTS

In the acute phase, treatments of stroke-induced damage have centered around restoration of blood flow and pharmacological interventions to target the primary causes of neuronal death.

5.3.2.1 Approaches to Enable Substances to Cross the Blood–Brain Barrier

Under normal conditions, only small hydrophilic molecules are able to cross the BBB via the paracellular pathway or through transporters expressed at the endothelial cells. Because of that, special strategies are needed for delivery of intravenously administered drugs to the brain (Gabathuler 2010; Patel et al. 2009; Rip, Schenk, and de Boer 2009; Rosenberg, Chen, and Prabhakaran 2010). Thus, conjugation to ligands that are recognized by receptors expressed at the BBB results in receptor-mediated transcytosis. The low-density lipoprotein receptor-related protein is the most suitable receptor for such usage. Furthermore, amidation or esterification of amino, carboxyl, and hydroxyl groups or coupling with fatty acids, glycerides, or phospholipids increases the permeability of drugs for biological membranes. Subsequent hydrolysis of these groups leads to the release of the active drug from the prodrug. Moreover, micelles can improve drug pharmacokinetics and biodistribution. They have been utilized for oral delivery. The chemical modifications often result in loss of the desired activity in the central nervous system or make the compounds a substrate for the efflux through the P-glycoprotein pump.

5.3.2.2 Hemodynamic Mechanisms

The cerebral blood flow can be increased physically (Mantz, Degos, and Laigle 2010; Medel et al. 2009; Rosenberg et al. 2010) and pharmacologically (Ginsberg 2008; Rosenberg et al. 2010). Physical approaches cover cavitation, endovascular thrombectomy, and stenting. Cavitation uses expanding gas bubbles produced by an ultrasound wave to create microdisruptions in the clot, endovascular thrombectomy leads to recanalization of vessels, and stenting allows the mechanical disruption of clots. Furthermore, mild hypothermia protects the brain, and normobaric hyperoxia increases the therapeutic window for thrombolysis and other neuroprotectant drugs without increasing oxidative stress.

Pharmacological approaches comprise intervention by intravenous or intraarterial administration of thrombolytic agents (e.g., tissue plasminogen activator, urokinase), antithrombotic agents (e.g., heparin), antiplatelet drugs (e.g., aspirin, dipyridamole, abciximab), and fibrinogen-depleting agents (e.g., ancrod).

5.3.2.3 Inhibition of Primary Causes of Neurodegeneration

Interventions affecting the primary causes of neuronal cell death mainly concern inhibition of excitotoxicity, reduction of overwhelming $[Ca^{2+}]_i$ load, and avoidance of free radicals. Indeed, different agonists, antagonists, and coagonists, which interfere with the excitotoxic pathway (mainly NMDA receptor antagonists), were shown to improve the outcome after ischemia in animals (Ginsberg 2008; Hardingham 2009). Similarly, pharmacological inhibition of ischemia-induced Ca^{2+} and ROS dysregulation was shown to reduce brain injury (Chen et al. 2011; Hardeland 2009; Zündorf and Reiser 2011).

5.3.3 TREATMENT OF SUBACUTE EFFECTS

5.3.3.1 Postconditioning

Endogenous processes similar to those identified in preconditioning are initiated within the first minutes of reperfusion. Thus, it has been found that brief repetitive interruption of the blood flow applied after the onset of reperfusion following an ischemic event in brain provides protection (Kaur, Jaggi, and Singh 2009; Sandu and Schaller 2010; Zhao 2009).

5.3.3.2 Inhibition of Apoptosis

The penumbra is salvageable from turning into infarcted tissue. Strategies for the induction of neuroprotection comprise both disruption of the detrimental cascades leading to apoptosis and activation of prosurvival pathways. Different candidate therapeutics are presented here as promising examples.

Erythropoietin and its receptor are expressed in the nervous system. Peripherally administered erythropoietin is able to cross the BBB and to stimulate neurogenesis, neuronal differentiation (see section 5.3.4.1.), and, furthermore, to induce antiapoptotic, antioxidant, and antiinflammatory signaling, as mediated via the transcription factor NF-κB (Siren et al. 2009). Erythropoietin also induces expression of BDNF and activation of its antiapoptotic receptor, TrkB, in primary hippocampal cells. In this way, administration of neurotrophins was shown to have neuroprotective effects after injury when they are administered as a chimeric protein to cross the BBB (Webster and Pirrung 2008).

In addition to their cholesterol-lowering properties, statins improve the outcome after neurodegeneration. Statin-induced inactivation of Rho and Rac upregulates endothelial NOS, inactivates NOX, and, thus, activates endothelial function, regulates progenitor cells, and inhibits inflammatory responses (Endres 2005).

Protein kinases represent network nodes, which transform various cellular stimuli into an integrated response (Chico, Van Eldik, and Watterson 2009). Treatment with low doses of protein kinase inhibitors preferentially modulates one pathway over another to attain the desired efficacy. As an example, glycogen synthase kinase 3 (GSK3) modulates different signaling pathways that are important for cell survival. In response to distinct input signals, GSK3 is inhibited by serine phosphorylation and activated by tyrosine phosphorylation. Upregulation of GSK3 activity has been linked to pathological conditions, including Alzheimer's disease, stroke, and neuropsychiatric disorders. Hence, GSK3 inhibitors have shown positive efficacy in models of these disorders. Similarly, death-associated protein kinase 1 (DAPK1) phosphorylates a set of substrate proteins by increased Ca^{2+} levels. DAPK1 inhibitors are able to inhibit stress- or injury-induced cascades within clinically relevant time frames.

Blockers of angiotensin receptors improve neurological outcome after ischemia. Angiotensin maintains blood flow that is mediated via angiotensin AT1 receptors and regulates apoptotic events via AT2 receptors, which are expressed both on neurons within the BBB and in cerebrovascular endothelial cells (Culman et al. 2002).

Pituitary adenylate cyclase-activating polypeptide suppresses neuronal cell death, even when administered several hours after ischemia. Activation of the underlying receptor modifies Bcl family members to produce an antiapoptotic effect by decreasing the release of cytochrome c and AIF as described earlier (Ohtaki et al. 2008).

5.3.3.3 Inflammation

The concept of reducing deleterious consequences and enhancing protective actions of specific types of inflammatory cells offers novel, promising neuroprotectant therapies (Jordan et al. 2008; Lakhan, Kirchgessner, and Hofer 2009). Thus, the inhibition of microglial activation by immunosuppression results in improved structural preservation and, mostly, in enhanced neuronal function (Hailer 2008; Kleinig and Vink 2009). The immunosuppressive drugs include antiinflammatory corticosteroids, cytotoxic drugs (e.g., azathioprine and cyclosphosphamide), fungal and bacterial derivatives inhibiting T-cell activation (e.g., cyclosporine A, rapamycin), and nonconventional immunosuppresive drugs (e.g., statins, G-CSF, cytosine-arabinoside, FK506, mycophenolate mofetil, minocycline). The effects of these substances are mainly exerted by inhibiting microglial proliferation or suppressing secretion of neurotoxic substances, such as proinflammatory cytokines and nitric oxide from microglia. However, immunosuppressants may have direct effects on neurons and on glial cells, such as astrocytes. Several approaches are under investigation for managing IL-1β in stroke. IL-1β acts via the membrane receptor IL-1R. The IL-1 receptor antagonist readily crosses the BBB and appears to afford some benefits, particularly for patients with cortical infarcts.

Reperfusion represents a highly vulnerable period for the brain, as it provides the potential benefits of restoring blood flow to an ischemic region and simultaneously opens the flood gates allowing a massive influx of activated leukocytes into the ischemic tissue. Hence, the subacute reperfusion period after a stroke is considered more amenable to treatment than the acute neurotoxicity. Inhibition of leukocyte infiltration into the ischemic brain via antiadhesion molecules (e.g., CD11b/CD18, ICAM-1, P-selectin) has been shown to reduce infarct size, edema, and neurological deficits in animal models. Furthermore, antileukocyte strategies extend the therapeutic time window of tPA reperfusion therapy. UK-279276, a recombinant glycoprotein, is a selective antagonist of the CD11b integrin of Mac-1 (CD11b/CD18) and has been shown to reduce neutrophil infiltration and ameliorate the infarct volume in rats.

The complement cascade is involved in triggering cell death and recruiting cells of the immune system to sites of inflammation in the brain. Activation of the complement cascade might be a potential approach for treatment of acute neurodegenerative diseases (Yanamadala and Friedlander 2010).

5.3.4 Restorative Therapies

Targeting neural stem cells, cerebral endothelial cells, astrocytes, oligodendrocytes, and neurons by stem cell activation (intrinsic) or stem cell transplantation (extrinsic) could lower the grade of disability in patients with neurodegenerative diseases.

5.3.4.1 Stem Cell Activation

Pharmacological targeting of natural repair mechanisms helps to establish new connections in the damaged brain area. Statins, erythropoietin, as well as trophic factors and growth factors promote neurogenesis, angiogenesis, synaptogenesis, and structural changes in the brain (Endres 2005; Endres et al. 2008; Zhang and Chopp 2009).

Erythropoietin activates the PI3K-Akt pathway by interaction with the erythropoietin receptor in neural progenitor cells, whereas phosphodiesterase-5 inhibitors and statins activate Akt via increased concentrations of cGMP. Akt regulates the assembly of mammalian achaete-scute homolog 1 (Mash1) and neurogenin 1 (Ngn1), which are basic helix-loop-helix transcription factors that promote the differentiation of neural progenitor cells into neurons. Accordingly, attenuation of expression of endogenous Mash1 and Ngn1 by small interfering RNA in neural progenitor cells reduces the rise in the neuronal population induced by erythropoietin and statins.

Furthermore, MMPs regulate the relation between erythropoietin-enhanced angiogenesis and neurogenesis. Erythropoietin stimulates cerebral endothelial cells to secrete active forms of MMP-2 and MMP-9. Coculture of cerebral endothelial cells, which were activated by erythropoietin, with neural progenitor cells promoted neuroblast migration and motility. Erythropoietin, statins, and phosphodiesterase-5 inhibitors increase the concentrations of VEGF in the ischemic boundary. Treatment with VEGF enhances angiogenesis by regulating expression of VEGF and VEGFR2, as well as angiopoietins 1 and 2, and Tie2. In mice, endothelial nitric oxide synthase mediated statin-induced angiogenesis. SDF-1α was shown to induce transmigration of hematopoietic progenitor cells, and monocyte chemotactic protein-1 directs the recruitment of cells to the damaged zone of the hippocampus.

Administration of endothelial growth factor increases the number of neurons in the olfactory bulb, whereas the infusion of fibroblast growth factor has the opposite effect in inducing differentiation of glial cells and reducing the production of neurons in the olfactory bulb.

5.3.4.2 Stem Cell Transplantation

In damaged brain tissue, stem cells administered locally by intraventricular, intraparenchymal, intravenous, or intraarterial injections can integrate themselves within the lesion area and induce self-repair by inhibition of inflammation or stimulation of growth of new blood vessels (Gutierrez et al. 2009; Hung et al. 2010; Urbaniak Hunter, Yarbrough, and Ciacci 2010).

Mesenchymal stem cells of bone marrow can differentiate into different cell types, including astrocytes, neurons, and endothelial cells, and improve the functional recovery in experimental models of neurodegenerative diseases and in clinical studies. Treatment with mesenchymal stem cells increases the concentration of SDF-1α, promotes migration of neuroblasts to the ischemic boundary, stimulates brain parenchymal cells to secrete basic fibroblast growth factor, BDNF, and to express VEGF, angiopoietin 1, and Tie2, leading to increased angiogenesis and maturation of newly formed vessels by reducing the vascular permeability and increasing the expression of tight-junction proteins. The administration or mobilization of endothelial progenitor cells, umbilical cord cells, fetal cells, and adult cerebral stem cells can improve the functional situation following a cerebral infarct. Further, implanting stem cells from an immortalized hippocampus cell line (MHP36 cells) improved the sensory-motor recovery following cerebral ischemia.

5.4 CONCLUSIONS AND OUTLOOK

Animal models provide mechanistic insights that have correlated quite well with clinical findings in terms of the pathophysiology of neurodegenerative disorders. However, clinical applicability of successful therapies, which can be gained in animals, is limited. This is because most of the studies are performed in young and healthy animals. A great number of variables come into play and result in a large heterogeneity of pathophysiological courses in different patients. Moreover, observations in one species are not necessarily valid in another species. More elaborate therapies, including combination therapies with pleiotropic beneficial effects, more sophisticated targeting and consideration of elevated inflammatory states, will increase the likelihood of successful clinical translation.

ACKNOWLEDGMENTS

The work from the authors' laboratory was supported by grants from Bundesministerium für Bildung und Forschung (BMBF), Europäischer Fonds für Regionalentwicklung (EFRE), Land Sachsen-Anhalt, and the German-Israeli Foundation.

REFERENCES

Amor, S., F. Puentes, D. Baker, and P. van der Valk. 2010. Inflammation in neurodegenerative diseases. *Immunology* 129:154–69.

Arumugam, T. V., D. N. Granger, and M. P. Mattson. 2005. Stroke and T-cells. *Neuromolecular Med* 7:229–42.

Azarashvili, T., R. Stricker, and G. Reiser. 2010. The mitochondria permeability transition pore complex in the brain with interacting proteins–promising targets for protection in neurodegenerative diseases. *Biol Chem* 391:619–29.

Candelario-Jalil, E. 2009. Injury and repair mechanisms in ischemic stroke: considerations for the development of novel neurotherapeutics. *Curr Opin Investig Drugs* 10:644–54.

Chen, H., H. Yoshioka, G. S. Kim, et al. 2011. Oxidative stress in ischemic brain damage: mechanisms of cell death and potential molecular targets for neuroprotection. *Antioxid Redox Signal* 14:1505–17.

Chico, L. K., L. J. Van Eldik, and D. M. Watterson. 2009. Targeting protein kinases in central nervous system disorders. *Nat Rev Drug Discov* 8:892–909.

Culman, J., A. Blume, P. Gohlke, and T. Unger. 2002. The renin-angiotensin system in the brain: possible therapeutic implications for AT(1)-receptor blockers. *J Hum Hypertens* 16(Suppl 3):S64–70.

del Zoppo, G. J., R. Milner, T. Mabuchi, et al. 2007. Microglial activation and matrix protease generation during focal cerebral ischemia. *Stroke* 38:646–51.

Dirnagl, U., K. Becker, and A. Meisel. 2009. Preconditioning and tolerance against cerebral ischaemia: from experimental strategies to clinical use. *Lancet Neurol* 8:398–412.

Dirnagl, U. and A. Meisel. 2008. Endogenous neuroprotection: mitochondria as gateways to cerebral preconditioning? *Neuropharmacology* 55:334–44.

Dirnagl, U., R. P. Simon, and J. M. Hallenbeck. 2003. Ischemic tolerance and endogenous neuroprotection. *Trends Neurosci* 26:248–54.

Doyle, K. P., R. P. Simon, and M. P. Stenzel-Poore. 2008. Mechanisms of ischemic brain damage. *Neuropharmacology* 55:310–18.

Duan, Y., R. A. Gross, and S. S. Sheu. 2007. Ca^{2+}-dependent generation of mitochondrial reactive oxygen species serves as a signal for poly(ADP-ribose) polymerase-1 activation during glutamate excitotoxicity. *J Physiol* 585:741–58.

Endres, M. 2005. Statins and stroke. *J Cereb Blood Flow Metab* 25:1093–110.

Endres, M., B. Engelhardt, J. Koistinaho, et al. 2008. Improving outcome after stroke: overcoming the translational roadblock. *Cerebrovasc Dis* 25:268–78.

Feissner, R. F., J. Skalska, W. E. Gaum, and S. S. Sheu. 2009. Crosstalk signaling between mitochondrial Ca^{2+} and ROS. *Front Biosci* 14:1197–218.

Gabathuler, R. 2010. Approaches to transport therapeutic drugs across the blood-brain barrier to treat brain diseases. *Neurobiol Dis* 37:48–57.

Gidday, J. M. 2006. Cerebral preconditioning and ischaemic tolerance. *Nat Rev Neurosci* 7:437–48.

Ginsberg, M. D. 2008. Neuroprotection for ischemic stroke: past, present and future. *Neuropharmacology* 55:363–89.

Gutierrez, M., J. J. Merino, M. A. de Lecinana, and E. Diez-Tejedor. 2009. Cerebral protection, brain repair, plasticity and cell therapy in ischemic stroke. *Cerebrovasc Dis* 27(Suppl 1):177–86.

Hailer, N. P. 2008. Immunosuppression after traumatic or ischemic CNS damage: it is neuroprotective and illuminates the role of microglial cells. *Prog Neurobiol* 84:211–33.

Hardeland, R. 2009. Neuroprotection by radical avoidance: search for suitable agents. *Molecules* 14:5054–102.

Hardingham, G. E. 2009. Coupling of the NMDA receptor to neuroprotective and neurodestructive events. *Biochem Soc Trans* 37:1147–60.

Hertz, L. 2008. Bioenergetics of cerebral ischemia: a cellular perspective. *Neuropharmacology* 55:289–309.

Huang, E. J. and L. F. Reichardt. 2003. Trk receptors: roles in neuronal signal transduction. *Annu Rev Biochem* 72:609–42.

Hung, C. W., Y. J. Liou, S. W. Lu, et al. 2010. Stem cell-based neuroprotective and neurorestorative strategies. *Int J Mol Sci* 11:2039–55.

Jin, R., G. Yang, and G. Li. 2010. Inflammatory mechanisms in ischemic stroke: role of inflammatory cells. *J Leukoc Biol* 87:779–89.

Jordan, J., T. Segura, D. Brea, M. F. Galindo, and J. Castillo. 2008. Inflammation as therapeutic objective in stroke. *Curr Pharm Des* 14:3549–64.

Kaur, S., A. S. Jaggi, and N. Singh. 2009. Molecular aspects of ischaemic postconditioning. *Fundam Clin Pharmacol* 23:521–36.

Kernie, S. G. and J. M. Parent. 2010. Forebrain neurogenesis after focal ischemic and traumatic brain injury. *Neurobiol Dis* 37:267–74.

Kleinig, T. J. and R. Vink. 2009. Suppression of inflammation in ischemic and hemorrhagic stroke: therapeutic options. *Curr Opin Neurol* 22:294–301.

Lakhan, S. E., A. Kirchgessner, and M. Hofer. 2009. Inflammatory mechanisms in ischemic stroke: therapeutic approaches. *J Transl Med* 7:97.

Lemasters, J. J., T. P. Theruvath, Z. Zhong, and A. L. Nieminen. 2009. Mitochondrial calcium and the permeability transition in cell death. *Biochim Biophys Acta* 1787:1395–401.

Llorens, F., V. Gil, and J. A. Del Rio. 2010. Emerging functions of myelin-associated proteins during development, neuronal plasticity, and neurodegeneration. *Faseb J* 25:463–75.

Lo, E. H., M. A. Moskowitz, and T. P. Jacobs. 2005. Exciting, radical, suicidal: how brain cells die after stroke. *Stroke* 36:189–92.

Mantz, J., V. Degos, and C. Laigle. 2010. Recent advances in pharmacologic neuroprotection. *Eur J Anaesthesiol* 27:6–10.

McColl, B. W., S. M. Allan, and N. J. Rothwell. 2009. Systemic infection, inflammation and acute ischemic stroke. *Neuroscience* 158:1049–61.

Medel, R., R. W. Crowley, M. S. McKisic, A. S. Dumont, and N. F. Kassell. 2009. Sonothrombolysis: an emerging modality for the management of stroke. *Neurosurgery* 65:979–93.

Mehta, S. L., N. Manhas, and R. Raghubir. 2007. Molecular targets in cerebral ischemia for developing novel therapeutics. *Brain Res Rev* 54:34–66.

Moroni, F. 2008. Poly(ADP-ribose)polymerase 1 (PARP-1) and postischemic brain damage. *Curr Opin Pharmacol* 8:96–103.

Murphy, M. P. 2009. How mitochondria produce reactive oxygen species. *Biochem J* 417:1–13.

Ohtaki, H., T. Nakamachi, K. Dohi, and S. Shioda. 2008. Role of PACAP in ischemic neural death. *J Mol Neurosci* 36:16–25.

Okouchi, M., O. Ekshyyan, M. Maracine, and T. Y. Aw. 2007. Neuronal apoptosis in neurodegeneration. *Antioxid Redox Signal* 9:1059–96.

Patel, M. M., B. R. Goyal, S. V. Bhadada, J. S. Bhatt, and A. F. Amin. 2009. Getting into the brain: approaches to enhance brain drug delivery. *CNS Drugs* 23:35–58.

Pignataro, G., R. P. Simon, and Z. G. Xiong. 2007. Prolonged activation of ASIC1a and the time window for neuroprotection in cerebral ischaemia. *Brain* 130:151–8.

Pradelli, L. A., M. Beneteau, and J. E. Ricci. 2010. Mitochondrial control of caspase-dependent and -independent cell death. *Cell Mol Life Sci* 67:1589–97.

Pun, P. B., J. Lu, and S. Moochhala. 2009. Involvement of ROS in BBB dysfunction. *Free Radic Res* 43:348–64.

Quarles, R. H. 2007. Myelin-associated glycoprotein (MAG): past, present and beyond. *J Neurochem* 100:1431–48.

Reichardt, L. F. 2006. Neurotrophin-regulated signalling pathways. *Philos Trans R Soc Lond B Biol Sci* 361:1545–64.

Rip, J., G. J. Schenk, and A. G. de Boer. 2009. Differential receptor-mediated drug targeting to the diseased brain. *Expert Opin Drug Deliv* 6:227–37.

Rosenberg, N., M. Chen, and S. Prabhakaran. 2010. New devices for treating acute ischemic stroke. *Recent Pat CNS Drug Discov* 5:118–34.

Sandu, N. and B. Schaller. 2010. Postconditioning: a new or old option after ischemic stroke? *Expert Rev Cardiovasc Ther* 8:479–82.

Siren, A. L., T. Fasshauer, C. Bartels, and H. Ehrenreich. 2009. Therapeutic potential of erythropoietin and its structural or functional variants in the nervous system. *Neurotherapeutics* 6:108–27.

Strbian, D., P. T. Kovanen, M. L. Karjalainen-Lindsberg, T. Tatlisumak, and P. J. Lindsberg. 2009. An emerging role of mast cells in cerebral ischemia and hemorrhage. *Ann Med* 41:438–50.

Urbaniak Hunter, K., C. Yarbrough, and J. Ciacci. 2010. Stem cells in the treatment of stroke. *Adv Exp Med Biol* 671:105–16.

Webster, N. J. and M. C. Pirrung. 2008. Small molecule activators of the Trk receptors for neuroprotection. *BMC Neurosci* 9(Suppl 2):S1.

Yanamadala, V. and R. M. Friedlander. 2010. Complement in neuroprotection and neurodegeneration. *Trends Mol Med* 16:69–76.

Yilmaz, G. and D. N. Granger. 2010. Leukocyte recruitment and ischemic brain injury. *Neuromolecular Med* 12:193–204.

Zhang, Z. G. and M. Chopp. 2009. Neurorestorative therapies for stroke: underlying mechanisms and translation to the clinic. *Lancet Neurol* 8:491–500.

Zhao, H. 2009. Ischemic postconditioning as a novel avenue to protect against brain injury after stroke. *J Cereb Blood Flow Metab* 29:873–85.

Zündorf, G., S. Kahlert, V. I. Bunik, and G. Reiser. 2009. α-Ketoglutarate dehydrogenase contributes to production of reactive oxygen species in glutamate-stimulated hippocampal neurons in situ. *Neuroscience* 158:610–16.

Zündorf, G. and G. Reiser. 2011. Calcium dysregulation and homeostasis of neural calcium in the molecular mechanisms of neurodegenerative diseases provide multiple targets for neuroprotection. *Antioxid Redox Signal* 14:1275–88.

6 Dynorphins in Central Nervous System Pathology

Kurt F. Hauser, Pamela E. Knapp,
Tatiana Yakovleva, Dineke S. Verbeek,
and Georgy Bakalkin

CONTENTS

6.1 INTRODUCTION

The preprodynorphin gene encodes several key peptides, which have emergent functions in the normal and pathophysiological processes of neurons and glia (Hauser et al. 2005). Major posttranslational products of the preprodynorphin gene include dynorphin A [dynorphin A(1-17)] and dynorphin B (collectively termed *dynorphins*) and α-neoendorphin (Goldstein et al. 1979; Chavkin, James, and Goldstein 1982; Kakidani et al. 1982; Evans, Hammond, and Frederickson 1988). Dynorphin A is especially unique in its dual ability to bind with high affinity κ-opioid receptors at physiological concentrations and activate glutamatergic receptors at higher, supraphysiological concentrations that perhaps only accompany injury or disease. The amino acid sequence of dynorphin A (YGGFLRRIRPKLKWDNQ) is highly conserved and is identical in humans, rodents, bovine, and porcine species, and in amphibians (Danielson et al. 2002; Pattee et al. 2003; Sundstrom, Dreborg, and Larhammar 2010). Accumulating evidence suggests that dynorphins can exert protective or proapoptotic effects depending on whether they activate opioid receptors or nonopioid mechanisms (glutamate receptors or protein–protein interactions). Irrespective of the mechanism involved, it is evident that dynorphins can contribute to nervous system pathology through complex interactions involving multiple receptors and signaling pathways.

The dualistic nature of dynorphin A is revealed in structure–activity studies in which NH_2-terminal and COOH-terminal peptide fragments are compared. The NH_2-terminal tyrosine of dynorphin A is essential for opioid activity, where dynorphins preferentially activate κ-opioid receptors (Chavkin and Goldstein 1981; Day et al. 1991). Initial insight into the bipotent nature of dynorphin A originated from findings that "des-tyrosine" cleavage products of dynorphin A, which lacked the *N*-terminal tyrosine, retained biological activity at nonopioid receptors (Walker, Moises, Coy, Young, et al. 1982; Walker, Tucker, et al. 1982; Faden and Jacobs 1983, 1984; Long, Kinney, et al. 1986; Hauser, Knapp, and Turbek 2001). It was soon realized that the nonopioid activity was principally glutamatergic and mediated in large part through *N*-methyl-D-aspartic acid (NMDA) receptors (Faden and Jacobs 1984; Caudle and Isaac 1988; Bakshi, Ni, and Faden 1992; Dubner and Ruda 1992; Faden and Salzman 1992; Gentile and McIntosh 1993; Shukla and Lemaire 1994; Chen, Gu, and Huang 1995a, 1995b; Vanderah et al. 1996; Laughlin et al. 1997; Chen and Huang 1998; Hauser, Foldes, and Turbek 1999; Caudle and Mannes 2000; Hauser et al. 2001; Laughlin, Larson, and Wilcox 2001; Tan-No et al. 2002; Koetzner et al. 2004). The relatively lower affinity of dynorphin A at NMDA, as well as α-amino-3-hydroxy-5-methyl-4-isoxazolepropionic acid (AMPA) (Singh et al. 2003), receptors compared to κ-opioid receptors assures that at physiological concentrations, dynorphin is acting largely through κ-opioid receptors. There are excellent reviews on the dynorphin–κ-opioid receptor system in addiction and behavior (Shippenberg, Zapata, and Chefer 2007; Bruijnzeel 2009; Mysels and Sullivan 2009; Schwarzer 2009; Knoll and Carlezon 2010; Wee and Koob 2010), as well as review on differential, agonist-selective signaling through the κ-opioid receptor (Aldrich and McLaughlin 2009; Bruchas and Chavkin 2010). The present review

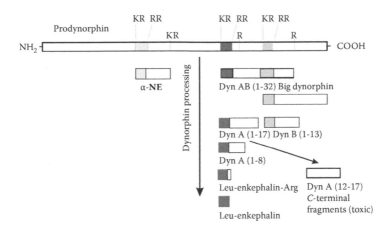

FIGURE 6.1 Prodynorphin processing. Prodynorphin is converted into multiple bioactive peptide fragments, including alpha-neoendorphin (α-NE), and the "dynorphins" big dynorphin (Dyn AB 1-32), leumorphin (Dyn B 1-29), dynorphin A (Dyn A 1-17), dynorphin B (Dyn B 1-16), leucine-enkephalin-arginine (Leu-enkephalin-Arg), which contain a leucine-enkephalin (Leu-enkephalin) sequence at their *N*-termini. The *N*-terminal tyrosine is essential for opioid activity at physiologic concentrations of the above peptides, while the *C*-terminal fragments of dynorphin A can be neurotoxic at high concentrations. Basic lysine (K) and arginine (R) residues are designated. (From Evans, C.J., et al., The Opiate Receptors, Humana, Clifton, NJ. 23–71, 1988; Day, R., et al., *J. Biol. Chem.*, 273, 829–836, 1998; Berman, Y., et al., *J. Neurochem.*, 75, 1763–1770, 2000.)

will focus on the pathobiology of dynorphins and the molecular basis for the pathophysiological effects (Figure 6.1).

6.2 PATHOLOGICAL MECHANISMS OF DYNORPHIN ACTION

The normal volume of the extracellular space in the central nervous system (CNS) is estimated to be about 10–15% of its total volume (Zoli et al. 1999; Sykova 2005; Sykova and Vargova 2008). Electron microscopy, through the use of hypertonic fixatives, compresses the extracellular space to about 2–3% of the total CNS volume (Patel, Hartmann, and Cohen 1971; Palay and Chan-Palay 1974). The volume and geometry of the extracellular space is highly dynamic. Following excitotoxic injury, trauma, or disease, the volume and tortuosity of the extracellular space may decrease by 10-fold, effectively concentrating substances and ions within this compartment, while simultaneously disrupting the kinetics of enzymes that degrade dynorphins (Sykova 1997; Ziak, Chvatal, and Sykova 1998; Sykova et al. 1999; Zoli et al. 1999). Experimentally increasing the extracellular ion concentration, for example, K^+-induced depolarization of cortical and striatal neurons, causes a large increase in the release of "processed" dynorphins (dynorphin B and Leu-enkephalin-Arg[6]) compared to nonprocessed or partially processed dynorphins (Yakovleva et al. 2006). This suggests that activity-dependent production of dynorphins and their release into the interstitial space of the CNS may be greatly exaggerated following excitotoxic conditions accompanying trauma or encephalitis.

6.2.1 Glutamatergic Effects of Dynorphins

Dynorphins appear to contribute to neuropathophysiological processes through several independent mechanisms. First, there is considerable evidence that CNS injury and disease can lead to pathophysiological changes in the biogenesis and processing of dynorphins (Cox et al. 1985; McIntosh, Head, and Faden 1987; Bakshi et al. 1992; Chen and Huang 1992; Dubner and Ruda 1992; Faden and Salzman 1992; Sharma, Nyberg, and Olsson 1992; Shukla and Lemaire 1992; Gentile and McIntosh 1993; McIntosh et al. 1994; Shukla and Lemaire 1994; Chen et al. 1995a, 1995b; Vanderah et al. 1996; Laughlin et al. 1997; Bian et al. 1999; Malan et al. 2000; Redell, Moore, and Dash 2003). Trauma or inflammation can cause large increases in cytotoxic, *C*-terminal fragments of dynorphin A (Hauser et al. 1999, 2001, 2005) and big dynorphin (a 32-amino acid fragment of prodynorphin that includes dynorphin A and dynorphin B) (Tan-No et al. 2001; Kuzmin et al. 2006; Merg et al. 2006). Dynorphin fragments that lack the *N*-terminal opioid Tyr [e.g., dynorphin A(2-17) and dynorphin A(2-13)] (Walker, Moises, Coy, Baldrighi, et al. 1982; Long, Martinez-Arizala, et al. 1986; Shukla and Lemaire 1994; Caudle and Mannes 2000; Laughlin et al. 2001) do not bind well with opioid receptors, but do retain excitotoxic properties suggesting that the toxic effects are nonopioid. For example, traumatic brain injury increases dynorphin release (Sirinathsinghji et al. 1990) despite large reductions in extracellular volume caused by swelling (Vorisek et al. 2002; Sykova 2005), which dramatically increases the relative concentration of dynorphins in the extracellular compartment. *In vitro* evidence shows *C*-terminal fragments of dynorphin A to be relatively stable (Singh et al. 2004, 2005) and intrinsically neurotoxic through the activation of NMDA and AMPA receptors (Hauser et al. 1999, 2001, 2005; Goody et al. 2003; Singh et al. 2003). Second, besides quantitative alterations in dynorphin expression and processing with injury or disease, recent observations by Bakalkin, Verbeek, and coworkers (2010) have found that specific mutations in the coding region of dynorphin A and in the nonopioid region of the prodynorphin gene cause human spinal cerebellar atrophy 23. This is a rare disorder in which dynorphin A mutants cause the degeneration of cerebellar Purkinje neurons, as well as, perhaps, other neuron types in affected individuals (Bakalkin et al. 2010). To our knowledge, the finding that prodynorphin mutations cause spinocerebellar ataxia type 23 (SCA23) is the first evidence that a mutation in a neuropeptide has been shown to be directly responsible for a human neurodegenerative disease (Bakalkin et al. 2010). Each of the general pathological mechanisms of dynorphin action noted earlier has important implications in translational medicine.

6.2.1.1 NMDA Receptors

Dynorphin–NMDA receptor interactions have long been discussed as a significant mechanism in aberrant dynorphin actions and have been reviewed before (Caudle and Isaac 1988; Shukla and Lemaire 1992; Caudle and Mannes 2000; Lai et al.

2001; Wollemann and Benyhe 2004; Hauser et al. 2005). Radioligand-binding assays show that dynorphin A(1-13) or the nonopioid dynorphin A(2-13) can bind with the NMDA receptor complex (Massardier and Hunt 1989; Tang et al. 1999), as well as displace the NMDA open channel blocker, MK-801 (Shukla and Lemaire 1992; Shukla et al. 1993; Shukla, Prasad, and Lemaire 1997; Dumont and Lemaire 1994; Tang et al. 1999). It is noteworthy that dynorphin A fragments that lack basic residues and possess opioid activity alone, such as Leu-enkephalin [dynorphin A(1-5)] or dynorphin A(1-11), are inherently nontoxic (Hauser et al. 2001). Dynorphin A-derived fragments containing Lys[13] and either the *N*-terminal [dynorphin A(3-13)] or *C*-terminal [dynorphin A(13-17)] residues flanking the Lys[13] are neurotoxic to striatal neurons *in vitro* (Hauser et al. 2001), although a somewhat different toxic profile is reported *in vivo* for dynorphin A fragments (Long, Martinez-Arizala, Echevarria, et al. 1988). Equimolar amounts of dynorphin A are more toxic than either the (3-13) or (13-17) fragments alone, suggesting a cumulative effect when both fragments are combined on the common precursor molecule. There is some indication that dynorphin can interact with NMDA receptor glycine sites (Dumont and Lemaire 1994; Zhang et al. 1997), whereas alterative studies infer that dynorphin interacts with the redox modulatory site of the NMDA receptor complex (Chen et al. 1995a; Laughlin, Kitto, and Wilcox 1999). Finally, the effects of dynorphin appear to differ markedly depending on the subunit composition of particular NMDA receptors. For example, in an artificial expression system, dynorphin-induced inhibition of NMDA receptor-evoked currents varied with subunit composition (Brauneis et al. 1996). In addition, NMDA receptor NR2B subunit-selective antagonists can reportedly differentiate antinociception from motor impairment (Boyce et al. 1999). We predict that future studies will increasingly unravel unique roles of dynorphins at different NMDA receptor subtypes and at NMDA receptor targets at "synaptic" versus "extrasynaptic" sites (Leveille et al. 2008), using a combination of pharmacological (subunit selective antagonists—when feasible), gene silencing, and genetic strategies. These approaches hold considerable promise in sorting the beneficial versus deleterious actions of dynorphins.

Although descriptive studies provided considerable insight into the existence and nature of dynorphin–NMDA receptor interactions, the ability to directly manipulate dynorphins *in vivo* has been critical in demonstrating their role in CNS injury. Initial attempts to selectively target dynorphin A involved the use of antidynorphin A immunoneutralizing antibodies to attenuate the consequences of spinal cord injury (Ossipov et al. 1996; Nichols et al. 1997; Winkler et al. 2002). Another novel strategy involves the use of "decoy" or "scavenger" peptides, which selectively bind dynorphin A, to limit ischemic injury accompanying stroke (Woods et al. 2006). The decoy peptide was conceived of by Woods, Shippenberg, and coworkers (2006) based on a highly conserved epitope of the common NR1 subunit of the NMDA receptor complex, which is a likely target for noncovalent interactions with dynorphin A. The hypothesis that the highly acidic KVNSEEEEEDA decoy would bind highly basic dynorphin A peptides via noncovalent interactions was confirmed by mass spectrometry. Importantly, the investigators then went on to show that KVNSEEEEEDA could prevent dynorphin

A-induced neurotoxicity, allodynia *in vivo*, and could block its cytotoxicity in cultured striatal neurons (Woods et al. 2006). Another strategy using knockout mice confirmed the role of dynorphins in secondary spinal cord injury (Adjan et al. 2007). In these studies, deletion of the prodynorphin gene was protective in spinal cord injury (Adjan et al. 2007). The spinal cords of prodynorphin knockout mice displayed significantly fewer cleaved caspase-3-positive cells compared to their wild-type counterparts. Importantly, activated caspase-3 was reduced in both oligodendroglia and astroglia of knockout mice, whereas expression in neurons was unchanged. By augmenting caspase-3 activation in glia, dynorphin peptides likely increase the probability of oligodendroglial and astroglial apoptosis after CNS trauma (Adjan et al. 2007). Collectively, the aforementioned findings suggest that prodynorphin or its peptide products, although normally beneficial, can become maladaptive and contribute to secondary injury following trauma.

6.2.1.2 AMPA Receptors

As previously reviewed (Hauser et al. 2005), there are some reports that dynorphins can activate AMPA/kainate receptors (Tortella and DeCoster 1994; Kolaj, Cerne, and Randic 1995), and this may impart neurotoxic signals at supraphysiological concentrations of dynorphin (Singh et al. 2003). This was revealed in striatal neurons, which, unlike many other neuron types, express AMPA receptors earlier than NMDA receptors during maturation (Martin, Furuta, and Blackstone 1998). Unlike spinal cord neurons in which dynorphin A toxicity is preferentially mediated by NMDA receptors (Hauser et al. 1999, 2001), in striatal neurons, dynorphin A toxicity can be reversed by the AMPA receptor antagonist CNQX and mimicked by the ampakine CX546 (Goody et al. 2003).

Dynorphin A-induced apoptosis in striatal neurons acting via NMDA and, to a lesser extent, AMPA receptors is accompanied by caspase-3 activation (Singh et al. 2003). This includes the activation of both p38-kinase and c-jun-*N*-terminal kinase (JNK) 1 and 2 mitogen-activated protein kinases (MAPKs). Phosphorylation of p38-kinase, but not JNK-1 and 2, has been speculated to be necessary for dynorphin to activate caspase-3 (Hauser et al. 2005). Similar patterns of striatal neurotoxicity are seen with HIV-1 Tat exposure, which are also mediated by mixed-lineage kinases, including p38 and JNK (Singh et al. 2004, 2005). We speculate that dynorphin acts in a similar manner to Tat because of the highly basic nature of both peptides and because both have been shown to activate NMDA receptors through noncovalent protein–protein interactions (Woods et al. 2006; Li et al. 2008), although additional studies are necessary to confirm or deny this assumption.

Supraphysiological amounts of dynorphin A, when applied extracellularly, can act via a generalized excitotoxic mechanism in a manner similar to excess glutamate (Caudle and Isaac 1988; Massardier and Hunt 1989; Isaac et al. 1990; Johnson et al. 1991; Bakshi et al. 1992; Skilling et al. 1992; Chen and Huang 1998; Caudle, Chavkin, and Dubner 1994; Shukla and Lemaire 1994; Chen et al. 1995a, 1995b; Hanson et al. 1995; Vanderah et al. 1996; Shukla et al. 1997; Caudle and Dubner 1998; Lai, Gu, and Huang 1998; Hauser et al. 1999, 2001, 2005; Laughlin et al. 1999; Szeto, Soong, and Wu 1999; Tang et al. 1999; Tan-No et al.

2002; Koetzner et al. 2004; Wollemann and Benyhe 2004; Woods et al. 2006). The excitotoxic and electrophysiological effects of dynorphins can be partially ameliorated by NMDA receptor antagonists, suggesting a role for NMDA receptors (Caudle and Isaac 1988; Massardier and Hunt 1989; Isaac et al. 1990; Johnson et al. 1991; Bakshi et al. 1992; Skilling et al. 1992; Caudle et al. 1994; Laugh et al. 1994; Shukla and Lemaire 1994; Hanson et al. 1995; Vanderah et al. 1996; Shukla et al. 1997; Lai et al. 1998; Hauser et al. 1999, 2001; Laughlin et al. 1999, 2001; Tang et al. 1999; Woods and Zangen 2001). In addition, the excitotoxic effects of high concentrations of dynorphins can also be attenuated through AMPA receptor blockade, suggesting AMPA receptors can also contribute to the excitatory effects of dynorphins. However, despite marked actions of dynorphins via NMDA receptors, and to a lesser extent AMPA receptors, there remained indications that dynorphin could depolarize neuronal membranes by mechanisms independent of either NMDA or AMPA receptors (Hauser et al. 2001).

6.2.2 NONTRADITIONAL ACTIONS OF DYNORPHINS—PRODYNORPHIN, BIG DYNORPHIN, AND DIRECT PROTEIN–PROTEIN INTERACTIONS

Dynorphins and especially larger, intact, or partially processed dynorphin precursors, such as prodynorphin and big dynorphin, may act through protein–protein interactions that are independent of cell surface receptors (Tan-No et al. 2001, 2002; Yakovleva et al. 2001) or by disrupting membrane function (Marinova et al. 2005; Hugonin et al. 2006, 2008). Paradoxically, prodynorphin and big dynorphin appear to serve pronociceptive functions that differ markedly from "processed" dynorphin fragments, which are antinociceptive (Tan-No et al. 2009). Big dynorphin is cytotoxic at high femtomolar concentrations when expressed intracellularly (Tan-No et al. 2001) and is markedly more toxic than equimolar concentrations of dynorphin A, suggesting that big dynorphin's toxicity is not due to its conversion to dynorphin A (Tan-No et al. 2001). As the cytotoxic profiles of big dynorphin and dynorphin A are similar, dynorphin A retains some key structural sequences of big dynorphin that are inherently toxic (Tan-No et al. 2001). The potential intracellular targets of big dynorphin are not completely understood. As previously reviewed (Hauser et al. 2005), big dynorphin appears to induce toxicity through an apoptotic mechanism that may involve interactions with the tumor suppressor protein p53. Summarized briefly, big dynorphin is highly basic and tentatively induces cell death through noncovalent protein–protein interactions with a variety of acidic target proteins in a manner similar to dynorphin A (Woods et al. 2006). Novel N-terminus-truncated prodynorphin variants (Nikoshkov et al. 2005) may traffic to the cell nucleus or cytoplasm, suggesting that they have functions other than those of full-length prodynorphin. Truncated prodynorphin or big dynorphin may interfere with intracellular processes by binding to cytoplasmic proteins. Prodynorphin variants lacking the signal peptide are widely expressed in neoplastic cells. These prodynorphin variants fail to translocate into the endoplasmic reticulum (ER), and instead degrade in the cytoplasm and may potentially influence tumor cell fate (Hauser et al. 2005).

6.2.3   NONTRADITIONAL ACTIONS OF DYNORPHINS—BIOPHYSICAL INTERACTIONS WITH BIOLOGICAL MEMBRANES

Besides intracellular actions of intact prodynorphin or partially processed fragments, prodynorphin fragments also appear to have intrinsic effects at the cell membrane. Dynorphin-induced membrane depolarization can be attenuated by L-type voltage-dependent Ca^{2+} channel (VDCC) blockers (Simmons et al. 1995). The challenge was then to determine how dynorphin was able to alter the biophysical properties of neurons. Considerable insight was provided by studies of Gräslund, Bakalkin, and coworkers, who showed that dynorphins could disrupt lipid membranes, making them more "leaky," and in doing so permit the fluxing of nearby ions (Marinova et al. 2005; Hugonin et al. 2006, 2008). In fact, dynorphin sufficiently increased membrane permeability to allow Ca^{2+} outside ($[Ca^{2+}]_o$) to enter intact liposomes. Thus, dynorphin increased permeability sufficiently to permit ions to flux across biological membranes. Direct evidence for this was obtained using fluorescence correlation spectroscopy, which visualized the movement of dynorphin A into and across the plasmalemma of living cells (Marinova et al. 2005). These novel biophysical properties suggest that dynorphin *per se* has the potential to depolarize neuronal membranes directly and this alone may be sufficient to remove the Mg^{2+} block from NMDA receptors and open VDCCs, thereby promoting excitotoxic injury.

6.2.4   DELETERIOUS EFFECTS OF DYNORPHINS AT OPIOID RECEPTORS

As previously reviewed (Hauser et al. 2005), some of dynorphin's actions resulting in neural dysfunction *in vivo* can be attenuated by broad acting μ-, δ-, and κ-opioid receptor antagonists, including naloxone or selective κ-opioid receptor antagonists, such as nor-binaltorphimine (Faden, Takemori, and Portoghese 1987; Faden et al. 1990; Yum and Faden 1990; Puniak et al. 1991; Benzel, Khare, and Fowler 1992; McIntosh et al. 1994; Caudle and Mannes 2000; Laughlin et al. 2001; Xu et al. 2004). Similarly, opioid antagonists are reported to be beneficial in experimental models of spinal cord injury or traumatic brain injury, suggesting that endogenous dynorphins (or potentially other endogenous opioid peptides) contribute to secondary CNS injury through actions at opioid receptors (Faden et al. 1987; McIntosh et al. 1987; Vink et al. 1990; Behrmann, Bresnahan, and Beattie 1993; Baskin et al. 1993; DeWitt et al. 1997). The aspects of dynorphin-evoked secondary CNS injury or neuropathic pain that are attributable to opioid receptor activation are multifaceted and more controversial than neuropathologic effects of dynorphins that are not mediated via opioid receptors. Potentially deleterious dynorphin–opioid receptor interactions have been reviewed previously (Long, Martinez-Arizala, et al. 1986; Faden 1990; Dubner and Ruda 1992; Faden and Salzman 1992; Lyeth and Hayes 1992; Shukla and Lemaire 1992; Caudle and Mannes 2000; Laughlin et al. 2001; Hall and Springer 2004; Bruijnzeel 2009; Mysels and Sullivan 2009; Schwarzer 2009; Tan-No et al. 2009; Bruchas and Chavkin 2010; Knoll and Carlezon 2010; Wee and Koob 2010) and have provided the basis for clinical trials of high-dose naloxone treatment for spinal cord injury

(Flamm et al. 1985; Bracken et al. 1992; Bracken and Holford 1993), which proved to be largely nonefficacious (Bracken et al. 1992; Bracken 1993; Hall and Springer 2004).

6.3 BENEFICIAL ROLES OF DYNORPHIN AT κ-OPIOID RECEPTORS

At physiological concentrations, dynorphins bind with κ-opioid receptors with higher affinity than either with μ- or δ-opioid receptors and are generally classified as the preferential endogenous ligands for κ-opioid receptors (Chavkin et al. 1982). Importantly, dynorphin A fragments lacking the C-terminal basic residues and possessing opioid activity alone, such as Leu-enkephalin [dynorphin A(1-5)] or [dynorphin A(1-11)], are nontoxic (Hauser et al. 2001). Consistent with a preferential action at κ-opioid receptors, dynorphin A can alter excitatory amino acid levels in the CNS, and these changes can be attenuated by the opioid receptor antagonist nalmefene or the selective κ-opioid receptor antagonist, nor-binaltorphimine (Bakshi, Newman, and Faden 1990; Long et al. 1994). The activation of opioid receptors by dynorphins at physiological concentrations serves normative functions, and, in some instances, appears to be neuroprotective.

6.3.1 BENEFICIAL ACTIONS OF DYNORPHIN A VIA κ-OPIOID RECEPTORS MAY BE SUBORDINATE TO NEUROTOXIC ACTIONS AT NMDA RECEPTORS

When dynorphin A is added at high concentrations to dorsal spinal cord neurons expressing both κ-opioid and NMDA receptors, the toxic actions through NMDA receptors override the protective effects at κ-receptors (Hauser et al. 1999). At lower dynorphin A concentrations, neurotoxicity is not an issue and any potential protective effects at κ-opioid receptors would predominate (Hauser et al. 1999). In neuronal populations possessing both AMPA and NMDA receptors, the cytotoxic effects of abnormally high levels of dynorphin A at NMDA receptors appear to be generally more potent than at other glutamate receptors, potentially masking negative effects at AMPA receptors (Hauser unpublished). Again, it is important to point out that when NMDA, AMPA, and κ-opioid receptors function in a coordinated manner, such as with excitatory transmission in the dentate gyrus, endogenous dynorphin A blocks excitatory neurotransmission, and the induction of long-term potentiation (Wagner, Terman, and Chavkin 1993). Again, owing to the lower affinity of dynorphins for glutamate versus opioid receptors, supraphysiological levels or aberrant processing of dynorphins are seemingly necessary before the more potent excitotoxic signaling pathways become operative.

Although there are a few reports that opioids can have deleterious effects following CNS injury, opioid receptor activation is generally perceived as inconsequential to injury outcome, or even beneficial depending on the particular opioid receptor type involved, the nature of the insult, and the outcome measure that is assessed (Baskin et al. 1984; Long, Martinez-Arizala, et al. 1986; Long, Petras, et al. 1988; Long et al. 1989; Silvia et al. 1987; Hall and Pazara 1988; Itoh, Ukai, and Kameyama

1993a, 1993b; Caudle and Dubner 1998; Tegeder and Geisslinger 2004). κ-Opioid receptor activation can be neuroprotective (Hudson et al. 1991; Behrmann et al. 1993; Genovese, Moreton, and Tortella 1994; Przewlocka, Machelska, and Lason 1994; Tortella and DeCoster 1994; Ossipov et al. 1996; Nichols et al. 1997; Sheng et al. 1997; Kong et al. 2000; Solbrig and Koobs 2004). Synthetic κ-opioid receptor agonists, such as U54,488H or U69,593, as well as dynorphin A(1-13), can improve aspects of neural injury or ischemic injury (Baskin et al. 1984; Silvia et al. 1987; Hall and Pazara 1988; Itoh Ukai, and Kameyama 1993a, 1993b; Caudle and Dubner 1998). Nerve damage from trauma is also reduced by κ-opioid receptor activation (Hall et al. 1987). Some negative effects of dynorphin may arise from κ-opioid receptor-mediated alterations in cerebrovascular functions following injury (La Torre et al. 1991; Lyeth and Hayes 1992; Armstead and Kurth 1994; Armstead 1997; DeWitt et al. 1997) or local opioid-dependent alterations in glutamate or aspartate release (Bakshi et al. 1990; Faden et al. 1990; Faden 1992; Graham et al. 1993). Opioid receptor blockade significantly increases the neurotoxic effects of dynorphin A(1-13) through glutamate receptors, providing indirect evidence that κ-opioid receptor stimulation is protective (Hauser et al. 1999). Finally, dynorphin A acting via κ-opioid receptors tends to attenuate macrophage function (Gabrilovac, Balog, and Andreis 2003).

The activation of postsynaptic κ-opioid receptors typically hyperpolarizes neurons by opening K^+ channels and closing Ca^{2+} channels (Macdonald and Werz 1986; Surprenant, Shen, and North 1990; Xiang et al. 1990; Huang 1995; Rusin et al. 1997), which should counter excitotoxic effects at glutamate receptors (Wagner, Caudle, and Chavkin 1992). Selective κ-opioid receptor agonists also limit the presynaptic release of glutamate (Gannon and Terrian 1991, 1992; Simmons and Chavkin 1996; Rusin et al. 1997; Herrera-Marschitz et al. 1998). Opioid receptors can couple with signaling pathways that promote cell survival (Mangoura and Dawson 1993; Hauser and Mangoura 1998; Polakiewicz et al. 1998; Bohn, Belcheva, and Coscia 2000b; Belcheva et al. 2001; Tegeder and Geisslinger 2004). Opioid-dependent activation of MAPK is likely to have complex effects on cell survival (Law and Bergsbaken 1995; Polakiewicz et al. 1998; Hauser et al. 2001; Knapp et al. 2001; Wang et al. 2003). κ-Opioid receptor agonists, via Gβγ-protein subunits and perhaps other agonist-selective pathways, have been shown to activate a variety of protective pathways in cells (Charron, Messier, and Plamondon 2008; Lin et al. 2008; McLennan et al. 2008; Peart et al. 2008; Husain, Potter, and Crosson 2009; Jaiswal et al. 2010). Coscia and coworkers have extended the aforementioned findings to show that opioid signaling can converge at both epidermal growth factor and fibroblast growth factor (FGF) neurotrophic pathways (Belcheva et al. 2001, 2002, 2003; Miyatake et al. 2009). Stimulating κ-opioid receptors can also direct the fate and survival of neural stem cells (Kim et al. 2006). Similar opioid and FGF-2 signaling convergence is likely to occur in adult hippocampal, neural progenitors (Persson, Thorlin, Bull, and Eriksson 2003; Persson, Thorlin, Bull, Zarnegar, et al. 2003), as well as glial progenitors, which can express μ-, δ-, and κ-opioid receptors (Hauser et al. 2009). Thus, κ-opioid receptors appear to couple with MAPKs in glia (Bohn, Belcheva, and Coscia 2000a; Belcheva and Coscia 2002; Belcheva et al.

2005; Kim et al. 2006; McLennan et al. 2008; Miyatake et al. 2009) and are likely to be intrinsically beneficial to glia.

Finally, recently discovered missense mutations in prodynorphin and, in particular, in the dynorphin A coding region, have been shown to underlie the etiology of at least one form of spinocerebellar ataxia (Bakalkin et al. 2010). Because of unique biological differences among mutant dynorphin peptides in SCA23 compared to the native dynorphin A peptide (Bakalkin et al. 2010), and the potential of disrupted prodynorphin biosynthesis intracellularly, the role of prodynorphin mutations in the pathogenesis of diseases is likely to be somewhat different from previously mentioned mechanisms of excitotoxic neuronal injury and death (Hauser et al. 2005).

6.4 DYNORPHIN IN THE ETIOLOGY OF CNS DISEASE

6.4.1 Pathobiology of Dynorphins in Drug Abuse

The interrelationship of drug abuse and the dynorphin–κ-opioid receptor system has been the topic of several recent reviews (Aldrich and McLaughlin 2009; Bruijnzeel 2009; Bruchas and Chavkin 2010; Wee and Koob 2010). We previously reviewed whether drug abuse might alter CNS function by causing pathological changes in dynorphins, and readers are referred to that review (Hauser et al. 2005). Briefly, dynorphins acting presynaptically via κ-opioid receptors decrease dopamine release from terminals in the nucleus accumbens and dorsal striatum (Di Chiara and Imperato 1988; Spanagel, Herz, and Shippenberg 1990, 1992; Chefer et al. 2000; Thompson et al. 2000). These actions are thought to mediate the rewarding effects of drugs with abuse liability and to contribute to the compulsive drug-seeking behavior that characterizes drug addiction (Wise 1998; Ito et al. 2002) (see reviews, Weiss and Porrino 2002; Shippenberg et al. 2007; Wee and Koob 2010). Glutamate-driven neuroplasticity is also important for psychostimulant addiction and agrees with the concept that glutamatergic signaling contributes to plasticity and learning (Cornish and Kalivas 2000; Hotsenpiller, Giorgetti, and Wolf 2001; Shippenberg and Chefer 2002; Everitt and Wolf 2002; McFarland, Lapish, and Kalivas 2003; Hauser et al. 2005).

6.4.2 Pathobiology of Dynorphins in Spinal Cord Injury and Neuropathic Pain

The interactive role of dynorphins in spinal cord injury and neuropathic pain is an important area of study and has been extensively reviewed (Shukla and Lemaire 1994; Caudle and Mannes 2000; Lai et al. 2001; Hauser et al. 2005; Tan-No et al. 2009). This review focuses on recent, novel potential areas where the dynorphin–κ-opioid system appears to influence disease processes. Dynorphins have long been implicated in the secondary pathophysiology of spinal cord or traumatic brain injury (Faden and Jacobs 1984; Caudle and Isaac 1988; Bakshi et al. 1992; Chen and Huang 1992, 1998; Dubner and Ruda 1992; Faden 1992; Faden and Salzman 1992; Gentile and McIntosh 1993; McIntosh et al. 1994; Shukla and Lemaire 1994; Chen et al.

1995a, 1995b; Vanderah et al. 1996; Laughlin et al. 1997, 2001; Hauser et al. 1999, 2001; Caudle and Mannes 2000; Tan-No et al. 2002). Intrathecal injections of high concentrations of dynorphin A result in a variety of problems, including paralysis, vascular dysfunction, and neurodegeneration (Long, Martinez-Arizala, et al. 1986; Long, Martinez-Arizala, Rigamonti, et al. 1988). As noted in the previous sections, the deleterious effects are markedly reduced by MK-801 and competitive NMDA receptor antagonists, suggesting that NMDA receptors mediate many of these maladaptive effects (Faden and Jacobs 1983; Long, Martinez-Arizala, Echevarria, et al. 1988; Bakshi et al. 1990, 1992; Isaac et al. 1990; Faden 1992; Skilling et al. 1992; Shukla and Lemaire 1994) More recent strategies that selectively bind or eliminate dynorphins confirm their role in secondary injury through excitotoxic modes of action (Woods et al. 2006; Adjan et al. 2007). However, because of the complexities of spinal cord injury, it is unlikely that targeting any one molecule or neurochemical system will be therapeutically successful. It has been suggested that therapeutic strategies be broadened to include the use of stem cells or engineered Schwann cells to replace neurons and glia lost by the ravages of primary and secondary traumatic injury (Thuret, Moon, and Gage 2006). The intricacies of spinal cord injury may necessitate targeting a large number of molecules that contribute to secondary injury in addition to dynorphins.

6.4.3 PRODYNORPHIN MUTATIONS AS THE ETIOLOGICAL BASIS OF SPINOCEREBELLAR ATAXIA TYPE 23

As mentioned earlier, Bakalkin, Verbeek, and coworkers (2010) recently found that several missense mutations in the prodynorphin gene, including the coding region of dynorphin A, underlie the neurodegenerative changes in SCA23. When several of the mutant dynorphin A peptides were bath-applied extracellularly, they were significantly more neurotoxic than wild-type dynorphin A. A logical assumption is that the toxic mutant variants of dynorphin A are triggering neuron death via a traditional excitotoxic, caspase-dependent apoptotic pathway (Singh et al. 2003). Importantly, neurodegeneration occurs in Purkinje cells (Bakalkin et al. 2010), as well as in the dentate and olivary nuclei (Verbeek et al. 2004), which contain dynorphinergic neurons (Zamir, Palkovits, and Brownstein 1984; Fallon and Leslie 1986; Altschuler et al. 1988; Sahley, Nodar, and Musiek 1999), but degeneration was not reported in the surrounding nondynorphinergic neuron populations, and there are limited accompanying gliotic changes (Verbeek et al. 2004), suggesting bystander effects are not operative. Other evidence leads us to question whether the neurodegeneration is SCA23 is exclusively caused via excitotoxic mechanisms. For example, rat insulinoma (RINm-5F) cells transfected with wild-type and SCA23 mutant prodynorphin variants, showed ~19-fold increase in dynorphin A accumulation with the prodynorphin variant p.R212W and ~10-fold increase with the p.L211S variant compared to expression of wild-type prodynorphin (Bakalkin et al. 2010). Although excitotoxic mechanisms (possible via an autocrine feedback mechanism) might be partially responsible, the massive buildup of mutant dynorphins suggests that another mode of neuronal injury may be active. An accumulation of unfolded proteins in the lumen of the endoplasmic reticulum is caused by a variety of mechanisms and initiates an ER stress response (Xu, Bailly-Maitre, and Reed 2005; Maiuri et al. 2007). The ER stress reaction attempts to restore homeostasis through the unfolded protein response (UPR) by (1) halting translation, (2) increasing the production of chaperones that fold proteins, and (3) signaling to destroy misfolded proteins by the ER-associated protein degradation system. If ER stress is severe or prolonged, then regulated cell death is triggered. Autophagy, the UPR, and apoptosis are interrelated. Autophagy counterbalances the rapid expansion of the ER during the UPR and may be obligatory in establishing the homeostatic equilibrium between ER genesis and degradation (Bernales, McDonald, and Walter 2006; Hoyer-Hansen and Jaattela 2007). In severely compromised cells, autophagy and the UPR can trigger death through apoptotic mechanisms. Clearly, additional studies are needed to assess this, as well as other potential prodynorphin-dependent neurodegenerative mechanisms, in SCA23 and perhaps other neurological disorders; however, the finding that prodynorphin mutations drive SCA23 provides new insight into possible mechanisms by which neuropeptides contribute to normal and abnormal CNS function.

6.4.4 Putative Role for the Dynorphin–κ-Opioid Receptor System in Oligodendrocyte Function and Dymyelinating Disease

κ-Opioid receptor activation appears to enhance the survival of oligodendroglia. The expression of κ-opioid receptors by oligodendrocytes is developmentally regulated and occurs in differentiating, myelin basic protein-expressing cells (Knapp, Maderspach, and Hauser 1998). Interestingly, oligodendroglia also express prodynorphin-derived peptides transiently during maturation (Knapp et al. 2001), and we speculate that the prodynorphin-derived peptides markedly enhance oligodendroglial survival through local (autocrine or paracrine) feedback mechanisms at κ-opioid receptors (Knapp et al. 2001). Thus far, however, the evidence for κ-opioid receptor protection in immature oligodendroglia is largely circumstantial and supported by two lines of evidence. In isolated oligodendrocytes *in vitro,* κ-opioid receptor blockade using nor-binaltorphimine significantly exacerbates the cytotoxicity seen following exposure to excess glutamate (Knapp et al. 2001). The assumption here is that endogenous dynorphins (or perhaps partially processed proenkephalin peptides, such as BAM 18, that possess high affinity for κ-opioid receptors) are tonically activating κ-opioid receptors on oligodendroglia (Knapp et al. 2001). In related studies, we examined *jimpy* mutant mice, which are a model of the rare, X-linked, Pelizaeus–Merzbacher disease. *Jimpy* mice possess a mutation in myelin proteolipid protein that results in severe hypomyelination and oligodendrocyte death and is lethal to young mice. Interestingly, we found oligodendrocytes in *jimpy* mice to be completely devoid of κ-opioid (but not μ-opioid) receptors, despite the normal presence of κ-receptors on neurons and other cell types (Knapp, Adjan, and Hauser 2009). Although additional study is warranted, this provides independent evidence linking κ-opioid receptor expression to oligodendrocyte maintenance and survival. Thus, κ-opioid receptor-mediated protective mechanisms might be operative in oligodendroglia, whereas pathophysiological changes in dynorphins have the potential to affect oligodendroglial function and even myelination.

In addition to κ-opioid receptors (Knapp and Hauser 1996; Knapp, Maderspach, and Hauser 1998; Knapp et al. 2001; Stiene-Martin et al. 2001), oligodendrocytes also can express NMDA (Karadottir et al. 2005; Salter and Fern 2005) and AMPA (Gallo, Wright, and McKinnon 1994; Gallo and Russell 1995) receptors, suggesting that dynorphins may serve dichotomous roles in oligodendroglia in a manner similar to spinal cord or striatal neurons. The finding that white-matter injury attenuates the NMDA receptor antagonist memantine (Manning et al. 2008) suggests that excess glutamate and, perhaps, dynorphins normally contribute to white-matter injury. The fact that elevations in cleaved caspase-3 in Oligodendrocytes are markedly reduced after spinal cord trauma in prodynorphin knockout mice supports this assumption (Adjan et al. 2007).

6.4.5 DYNORPHIN–κ-OPIOID SYSTEM IN THE PATHOGENESIS OF NEURO-ACQUIRED IMMUNODEFICIENCY SYNDROME

Interestingly, in a promonocytic U1 cell line, Peterson, Chao, Lokensgard, and coworkers demonstrated that dynorphin or U50,488H caused inverted U-shaped, concentration-dependent increases in HIV-1 expression that were antagonized by nor-binaltorphimine, suggesting a complex relationship between κ-opioid receptor activation and HIV-1 expression (Chao et al. 1995). Later studies by these investigators confirmed that κ-opioid receptor activation caused an initial inhibition, followed by a time-dependent increase in HIV-1 replication in T-lymphocytes and monocytes (Chao et al. 1996, 1998, 2000, 2001; Peterson et al. 2001). Importantly, the initial attenuation in viral production coincided with (1) a significant reduction in the fusion of HIV-1 IIIB Env glycoprotein-expressing HeLa cells with CD4+ lymphocytes and (2) a κ-opioid-receptor-dependent downregulation of CXCR4 (Lokensgard, Gekker, and Peterson 2002). CXCR4 is the major coreceptor that, along with CD4, is responsible for HIV-1 entry into T-lymphocytes (Feng et al. 1996; Carroll et al. 1997). Additional evidence, provided by Rogers and coworkers, demonstrates that κ-opioid receptors undergo heterologous cross desensitization with CXCR4 (Finley et al. 2008). Collectively, the findings that the dynorphin–κ-opioid receptor system can modulate the function of the coreceptor have important implications for the pathogenesis of HIV-1 (Peterson et al. 2001; Lokensgard et al. 2002; Finley et al. 2008).

Alternatively, HIV-1 may affect the endogenous opioid peptides and receptors, including dynorphin–κ-opioid receptor interfacing. Interestingly, in an inducible Tat-expressing transgenic mouse model of HIV-1, prodynorphin and κ-opioid receptor gene transcripts were significantly altered compared to control mice lacking the Tat transgene (Fitting et al. 2010). Moreover, changes in the κ-receptors differed more greatly than μ- or δ-opioid receptor mRNA levels. Most notably, κ-opioid receptor mRNA levels increased dramatically in the whole brain and cerebral cortex, while declining in the striatum. In contrast, prodynorphin transcripts in the hippocampus and striatum declined with high levels of Tat induction, although more modest levels of sustained Tat expression increased prodynorphin transcripts in the cerebral cortex (Fitting et al. 2010). The general pattern suggests that HIV-1 Tat exposure is accompanied by marked reductions in prodynorphin–κ-opioid receptor signaling in the striatum, and to a lesser extent in the hippocampus, compared to wild-type controls, whereas adaptive increases in dynorphin–κ-opioid receptor signaling were present in the cerebral cortex depending on the level of Tat induction and perhaps the duration of Tat exposure. These findings suggest that while the opioid system can influence HIV-1 pathogenesis, the converse is also true—HIV-1 can widely affect the opioid system. Further study is needed to determine whether the changes in the opioid system are contributing to pathology or to an adaptive response to the insult. The findings suggest that the opioid system is fundamentally involved in the pathogenesis of HIV-1 in the CNS.

6.5 CONCLUSIONS

Prodynorphin and dynorphin A (Sundstrom et al. 2010), which can bind κ-opioid (Dreborg et al. 2008) and ionotropic glutamate (Tikhonov and Magazanik 2009) receptors, are highly conserved evolutionarily, suggesting that dynorphin's roles as an agonist at both opioid and glutamate receptors may be important. The bipotent nature of dynorphin underscores the well-established "near-symbiotic" interrelatedness of the opioid and glutamatergic systems. For this reason, dynorphin's role at glutamate receptors may not be entirely maladaptive and is likely to be biologically important. The interdependence of opioidergic and glutamatergic systems is revealed in the pathophysiology of neurotrauma, drug addiction, and neuropathic pain, as well as in normal CNS function (Zhang et al. 1991; Bonham 1995; McGinty 1995; Koob 2000; Ossipov et al. 2000; Solbrig and Koobs 2004). Accordingly, aberrant changes in dynorphin biosynthesis and processing continue to be implicated in the pathogenesis of injury and disease in the CNS. Pathophysiological changes in dynorphin are likely to be important therapeutic targets in specific diseases, such as SCA23, which is caused by prodynorphin mutations, as well as an adjunctive therapy in managing secondary excitotoxic injury caused by CNS trauma.

ACKNOWLEDGMENTS

This work was supported by the National Institute on Drug Abuse grant K02 DA027374 and by grants from the Swedish Council for Working Life and Social Research (FAS), Research Foundation of the Swedish Alcohol Retail Monopoly (SRA), and the Swedish Science Research Council.

REFERENCES

Adjan, V. V., K. F. Hauser, G. Bakalkin, T. Yakovleva, A. Gharibyan, S. W. Scheff & P. E. Knapp. 2007. Caspase-3 activity is reduced after spinal cord injury in mice lacking dynorphin: differential effects on glia and neurons. *Neurosci.* 148: 724–736.
Aldrich, J. V. & J. P. McLaughlin. 2009. Peptide kappa opioid receptor ligands: potential for drug development. *AAPS J.* 11: 312–322.
Altschuler, R. A., K. A. Reeks, J. Fex & D. W. Hoffman. 1988. Lateral olivocochlear neurons contain both enkephalin and dynorphin immunoreactivities: immunocytochemical co-localization studies. *J. Histochem. Cytochem.* 36: 797–801.
Armstead, W. M. 1997. Role of opioids in the physiologic and pathophysiologic control of the cerebral circulation. *Proc. Soc. Exp. Biol. Med.* 214: 210–221.

Armstead, W. M. & C. D. Kurth. 1994. The role of opioids in newborn pig fluid percussion brain injury. *Brain Res.* 660: 19–26.

Bakalkin, G., H. Watanabe, C. Jezierska, C. Depoorter, C. Verschuuren-Bemelmans, I. Bazov, K. A. Artemenko, T. Yakovleva, D. Dooijes, B. P. C. Van de Warrenburg, R. A. Zubarev, B. Kremer, P. E. Knapp, K. F. Hauser, C. Wijmenga, F. Nyberg, R. J. Sinke & D. S. Verbeek. 2010. Prodynorphin mutations cause the neurodegenerative disorder spinocerebellar ataxia type 23. *Am. J. Hum. Genet.* 87: 593–603.

Bakshi, R., A. H. Newman & A. I. Faden. 1990. Dynorphin A-(1-17) induces alterations in free fatty acids, excitatory amino acids, and motor function through an opiate- receptor-mediated mechanism. *J. Neurosci.* 10: 3793–3800.

Bakshi, R., R.-X. Ni & A. I. Faden. 1992. *N*-Methyl-D-aspartate (NMDA) and opioid receptors mediate dynorphin-induced spinal cord injury: behavioral and histological studies. *Brain Res.* 580: 255–264.

Baskin, D. S., Y. Hosobuchi, H. H. Loh & N. M. Lee. 1984. Dynorphin(1-13) improves survival in cats with focal cerebral ischaemia. *Nature.* 312: 551–552.

Baskin, D. S., R. K. Simpson, Jr., J. L. Browning, A. W. Dudley, F. Rothenberg & L. Bogue. 1993. The effect of long-term high-dose naloxone infusion in experimental blunt spinal cord injury. *J. Spinal Disord.* 6: 38–43.

Behrmann, D. L., J. C. Bresnahan & M. S. Beattie. 1993. A comparison of YM-14673, U-50488H, and nalmefene after spinal cord injury in the rat. *Exp. Neurol.* 119: 258–267.

Belcheva, M. M., A. L. Clark, P. D. Haas, J. S. Serna, J. W. Hahn, A. Kiss & C. J. Coscia. 2005. Mu and kappa opioid receptors activate ERK/MAPK via different protein kinase C isoforms and secondary messengers in astrocytes. *J. Biol. Chem.* 280: 27662–27669.

Belcheva, M. M. & C. J. Coscia. 2002. Diversity of G protein-coupled receptor signaling pathways to ERK/MAP kinase. *Neurosignals.* 11: 34–44.

Belcheva, M. M., P. D. Haas, Y. Tan, V. M. Heaton & C. J. Coscia. 2002. The fibroblast growth factor receptor is at the site of convergence between mu-opioid receptor and growth factor signaling pathways in rat C6 glioma cells. *J. Pharmacol. Exp. Ther.* 303: 909–918.

Belcheva, M. M., M. Szucs, D. Wang, W. Sadee & C. J. Coscia. 2001. mu-Opioid receptor-mediated ERK activation involves calmodulin-dependent epidermal growth factor receptor transactivation. *J. Biol. Chem.* 276: 33847–33853.

Belcheva, M. M., Y. Tan, V. M. Heaton, A. L. Clark & C. J. Coscia. 2003. μ Opioid transactivation and down-regulation of the epidermal growth factor receptor in astrocytes: implications for mitogen-activated protein kinase signaling. *Mol. Pharmacol.* 64: 1391–1401.

Benzel, E. C., V. Khare & M. R. Fowler. 1992. Effects of naloxone and nalmefene in rat spinal cord injury induced by the ventral compression technique. *J. Spinal Disord.* 5: 75–77.

Berman, Y., N. Mzhavia, A. Polonskaia, M. Furuta, D. F. Steiner, J. E. Pintar & L. A. Devi. 2000. Defective prodynorphin processing in mice lacking prohormone convertase PC2. *J. Neurochem.* 75: 1763–1770.

Bernales, S., K. L. McDonald & P. Walter. 2006. Autophagy counterbalances endoplasmic reticulum expansion during the unfolded protein response. *PLoS. Biol.* 4: e423.

Bian, D., M. H. Ossipov, M. Ibrahim, R. B. Raffa, R. J. Tallarida, T. P. Malan, Jr., J. Lai & F. Porreca. 1999. Loss of antiallodynic and antinociceptive spinal/supraspinal morphine synergy in nerve-injured rats: restoration by MK-801 or dynorphin antiserum. *Brain Res.* 831: 55–63.

Bohn, L. M., M. M. Belcheva & C. J. Coscia. 2000a. Mitogenic signaling via endogenous kappa-opioid receptors in C6 glioma cells: evidence for the involvement of protein kinase C and the mitogen-activated protein kinase signaling cascade. *J. Neurochem.* 74: 564–573.

Bohn, L. M., M. M. Belcheva & C. J. Coscia. 2000b. Mu-opioid agonist inhibition of kappa-opioid receptor-stimulated extracellular signal-regulated kinase phosphorylation is dynamin-dependent in C6 glioma cells. *J. Neurochem.* 74: 574–581.

Bonham, A. C. 1995. Neurotransmitters in the CNS control of breathing. *Respir. Physiol.* 101: 219–230.

Boyce, S., A. Wyatt, J. K. Webb, R. O'Donnell, G. Mason, M. Rigby, D. Sirinathsinghji, R. G. Hill & N. M. Rupniak. 1999. Selective NMDA NR2B antagonists induce antinociception without motor dysfunction: correlation with restricted localisation of NR2B subunit in dorsal horn. *Neuropharmacology.* 38: 611–623.

Bracken, M. B. 1993. Pharmacological treatment of acute spinal cord injury: current status and future projects. *J. Emerg. Med.* 11 (Suppl 1): 43–48.

Bracken, M. B. & T. R. Holford. 1993. Effects of timing of methylprednisolone or naloxone administration on recovery of segmental and long-tract neurological function in NASCIS 2. *J. Neurosurg.* 79: 500–507.

Bracken, M. B., M. J. Shepard, W. F. Collins, Jr., T. R. Holford, D. S. Baskin, H. M. Eisenberg, E. Flamm, L. Leo-Summers, J. C. Maroon, L. F. Marshall, P. L. Perot, Jr., J. Piepmeier, V. K. H. Sonntag, F. C. Wagner, Jr., J. L. Wilberger, H. R. Winn & W. Young. 1992. Methylprednisolone or naloxone treatment after acute spinal cord injury: 1-year follow-up data. Results of the second National Acute Spinal Cord Injury Study. *J. Neurosurg.* 76: 23–31.

Brauneis, U., M. Oz, R. W. Peoples, F. F. Weight & L. Zhang. 1996. Differential sensitivity of recombinant *N*-methyl-D-aspartate receptor subunits to inhibition by dynorphin. *J. Pharmacol. Exp. Ther.* 279: 1063–1068.

Bruchas, M. R. & C. Chavkin. 2010. Kinase cascades and ligand-directed signaling at the kappa opioid receptor. *Psychopharmacology (Berl).* 210: 137–147.

Bruijnzeel, A. W. 2009. Kappa-opioid receptor signaling and brain reward function. *Brain Res. Rev.* 62: 127–146.

Carroll, R. G., J. L. Riley, B. L. Levine, Y. Feng, S. Kaushal, D. W. Ritchey, W. Bernstein, O. S. Weislow, C. R. Brown, E. A. Berger, C. H. June & D. C. St Louis. 1997. Differential regulation of HIV-1 fusion cofactor expression by CD28 costimulation of CD4+ T cells. *Science.* 276: 273–276.

Caudle, R. M., C. Chavkin & R. Dubner. 1994. Kappa 2 opioid receptors inhibit NMDA receptor-mediated synaptic currents in guinea pig CA3 pyramidal cells. *J. Neurosci.* 14: 5580–5589.

Caudle, R. M. & R. Dubner. 1998. Ifenprodil blocks the excitatory effects of the opioid peptide dynorphin 1-17 on NMDA receptor-mediated currents in the CA3 region of the guinea pig hippocampus. *Neuropeptides.* 32: 87–95.

Caudle, R. M. & L. Isaac. 1988. A novel interaction between dynorphin(1-13) and an N-methyl-D-aspartate site. *Brain Res.* 443: 329–332.

Caudle, R. M. & A. J. Mannes. 2000. Dynorphin: friend or foe? *Pain.* 87: 235–239.

Chao, C. C., G. Gekker, S. Hu, F. Kravitz & P. K. Peterson. 1998. Kappa-opioid potentiation of tumor necrosis factor-alpha-induced anti- HIV-1 activity in acutely infected human brain cell cultures. *Biochem. Pharmacol.* 56: 397–404.

Chao, C. C., G. Gekker, S. Hu, W. S. Sheng, P. S. Portoghese & P. K. Peterson. 1995. Upregulation of HIV-1 expression in cocultures of chronically infected promonocytes and human brain cells by dynorphin. *Biochem. Pharmacol.* 50: 715–722.

Chao, C. C., G. Gekker, S. Hu, W. S. Sheng, K. B. Shark, D. F. Bu, S. Archer, J. M. Bidlack & P. K. Peterson. 1996. Kappa opioid receptors in human microglia downregulate human immunodeficiency virus 1 expression. *Proc. Natl. Acad. Sci. U. S. A.* 93: 8051–8056.

Chao, C. C., G. Gekker, W. S. Sheng, S. Hu & P. K. Peterson. 2001. U50488 inhibits HIV-1 expression in acutely infected monocyte-derived macrophages. *Drug Alcohol Depend.* 62: 149–154.

Chao, C. C., S. Hu, G. Gekker, J. R. Lokensgard, M. P. Heyes & P. K. Peterson. 2000. U50,488 protection against HIV-1-related neurotoxicity: involvement of quinolinic acid suppression. *Neuropharmacology.* 39: 150–160.

Charron, C., C. Messier & H. Plamondon. 2008. Neuroprotection and functional recovery conferred by administration of kappa- and delta 1-opioid agonists in a rat model of global ischemia. *Physiol. Behav.* 93: 502–511.

Chavkin, C. & A. Goldstein. 1981. Specific receptor for the opioid peptide dynorphin: structure—activity relationships. *Proc. Natl. Acad. Sci. U. S. A.* 78: 6543–6547.

Chavkin, C., I. F. James & A. Goldstein. 1982. Dynorphin is a specific endogenous ligand of the kappa opioid receptor. *Science.* 215: 413–415.

Chefer, V. I., J. A. Moron, B. Hope, W. Rea & T. S. Shippenberg. 2000. Kappa-opioid receptor activation prevents alterations in mesocortical dopamine neurotransmission that occur during abstinence from cocaine. *Neurosci.* 101: 619–627.

Chen, L., Y. Gu & L.-Y. M. Huang. 1995a. The mechanism of action for the block of NMDA receptor channels by the opioid peptide dynorphin. *J. Neurosci.* 15: 4602–4611.

Chen, L., Y. Gu & L.-Y. M. Huang. 1995b. The opioid peptide dynorphin directly blocks NMDA receptor channels in the rat. *J. Physiol.* 482: 575–581.

Chen, L. & L. Y. Huang. 1998. Dynorphin block of N-methyl-D-aspartate channels increases with the peptide length. *J. Pharmacol. Exp. Ther.* 284: 826–831.

Chen, L. & L.-Y. M. Huang. 1992. Protein kinase C reduces Mg^{2+} block of NMDA-receptor channels as a mechanism of modulation. *Nature.* 356: 521–523.

Cornish, J. L. & P. W. Kalivas. 2000. Glutamate transmission in the nucleus accumbens mediates relapse in cocaine addiction. *J. Neurosci.* 20: RC89.

Cox, B. M., C. J. Molineaux, T. P. Jacobs, J. G. Rosenberger & A. I. Faden. 1985. Effects of traumatic injury on dynorphin immunoreactivity in spinal cord. *Neuropeptides.* 5: 571–574.

Danielson, P., D. Walker, J. Alrubaian & R. M. Dores. 2002. Identification of a fourth opioid core sequence in a prodynorphin cDNA cloned from the brain of the amphibian, *Bufo marinus*: deciphering the evolution of prodynorphin and proenkephalin. *Neuroendocrinology.* 76: 55–62.

Day, R., C. Lazure, A. Basak, A. Boudreault, P. Limperis, W. Dong & I. Lindberg. 1998. Prodynorphin processing by proprotein convertase 2. Cleavage at single basic residues and enhanced processing in the presence of carboxypeptidase activity. *J. Biol. Chem.* 273: 829–836.

Day, R., M. K. Schafer, S. J. Collard, S. J. Watson & H. Akil. 1991. Atypical prodynorphin gene expression in corticosteroid-producing cells of the rat adrenal gland. *Proc. Natl. Acad. Sci. U. S. A.* 88: 1320–1324.

DeWitt, D. S., D. S. Prough, T. Uchida, D. D. Deal & S. M. Vines. 1997. Effects of nalmefene, CG3703, tirilazad, or dopamine on cerebral blood flow, oxygen delivery, and electroencephalographic activity after traumatic brain injury and hemorrhage. *J. Neurotrauma.* 14: 931–941.

Di Chiara, G. & A. Imperato. 1988. Drugs abused by humans preferentially increase synaptic dopamine concentrations in the mesolimbic system of freely moving rats. *Proc. Natl. Acad. Sci. U. S. A.* 85: 5274–5278.

Dreborg, S., G. Sundstrom, T. A. Larsson & D. Larhammar. 2008. Evolution of vertebrate opioid receptors. *Proc. Natl. Acad. Sci. U. S. A.* 105: 15487–15492.

Dubner, R. & M. A. Ruda. 1992. Activity-dependent neuronal plasticity following tissue injury and inflammation. *Trends Neurosci.* 15: 96–103.

Dumont, M. & S. Lemaire. 1994. Dynorphin potentiation of [3H]CGP-39653 binding to rat brain membranes. *Eur. J. Pharmacol.* 271: 241–244.

Evans, C. J., D. L. Hammond & R. C. A. Frederickson. 1988. The opioid peptides. In Pasternak, G. W. (ed.), *The Opiate Receptors.* Clifton, NJ: Humana: 23–71.

Everitt, B. J. & M. E. Wolf. 2002. Psychomotor stimulant addiction: a neural systems perspective. *J. Neurosci.* 22: 3312–3320.

Faden, A. I. 1990. Opioid and nonopioid mechanisms may contribute to dynorphin's pathophysiological actions in spinal cord injury. *Ann. Neurol.* 27: 67–74.

Faden, A. I. 1992. Dynorphin increases extracellular levels of excitatory amino acids in the brain through a non-opioid mechanism. *J. Neurosci.* 12: 425–429.

Faden, A. I. 1996. Neurotoxic versus neuroprotective actions of endogenous opioid peptides: implications for treatment of CNS injury. *NIDA Res. Monogr.* 163: 318–330.

Faden, A. I. & T. P. Jacobs. 1983. Dynorphin induces partially reversible paraplegia in the rat. *Eur. J. Pharmacol.* 91: 321–324.

Faden, A. I. & T. P. Jacobs. 1984. Dynorphin-related peptides cause motor dysfunction in the rat through a non-opiate action. *Br. J. Pharmacol.* 81: 271–276.

Faden, A. I. & S. Salzman. 1992. Pharmacological strategies in CNS trauma. *Trends. Pharmacol. Sci.* 13: 29–35.

Faden, A. I., R. Shirane, L.-H. Chang, T. L. James, M. Lemke & P. R. Weinstein. 1990. Opiate-receptor antagonist improves metabolic recovery and limits neurochemical alterations associated with reperfusion after global brain ischemia in rats. *J. Pharmacol. Exp. Ther.* 255: 451–458.

Faden, A. I., A. E. Takemori & P. S. Portoghese. 1987. Kappa-selective opiate antagonist nor-binaltorphimine improves outcome after traumatic spinal cord injury in rats. *Cent. Nerv. Syst. Trauma.* 4: 227–237.

Fallon, J. H. & F. M. Leslie. 1986. Distribution of dynorphin and enkephalin peptides in the rat brain. *J. Comp. Neurol.* 249: 293–336.

Feng, Y., C. C. Broder, P. E. Kennedy & E. A. Berger. 1996. HIV-1 entry cofactor: functional cDNA cloning of a seven-transmembrane, G protein-coupled receptor. *Science.* 272: 872–877.

Finley, M. J., X. Chen, G. Bardi, P. Davey, E. B. Geller, L. Zhang, M. W. Adler & T. J. Rogers. 2008. Bi-directional heterologous desensitization between the major HIV-1 co-receptor CXCR4 and the kappa-opioid receptor. *J. Neuroimmunol.* 197: 114–123.

Fitting, S., R. Xu, C. M. Bull, S. K. Buch, N. El-Hage, A. Nath, P. E. Knapp & K. F. Hauser. 2010. Interactive comorbidity between opioid drug abuse and HIV-1 Tat: chronic exposure augments spine loss and sublethal dendritic pathology in striatal neurons. *Am. J. Pathol.* 177(3): 1397–1410.

Flamm, E. S., W. Young, W. F. Collins, J. Piepmeier, G. L. Clifton & B. Fischer. 1985. A phase I trial of naloxone treatment in acute spinal cord injury. *J. Neurosurg.* 63: 390–397.

Gabrilovac, J., T. Balog & A. Andreis. 2003. Dynorphin-A(1-17) decreases nitric oxide release and cytotoxicity induced with lipopolysaccharide plus interferon-gamma in murine macrophage cell line J774. *Biomed. Pharmacother.* 57: 351–358.

Gallo, V. & J. T. Russell. 1995. Excitatory amino acid receptors in glia: different subtypes for distinct functions? *J. Neurosci. Res.* 42: 1–8.

Gallo, V., P. Wright & R. D. McKinnon. 1994. Expression and regulation of a glutamate receptor subunit by bFGF in oligodendrocyte progenitors. *Glia.* 10: 149–153.

Gannon, R. L. & D. M. Terrian. 1991. Presynaptic modulation of glutamate and dynorphin release by excitatory amino acids in the guinea-pig hippocampus. *Neuroscience.* 41: 401–410.

Gannon, R. L. & D. M. Terrian. 1992. Kappa opioid agonists inhibit transmitter release from guinea pig hippocampal mossy fiber synaptosomes. *Neurochem. Res.* 17: 741–747.

Genovese, R. F., J. E. Moreton & F. C. Tortella. 1994. Evaluation of neuroprotection and behavioral recovery by the kappa-opioid, PD117302 following transient forebrain ischemia. *Brain Res. Bull.* 34: 111–116.

Gentile, N. T. & T. K. McIntosh. 1993. Antagonists of excitatory amino acids and endogenous opioid peptides in the treatment of experimental central nervous system injury. *Ann. Emer. Med.* 22: 1028–1034.

Goldstein, A., S. Tachibana, L. I. Lowney, M. Hunkapiller & L. Hood. 1979. Dynorphin-(1-13), an extraordinarily potent opioid peptide. *Proc. Natl. Acad. Sci. U. S. A.* 76: 6666–6670.

Goody, R. J., K. M. Martin, S. M. Goebel & K. F. Hauser. 2003. Dynorphin A toxicity in striatal neurons via an alpha-amino-3-hydroxy-5-methylisoxazole-4-propionate/Kainate receptor mechanism. *Neurosci.* 116: 807–816.

Graham, S. H., H. Shimizu, A. Newman, P. Weinstein & A. I. Faden. 1993. Opioid receptor antagonist nalmefene stereospecifically inhibits glutamate release during global cerebral ischemia. *Brain Res.* 632: 346–350.

Hall, E. D. & K. E. Pazara. 1988. Quantitative analysis of effects of kappa-opioid agonists on postischemic hippocampal CA1 neuronal necrosis in gerbils. *Stroke* 19: 1008–1012.

Hall, E. D. & J. E. Springer. 2004. Neuroprotection and acute spinal cord injury: a reappraisal. *NeuroRx.* 1: 80–100.

Hall, E. D., D. L. Wolf, J. S. Althaus & P. F. Von Voigtlander. 1987. Beneficial effects of the kappa opioid receptor agonist U-50488H in experimental acute brain and spinal cord injury. *Brain Res.* 435: 174–180.

Hanson, G. R., N. Singh, K. Merchant, M. Johnson & J. W. Gibb. 1995. The role of NMDA receptor systems in neuropeptide responses to stimulants of abuse. *Drug Alcohol Depend.* 37: 107–110.

Hauser, K. F., J. V. Aldrich, K. J. Anderson, G. Bakalkin, M. J. Christie, E. D. Hall, P. E. Knapp, S. W. Scheff, I. N. Singh, B. Vissel, A. S. Woods, T. Yakovleva & T. S. Shippenberg. 2005. Pathobiology of dynorphins in trauma and disease. *Front. Biosci.* 10: 216–235.

Hauser, K. F., J. K. Foldes & C. S. Turbek. 1999. Dynorphin A (1-13) neurotoxicity *in vitro*: opioid and non-opioid mechanisms in mouse spinal cord neurons. *Exp. Neurol.* 160: 361–375.

Hauser, K. F., Y. K. Hahn, V. V. Adjan, S. Zou, S. K. Buch, A. Nath, A. J. Bruce-Keller & P. E. Knapp. 2009. HIV-1 Tat and morphine have interactive effects on oligodendrocyte survival and morphology. *Glia.* 57: 194–206.

Hauser, K. F., P. E. Knapp & C. S. Turbek. 2001. Structure-activity analysis of dynorphin A toxicity in spinal cord neurons: intrinsic neurotoxicity of dynorphin A and its carboxyl-terminal, nonopioid metabolites. *Exp. Neurol.* 168: 78–87.

Hauser, K. F. & D. Mangoura. 1998. Diversity of the endogenous opioid system in development: novel signal transduction translates multiple extracellular signals into neural cell growth and differentiation. *Perspect. Dev. Neurobiol.* 5: 437–449.

Herrera-Marschitz, M., M. Goiny, Z. B. You, J. J. Meana, E. Engidawork, Y. Chen, R. Rodriguez-Puertas, C. Broberger, K. Andersson, L. Terenius, T. Hokfelt & U. Ungerstedt. 1998. Release of endogenous excitatory amino acids in the neostriatum of the rat under physiological and pharmacologically-induced conditions. *Amino Acids.* 14: 197–203.

Hotsenpiller, G., M. Giorgetti & M. E. Wolf. 2001. Alterations in behaviour and glutamate transmission following presentation of stimuli previously associated with cocaine exposure. *Eur. J. Neurosci.* 14: 1843–1855.

Hoyer-Hansen, M. & M. Jaattela. 2007. Connecting endoplasmic reticulum stress to autophagy by unfolded protein response and calcium. *Cell Death Differ.* 14: 1576–1582.

Huang, L.-Y. M. 1995. Cellular mechanisms of excitatory and inhibitory actions of opioids. In Tseng, L. F. (ed.), *The Pharmacology of Opioid Peptides*. Langhorn, PA: Harwood Academic Publishers: 131–149.

Hudson, C. J., P. F. Von Voigtlander, J. S. Althaus, H. M. Scherch & E. D. Means. 1991. The kappa opioid-related anticonvulsants U-50488H and U-54494A attenuate *N*-methyl-D-aspartate induced brain injury in the neonatal rat. *Brain Res.* 564: 261–267.

Hugonin, L., V. Vukojevic, G. Bakalkin & A. Graslund. 2006. Membrane leakage induced by dynorphins. *FEBS Lett.* 580: 3201–3205.

Hugonin, L., V. Vukojevic, G. Bakalkin & A. Graslund. 2008. Calcium influx into phospholipid vesicles caused by dynorphin neuropeptides. *Biochim. Biophys. Acta.* 1778: 1267–1273.

Husain, S., D. E. Potter & C. E. Crosson. 2009. Opioid receptor-activation: retina protected from ischemic injury. *Invest. Ophthalmol. Vis. Sci.* 50: 3853–3859.

Isaac, L., T. V. Z. O'Malley, H. Ristic & P. Stewart. 1990. MK-801 blocks dynorphin A (1-13)-induced loss of the tail-flick reflex in the rat. *Brain Res.* 531: 83–87.

Ito, R., J. W. Dalley, T. W. Robbins & B. J. Everitt. 2002. Dopamine release in the dorsal striatum during cocaine-seeking behavior under the control of a drug-associated cue. *J. Neurosci.* 22: 6247–6253.

Itoh, J., M. Ukai & T. Kameyama. 1993a. Dynorphin A-(1-13) potently prevents memory dysfunctions induced by transient cerebral ischemia in mice. *Eur. J. Pharmacol.* 234: 9–15.

Itoh, J., M. Ukai & T. Kameyama. 1993b. U-50,488H, a kappa-opioid receptor agonist, markedly prevents memory dysfunctions induced by transient cerebral ischemia in mice. *Brain Res.* 619: 223–228.

Jaiswal, A., S. Kumar, S. Seth, A. K. Dinda & S. K. Maulik. 2010. Effect of U50,488H, a kappa-opioid receptor agonist on myocardial alpha-and beta-myosin heavy chain expression and oxidative stress associated with isoproterenol-induced cardiac hypertrophy in rat. *Mol. Cell Biochem.* 345(1–2): 231–240.

Johnson, M., L. G. Bush, J. W. Gibb & G. R. Hanson. 1991. Role of N-methyl-D-aspartate (NMDA) receptors in the response of extrapyramidal neurotensin and dynorphin A systems to cocaine and GBR 12909. *Biochem. Pharmacol.* 41: 649–652.

Kakidani, H., Y. Furutani, H. Takahashi, M. Noda, Y. Morimoto, T. Hirose, M. Asai, S. Inayama, S. Nakanishi & S. Numa. 1982. Cloning and sequence analysis of cDNA for porcine beta-neo-endorphin/dynorphin precursor. *Nature.* 298: 245–249.

Karadottir, R., P. Cavelier, L. H. Bergersen & D. Attwell. 2005. NMDA receptors are expressed in oligodendrocytes and activated in ischaemia. *Nature.* 438: 1162–1166.

Kim, E., A. L. Clark, A. Kiss, J. W. Hahn, R. Wesselschmidt, C. J. Coscia & M. M. Belcheva. 2006. Mu and kappa opioids induce the differentiation of embryonic stem cells to neural progenitors. *J. Biol. Chem.* 281(44): 33749–33760.

Knapp, P. E., V. V. Adjan & K. F. Hauser. 2009. Cell-specific loss of κ opioid receptors in oligodendrocytes of the dysmyelinating jimpy mouse. *Neurosci. Lett.* 451(2): 114–118.

Knapp, P. E. & K. F. Hauser. 1996. μ-Opioid receptor activation enhances DNA synthesis in immature oligodendrocytes. *Brain Res.* 743: 341–345.

Knapp, P. E., O. S. Itkis, L. Zhang, B. A. Spruce, G. Bakalkin & K. F. Hauser. 2001. Endogenous opioids and oligodendroglial function: possible autocrine/paracrine effects on cell survival and development. *Glia.* 35: 156–165.

Knapp, P. E., K. Maderspach & K. F. Hauser. 1998. Endogenous opioid system in developing normal and jimpy oligodendrocytes: μ and κ opioid receptors mediate differential mitogenic and growth responses. *Glia.* 22: 189–201.

Knoll, A. T. & W. A. Carlezon, Jr. 2010. Dynorphin, stress, and depression. *Brain Res.* 1314: 56–73.

Koetzner, L., X. Y. Hua, J. Lai, F. Porreca & T. Yaksh. 2004. Nonopioid actions of intrathecal dynorphin evoke spinal excitatory amino acid and prostaglandin E2 release mediated by cyclooxygenase-1 and -2. *J. Neurosci.* 24: 1451–1458.

Kolaj, M., R. Cerne & M. Randic. 1995. The opioid peptide dynorphin modulates AMPA and kainate responses in acutely isolated neurons from the dorsal horn. *Brain Res.* 671: 227–244.

Kong, L., G. Jeohn, P. M. Hudson, L. Du, B. Liu & J. Hong. 2000. Reduction of lipopolysaccharide-induced neurotoxicity in mouse mixed cortical neuron/glia cultures by ultralow concentrations of dynorphins. *J. Biomed. Sci.* 7: 241–247.

Koob, G. F. 2000. Neurobiology of addiction. Toward the development of new therapies. *Ann. N. Y. Acad. Sci.* 909: 170–185.

Kuzmin, A., N. Madjid, L. Terenius, S. O. Ogren & G. Bakalkin. 2006. Big dynorphin, a prodynorphin-derived peptide produces NMDA receptor-mediated effects on memory, anxiolytic-like and locomotor behavior in mice. *Neuropsychopharmacology.* 31: 1928–1937.

La Torre, B. P., L. Favalli, A. Rozza, E. Lanza, C. Scavini, G. Racagni & F. Savoldi. 1991. Ischemic cerebral pathologies and K opioid receptors in rabbits. *Ital. J. Neurol. Sci.* 12: 7–10.

Lai, J., M. H. Ossipov, T. W. Vanderah, T. P. Malan, Jr. & F. Porreca. 2001. Neuropathic pain: the paradox of dynorphin. *Mol. Interv.* 1: 160–167.

Lai, S. L., Y. Gu & L. Y. Huang. 1998. Dynorphin uses a non-opioid mechanism to potentiate N-methyl-D-aspartate currents in single rat periaqueductal gray neurons. *Neurosci. Lett.* 247: 115–118.

Laughlin, T. M., K. F. Kitto & G. L. Wilcox. 1999. Redox manipulation of NMDA receptors in vivo: alteration of acute pain transmission and dynorphin-induced allodynia. *Pain.* 80: 37–43.

Laughlin, T. M., A. A. Larson & G. L. Wilcox. 2001. Mechanisms of induction of persistent nociception by dynorphin. *J. Pharmacol. Exp. Ther.* 299: 6–11.

Laughlin, T. M., T. W. Vanderah, J. Lashbrook, M. L. Nichols, M. Ossipov, F. Porreca & G. L. Wilcox. 1997. Spinally administered dynorphin A produces long-lasting allodynia: involvement of NMDA but not opioid receptors. *Pain.* 72: 253–260.

Law, P. Y. & C. Bergsbaken. 1995. Properties of delta opioid receptor in neuroblastoma NS20Y: receptor activation and neuroblastoma proliferation. *J. Pharmacol. Exp. Ther.* 272: 322–332.

Leveille, F., G. F. El, E. Gouix, M. Lecocq, D. Lobner, O. Nicole & A. Buisson. 2008. Neuronal viability is controlled by a functional relation between synaptic and extrasynaptic NMDA receptors. *FASEB J.* 22: 4258–4271.

Li, W., Y. Huang, R. Reid, J. Steiner, T. Malpica-Llanos, T. A. Darden, S. K. Shankar, A. Mahadevan, P. Satishchandra & A. Nath. 2008. NMDA receptor activation by HIV-Tat protein is clade dependent. *J. Neurosci.* 28: 12190–12198.

Lin, J. Y., L. M. Hung, L. Y. Lai & F. C. Wei. 2008. Kappa-opioid receptor agonist protects the microcirculation of skeletal muscle from ischemia reperfusion injury. *Ann. Plast. Surg.* 61: 330–336.

Lokensgard, J. R., G. Gekker & P. K. Peterson. 2002. Kappa-opioid receptor agonist inhibition of HIV-1 envelope glycoprotein-mediated membrane fusion and CXCR4 expression on CD4(+) lymphocytes. *Biochem. Pharmacol.* 63: 1037–1041.

Long, J. B., R. C. Kinney, D. S. Malcolm, G. M. Graeber & J. W. Holaday. 1986. Intrathecal dynorphin A (1-13) and (3-13) reduce spinal cord blood flow by non-opioid mechanisms. *NIDA Res. Monogr.* 75: 524–526.

Long, J. B., A. Martinez-Arizala, E. E. Echevarria, R. E. Tidwell & J. W. Holaday. 1988. Hindlimb paralytic effects of prodynorphin-derived peptides following spinal subarachnoid injection in rats. *Eur. J. Pharmacol.* 153: 45–54.

Long, J. B., A. Martinez-Arizala, J. M. Petras & J. W. Holaday. 1986. Endogenous opioids in spinal cord injury: a critical evaluation. *Cent. Nerv. Syst. Trauma.* 3: 295–315.

Long, J. B., A. Martinez-Arizala, D. D. Rigamonti & J. W. Holaday. 1988. Hindlimb paralytic effects of arginine vasopressin and related peptides following spinal subarachnoid injection in the rat. *Peptides.* 9: 1335–1344.

Long, J. B., J. M. Petras, W. C. Mobley & J. W. Holaday. 1988. Neurological dysfunction after intrathecal injection of dynorphin A (1-13) in the rat. II. Nonopioid mechanisms mediate loss of motor, sensory and autonomic function. *J. Pharmacol. Exp. Ther.* 246: 1167–1174.

Long, J. B., D. D. Rigamonti, B. de Costa, K. C. Rice & A. Martinez-Arizala. 1989. Dynorphin A-induced rat hindlimb paralysis and spinal cord injury are not altered by the kappa opioid antagonist nor- binaltorphimine. *Brain Res.* 497: 155–162.

Long, J. B., D. D. Rigamonti, M. A. Oleshansky, C. P. Wingfield & A. Martinez-Arizala. 1994. Dynorphin A-induced rat spinal cord injury: evidence for excitatory amino acid involvement in a pharmacological model of ischemic spinal cord injury. *J. Pharmacol. Exp. Ther.* 269: 358–366.

Lyeth, B. G. & R. L. Hayes. 1992. Cholinergic and opioid mediation of traumatic brain injury. *J. Neurotrauma.* 9 (Suppl 2): S463–S474.

Macdonald, R. L. & M. A. Werz. 1986. Dynorphin A decreases voltage-dependent calcium conductance of mouse dorsal root ganglion neurones. *J. Physiol.* 377: 237–249.

Maiuri, M. C., E. Zalckvar, A. Kimchi & G. Kroemer. 2007. Self-eating and self-killing: crosstalk between autophagy and apoptosis. *Nat. Rev. Mol. Cell Biol.* 8: 741–752.

Malan, T. P., M. H. Ossipov, L. R. Gardell, M. Ibrahim, D. Bian, J. Lai & F. Porreca. 2000. Extraterritorial neuropathic pain correlates with multisegmental elevation of spinal dynorphin in nerve-injured rats. *Pain.* 86: 185–194.

Mangoura, D. & G. Dawson. 1993. Opioid peptides activate phospholipase D and protein kinase C-ε in chicken embryo neuron cultures. *Proc. Natl. Acad. Sci. U. S. A.* 90: 2915–2919.

Manning, S. M., D. M. Talos, C. Zhou, D. B. Selip, H. K. Park, C. J. Park, J. J. Volpe & F. E. Jensen. 2008. NMDA receptor blockade with memantine attenuates white matter injury in a rat model of periventricular leukomalacia. *J. Neurosci.* 28: 6670–6678.

Marinova, Z., V. Vukojevic, S. Surcheva, T. Yakovleva, G. Cebers, N. Pasikova, I. Usynin, L. Hugonin, W. Fang, M. Hallberg, D. Hirschberg, T. Bergman, U. Langel, K. F. Hauser, A. Pramanik, J. V. Aldrich, A. Graslund, L. Terenius & G. Bakalkina. 2005. Translocation of dynorphin neuropeptides across the plasma membrane. A putative mechanism of signal transmission. *J. Biol. Chem.* 280: 26360–26370.

Martin, L. J., A. Furuta & C. D. Blackstone. 1998. AMPA receptor protein in developing rat brain: glutamate receptor-1 expression and localization change at regional, cellular, and subcellular levels with maturation. *Neuroscience.* 83: 917–928.

Massardier, D. & P. F. Hunt. 1989. A direct non-opiate interaction of dynorphin-(1-13) with the N-methyl-D-aspartate (NMDA) receptor. *Eur. J. Pharmacol.* 170: 125–126.

McFarland, K., C. C. Lapish & P. W. Kalivas. 2003. Prefrontal glutamate release into the core of the nucleus accumbens mediates cocaine-induced reinstatement of drug-seeking behavior. *J. Neurosci.* 23: 3531–3537.

McGinty, J. F. 1995. Introduction to the role of excitatory amino acids in the actions of abused drugs: a symposium presented at the 1993 annual meeting of the College on Problems of Drug Dependence. *Drug Alcohol Depend.* 37: 91–94.

McIntosh, T. K., S. Fernyak, I. Yamakami & A. I. Faden. 1994. Central and systemic kappa-opioid agonists exacerbate neurobehavioral response to brain injury in rats. *Am. J. Physiol.* 267: R665–672.

McIntosh, T. K., R. L. Hayes, D. S. DeWitt, V. Agura & A. I. Faden. 1987. Endogenous opioids may mediate secondary damage after experimental brain injury. *Am. J. Physiol.* 253: E565–E574.

McIntosh, T. K., V. A. Head & A. I. Faden. 1987. Alterations in regional concentrations of endogenous opioids following traumatic brain injury in the cat. *Brain Res.* 425: 225–233.

McLennan, G. P., A. Kiss, M. Miyatake, M. M. Belcheva, K. T. Chambers, J. J. Pozek, Y. Mohabbat, R. A. Moyer, L. M. Bohn & C. J. Coscia. 2008. Kappa opioids promote the proliferation of astrocytes via Gbetagamma and beta-arrestin 2-dependent MAPK-mediated pathways. *J. Neurochem.* 107: 1753–1765.

Merg, F., D. Filliol, I. Usynin, I. Bazov, N. Bark, Y. L. Hurd, T. Yakovleva, B. L. Kieffer & G. Bakalkin. 2006. Big dynorphin as a putative endogenous ligand for the kappa-opioid receptor. *J. Neurochem.* 97: 292–301.

Miyatake, M., T. J. Rubinstein, G. P. McLennan, M. M. Belcheva & C. J. Coscia. 2009. Inhibition of EGF-induced ERK/MAP kinase-mediated astrocyte proliferation by mu opioids: integration of G protein and beta-arrestin 2-dependent pathways. *J. Neurochem.* 110: 662–674.

Mysels, D. & M. A. Sullivan. 2009. The kappa-opiate receptor impacts the pathophysiology and behavior of substance use. *Am. J. Addict.* 18: 272–276.

Nichols, M. L., Y. Lopez, M. H. Ossipov, D. Bian & F. Porreca. 1997. Enhancement of the antiallodynic and antinociceptive efficacy of spinal morphine by antisera to dynorphin A (1-13) or MK-801 in a nerve-ligation model of peripheral neuropathy. *Pain.* 69: 317–322.

Nikoshkov, A., Y. L. Hurd, T. Yakovleva, I. Bazov, Z. Marinova, G. Cebers, N. Pasikova, A. Gharibyan, L. Terenius & G. Bakalkin, 2005. Prodynorphin transcripts and proteins differentially expressed and regulated in the adult human brain, *FASEB J.* 19: 1543–1545.

Ossipov, M. H., C. J. Kovelowski, H. Wheeler-Aceto, A. Cowan, J. C. Hunter, J. Lai, T. P. Malan, Jr. & F. Porreca. 1996. Opioid antagonists and antisera to endogenous opioids increase the nociceptive response to formalin: demonstration of an opioid kappa and delta inhibitory tone. *J. Pharmacol. Exp. Ther.* 277: 784–788.

Ossipov, M. H., J. Lai, T. P. Malan, Jr. & F. Porreca. 2000. Spinal and supraspinal mechanisms of neuropathic pain. *Ann. N. Y. Acad. Sci.* 909: 12–24.

Palay, S. L. & V. Chan-Palay. 1974. *The Cerebellar Cortex, Cytology and Organization.* New York: Springer-Verlag.

Patel, K. K., J. F. Hartmann & M. M. Cohen. 1971. Ultrastructural estimation of relative volume of extracellular space in brain slices. *J. Neurol. Sci.* 12: 275–288.

Pattee, P., A. E. Ilie, S. Benyhe, G. Toth, A. Borsodi & S. R. Nagalla. 2003. Cloning and characterization of Xen-dorphin prohormone from Xenopus laevis: a new opioid-like prohormone distinct from proenkephalin and prodynorphin. *J. Biol. Chem.* 278(52): 53098–53104.

Peart, J. N., E. R. Gross, M. E. Reichelt, A. Hsu, J. P. Headrick & G. J. Gross. 2008. Activation of kappa-opioid receptors at reperfusion affords cardioprotection in both rat and mouse hearts. *Basic Res. Cardiol.* 103: 454–463.

Persson, A. I., T. Thorlin, C. Bull & P. S. Eriksson. 2003. Opioid-induced proliferation through the MAPK pathway in cultures of adult hippocampal progenitors. *Mol. Cell Neurosci.* 23: 360–372.

Persson, A. I., T. Thorlin, C. Bull, P. Zarnegar, R. Ekman, L. Terenius & P. S. Eriksson. 2003. Mu- and delta-opioid receptor antagonists decrease proliferation and increase neurogenesis in cultures of rat adult hippocampal progenitors. *Eur. J. Neurosci.* 17: 1159–1172.

Peterson, P. K., G. Gekker, J. R. Lokensgard, J. M. Bidlack, A. Chang, X. Fang & P. S. Portoghese. 2001. Kappa-opioid receptor agonist suppression of HIV-1 expression in CD4(+) lymphocytes. *Biochem. Pharmacol.* 61: 1145–1151.

Polakiewicz, R. D., S. M. Schieferl, A. C. Gingras, N. Sonenberg & M. J. Comb. 1998. mu-Opioid receptor activates signaling pathways implicated in cell survival and translational control. *J. Biol. Chem.* 273: 23534–23541.

Przewlocka, B., H. Machelska & W. Lason. 1994. Kappa opioid receptor agonists inhibit the pilocarpine-induced seizures and toxicity in the mouse. *Eur. Neuropsychopharmacol.* 4: 527–533.

Puniak, M. A., G. M. Freeman, C. A. Agresta, L. Van Newkirk, C. A. Barone & S. K. Salzman. 1991. Comparison of a serotonin antagonist, opioid antagonist, and TRH analog for the acute treatment of experimental spinal trauma. *J. Neurotrauma.* 8: 193–203.

Redell, J. B., A. N. Moore & P. K. Dash. 2003. Expression of the prodynorphin gene after experimental brain injury and its role in behavioral dysfunction. *Exp. Biol. Med.* 228: 261–269.

Rusin, K. I., D. R. Giovannucci, E. L. Stuenkel & H. C. Moises. 1997. Kappa-opioid receptor activation modulates Ca^{2+} currents and secretion in isolated neuroendocrine nerve terminals. *J. Neurosci.* 17: 6565–6574.

Sahley, T. L., R. H. Nodar & F. E. Musiek. 1999. Endogenous dynorphins: possible role in peripheral tinnitus. *Int. Tinnitus. J.* 5: 76–91.

Salter, M. G. & R. Fern. 2005. NMDA receptors are expressed in developing oligodendrocyte processes and mediate injury. *Nature.* 438: 1167–1171.

Schwarzer, C. 2009. 30 years of dynorphins—new insights on their functions in neuropsychiatric diseases. *Pharmacol. Ther.* 123: 353–370.

Sharma, H. S., F. Nyberg & Y. Olsson. 1992. Dynorphin A content in the rat brain and spinal cord after a localized trauma to the spinal cord and its modification with p-chlorophenylalanine. An experimental study using radioimmunoassay technique. *Neurosci. Res.* 14: 195–203.

Sheng, W. S., S. Hu, G. Gekker, S. Zhu, P. K. Peterson & C. C. Chao. 1997. Immunomodulatory role of opioids in the central nervous system. *Arch. Immunol. Ther. Exp.* 45: 359–366.

Shippenberg, T. S. & V. Chefer. 2002. Opioid modulation of psychomotor stimulant effects. In Maldonado, R. (ed.), *Molecular Biology of Drug Addiction.* Clifton, NJ: Humana Press: 107–133.

Shippenberg, T. S., A. Zapata & V. I. Chefer. 2007. Dynorphin and the pathophysiology of drug addiction. *Pharmacol. Ther.* 116: 306–321.

Shukla, V. K. & S. Lemaire. 1992. Central non-opioid physiological and pathophysiological effects of dynorphin A and related peptides. *J. Psychiatry Neurosci.* 17: 106–119.

Shukla, V. K. & S. Lemaire. 1994. Non-opioid effects of dynorphins: possible role of the NMDA receptor. *Trends Pharmacol. Sci.* 15: 420–424.

Shukla, V. K., S. Lemaire, I. H. Ibrahim, T. D. Cyr, Y. Chen & R. Michelot. 1993. Design of potent and selective dynorphin A related peptides devoid of supraspinal motor effects in mice. *Can. J. Physiol. Pharmacol.* 71: 211–216.

Shukla, V. K., J. A. Prasad & S. Lemaire. 1997. Nonopioid motor effects of dynorphin A and related peptides: structure dependence and role of the N-methyl-D-aspartate receptor. *J. Pharmacol. Exp. Ther.* 283: 604–610.

Silvia, R. C., G. R. Slizgi, J. H. Ludens & A. H. Tang. 1987. Protection from ischemia-induced cerebral edema in the rat by U-50488H, a kappa opioid receptor agonist. *Brain Res.* 403: 52–57.

Simmons, M. L. & C. Chavkin. 1996. Endogenous opioid regulation of hippocampal function. *Int. Rev. Neurobiol.* 39: 145–196.

Simmons, M. L., G. W. Terman, S. M. Gibbs & C. Chavkin. 1995. L-type calcium channels mediate dynorphin neuropeptide release from dendrites but not axons of hippocampal granule cells. *Neuron.* 14: 1265–1272.

Singh, I. N., N. El-Hage, M. E. Campbell, S. E. Lutz, P. E. Knapp, A. Nath & K. F. Hauser. 2005. Differential involvement of p38 and JNK MAP kinases in HIV-1 Tat and gp120-induced apoptosis and neurite degeneration in striatal neurons. *Neuroscience.* 135: 781–790.

Singh, I. N., R. J. Goody, C. Dean, N. M. Ahmad, S. E. Lutz, P. E. Knapp, A. Nath & K. F. Hauser. 2004. Apoptotic death of striatal neurons induced by HIV-1 Tat and gp120: differential involvement of caspase-3 and endonuclease G. *J. Neurovirol.* 10: 141–151.

Singh, I. N., R. J. Goody, S. M. Goebel, K. M. Martin, Z. Marinova, D. Hirschberg, T. Yakovleva, T. Bergman, G. Bakalkin & K. F. Hauser. 2003. Dynorphin A (1-17) induces apoptosis in striatal neurons through alpha-amino-3-hydroxy-5-methylisoxazole-4-propionate/kainate receptor-mediated cytochrome c release and caspase-3 activation. *Neurosci.* 122: 1013–1023.

Sirinathsinghji, D. J. S., K. E. Nikolarakis, S. Reimer & A. Herz. 1990. Nigrostriatal dopamine mediates the stimulatory effects of corticotropin releasing factor on methionine-enkephalin and dynorphin release from the rat neostriatum. *Brain Res.* 526: 173–176.

Skilling, S. R., X. Sun, H. J. Kurtz & A. A. Larson. 1992. Selective potentiation of NMDA-induced activity and release of excitatory amino acids by dynorphin: possible roles in paralysis and neurotoxicity. *Brain Res.* 575: 272–278.

Solbrig, M. V. & G. F. Koob. 2004. Epilepsy, CNS viral injury and dynorphin. *Trends Pharmacol. Sci.* 25: 98–104.

Spanagel, R., A. Herz & T. S. Shippenberg. 1990. The effects of opioid peptides on dopamine release in the nucleus accumbens: an in vivo microdialysis study. *J. Neurochem.* 55: 1734–1740.

Spanagel, R., A. Herz & T. S. Shippenberg. 1992. Opposing tonically active endogenous opioid systems modulate the mesolimbic dopaminergic pathway. *Proc. Natl. Acad. Sci. U. S. A.* 89: 2046–2050.

Stiene-Martin, A., P. E. Knapp, K. M. Martin, J. A. Gurwell, S. Ryan, S. R. Thornton, F. L. Smith & K. F. Hauser. 2001. Opioid system diversity in developing neurons, astroglia, and oligodendroglia in the subventricular zone and striatum: impact on gliogenesis *in vivo. Glia.* 36: 78–88.

Sundstrom, G., S. Dreborg & D. Larhammar. 2010. Concomitant duplications of opioid peptide and receptor genes before the origin of jawed vertebrates. *PLoS. One.* 5: e10512.

Surprenant, A., K.-Z. Shen & R. A. North. 1990. Inhibiton of calcium currents by noradrenaline, somatostatin and opioids in guinea-pig submucosal neurons. *J. Physiol.* 431: 585–608.

Sykova, E. 1997. Extracellular space volume and geometry of the rat brain after ischemia and central injury. *Adv. Neurol.* 73: 121–135.

Sykova, E. 2005. Glia and volume transmission during physiological and pathological states. *J. Neural. Transm.* 112: 137–147.

Sykova, E. & L. Vargova. 2008. Extrasynaptic transmission and the diffusion parameters of the extracellular space. *Neurochem. Int.* 52: 5–13.

Sykova, E., L. Vargova, S. Prokopova & Z. Simonova. 1999. Glial swelling and astrogliosis produce diffusion barriers in the rat spinal cord. *Glia.* 25: 56–70.

Szeto, H. H., Y. Soong & D. Wu. 1999. The role of N-methyl-D-aspartate receptors in the release of adrenocorticotropin by dynorphin A1-13. *Neuroendocrinology.* 69: 28–33.

Tan-No, K., G. Cebers, T. Yakovleva, G. B. Hoon, I. Gileva, K. Reznikov, M. Aguilar-Santelises, K. F. Hauser, L. Terenius & G. Bakalkin. 2001. Cytotoxic effects of dynorphins through nonopioid intracellular mechanisms. *Exp. Cell Res.* 269: 54–63.

Tan-No, K., A. Esashi, O. Nakagawasai, F. Niijima, T. Tadano, C. Sakurada, T. Sakurada, G. Bakalkin, L. Terenius & K. Kisara. 2002. Intrathecally administered big dynorphin, a prodynorphin-derived peptide, produces nociceptive behavior through an N-methyl-D-aspartate receptor mechanism. *Brain Res.* 952: 7–14.

Tan-No, K., H. Takahashi, O. Nakagawasai, F. Niijima, S. Sakurada, G. Bakalkin, L. Terenius & T. Tadano. 2009. Nociceptive behavior induced by the endogenous opioid peptides dynorphins in uninjured mice: evidence with intrathecal N-ethylmaleimide inhibiting dynorphin degradation. *Int. Rev. Neurobiol.* 85: 191–205.

Tang, Q., R. Gandhoke, A. Burritt, V. J. Hruby, F. Porreca & J. Lai. 1999. High-affinity interaction of (des-Tyrosyl)dynorphin A(2-17) with NMDA receptors. *J. Pharmacol. Exp. Ther.* 291: 760–765.

Tegeder, I. & G. Geisslinger. 2004. Opioids as modulators of cell death and survival— unraveling mechanisms and revealing new indications. *Pharmacol. Rev.* 56: 351–369.

Thompson, A. C., A. Zapata, J. B. Justice, Jr., R. A. Vaughan, L. G. Sharpe & T. S. Shippenberg. 2000. Kappa-opioid receptor activation modifies dopamine uptake in the nucleus accumbens and opposes the effects of cocaine. *J. Neurosci.* 20: 9333–9340.

Thuret, S., L. D. Moon & F. H. Gage. 2006. Therapeutic interventions after spinal cord injury. *Nat. Rev. Neurosci.* 7: 628–643.

Tikhonov, D. B. & L. G. Magazanik. 2009. Origin and molecular evolution of ionotropic glutamate receptors. *Neurosci. Behav. Physiol.* 39: 763–773.

Tortella, F. C. & M. A. DeCoster. 1994. Kappa opioids: therapeutic considerations in epilepsy and CNS injury. *Clin. Neuropharmacol.* 17: 403–416.

Vanderah, T. W., T. Laughlin, J. M. Lashbrook, M. L. Nichols, G. L. Wilcox, M. H. Ossipov, T. P. J. Malan & F. Porreca. 1996. Single intrathecal injections of dynorphin A or des-Tyr-dynorphins produce long-lasting allodynia in rats: blockade by MK-801 but not naloxone. *Pain.* 68: 275–281.

Verbeek, D. S., B. P. van de Warrenburg, P. Wesseling, P. L. Pearson, H. P. Kremer & R. J. Sinke. 2004. Mapping of the SCA23 locus involved in autosomal dominant cerebellar ataxia to chromosome region 20p13-12.3. *Brain.* 127: 2551–2557.

Vink, R., T. K. McIntosh, R. Rhomhanyi & A. I. Faden. 1990. Opiate antagonist nalmefene improves intracellular free Mg^{2+}, bioenergetic state, and neurologic outcome following traumatic brain injury in rats. *J. Neurosci.* 10: 3524–3530.

Vorisek, I., M. Hajek, J. Tintera, K. Nicolay & E. Sykova. 2002. Water ADC, extracellular space volume, and tortuosity in the rat cortex after traumatic injury. *Magn. Reson. Med.* 48: 994–1003.

Wagner, J. J., R. M. Caudle & C. Chavkin. 1992. Kappa-opioids decrease excitatory transmission in the dentate gyrus of the guinea pig hippocampus. *J. Neurosci.* 12: 132–141.

Wagner, J. J., G. W. Terman & C. Chavkin. 1993. Endogenous dynorphins inhibit excitatory neurotransmission and block LTP induction in the hippocampus. *Nature.* 363: 451–454.

Walker, J. M., H. C. Moises, D. H. Coy, G. Baldrighi & H. Akil. 1982. Nonopiate effects of dynorphin and des-Tyr-dynorphin. *Science.* 218: 1136–1138.

Walker, J. M., H. C. Moises, D. H. Coy, E. A. Young, S. J. Watson & H. Akil. 1982. Dynorphin (1-17): lack of analgesia but evidence for non-opiate electrophysiological and motor effects. *Life Sci.* 31: 1821–1824.

Walker, J. M., D. E. Tucker, D. H. Coy, B. B. Walker & H. Akil. 1982. Des-tyrosine-dynorphin antagonizes morphine analgesia. *Eur. J. Pharmacol.* 85: 121–122.

Wang, P., H. Gao, Y. Ni, B. Wang, Y. Wu, L. Ji, L. Qin, L. Ma & G. Pei. 2003. Beta-arrestin 2 functions as a G-protein-coupled receptor-activated regulator of oncoprotein Mdm2. *J. Biol. Chem.* 278: 6363–6370.

Wee, S. & G. F. Koob. 2010. The role of the dynorphin-kappa opioid system in the reinforcing effects of drugs of abuse. *Psychopharmacology.* 210: 121–135.

Weiss, F. & L. J. Porrino. 2002. Behavioral neurobiology of alcohol addiction: recent advances and challenges. *J. Neurosci.* 22: 3332–3337.

Winkler, T., H. S. Sharma, T. Gordh, R. D. Badgaiyan, E. Stalberg & J. Westman. 2002. Topical application of dynorphin A (1-17) antiserum attenuates trauma induced alterations in spinal cord evoked potentials, microvascular permeability disturbances, edema formation and cell injury: an experimental study in the rat using electrophysiological and morphological approaches. *Amino Acids.* 23: 273–281.

Wise, R. A. 1998. Drug-activation of brain reward pathways. *Drug Alcohol Depend.* 51: 13–22.

Wollemann, M. & S. Benyhe. 2004. Non-opioid actions of opioid peptides. *Life Sci.* 75: 257–270.

Woods, A. S., R. Kaminski, Y. Wang, M. Oz, K. Hauser, R. Goody, H.-Y. J. Wang, P. Zeitz, K. P. Zeitz, D. Zolkowska, R. Schepers, C. Chang, H. Shen, M. Nold, J. Danielson, A. Gräslund, V. Vukojevic, G. Bakalkin, A. Basbaum & T. Shippenberg. 2006. Decoy peptides that bind dynorphin noncovalently prevent NMDA receptor-mediated neurotoxicity. *J. Proteome Res.* 5: 1017–1023.

Woods, A. & A. Zangen. 2001. A direct chemical interaction between dynorphin and excitatory amino acids. *Neurochem. Res.* 26: 395–400.

Xiang, J. Z., P. Adamson, M. J. Brammer & I. C. Campbell. 1990. The kappa-opiate agonist U50488H decreases the entry of 45Ca into rat cortical synaptosomes by inhibiting N- but not L-type calcium channels. *Neuropharmacology.* 29: 439–444.

Xu, C., B. Bailly-Maitre & J. C. Reed. 2005. Endoplasmic reticulum stress: cell life and death decisions. *J. Clin. Invest.* 115: 2656–2664.

Xu, M., M. Petraschka, J. P. McLaughlin, R. E. Westenbroek, M. G. Caron, R. J. Lefkowitz, T. A. Czyzyk, J. E. Pintar, G. W. Terman & C. Chavkin. 2004. Neuropathic pain activates the endogenous kappa opioid system in mouse spinal cord and induces opioid receptor tolerance. *J. Neurosci.* 24: 4576–4584.

Yakovleva, T., I. Bazov, G. Cebers, Z. Marinova, Y. Hara, A. Ahmed, M. Vlaskovska, B. Johansson, U. Hochgeschwender, I. N. Singh, A. J. Bruce-Keller, Y. L. Hurd, T. Kaneko, L. Terenius, T. J. Ekstrom, K. F. Hauser, V. M. Pickel & G. Bakalkin. 2006. Prodynorphin storage and processing in axon terminals and dendrites. *FASEB J.* 20: 2124–2126.

Yakovleva, T., A. Pramanik, T. Kawasaki, K. Tan-No, I. Gileva, H. Lindegren, U. Langel, T. J. Ekstrom, R. Rigler, L. Terenius & G. Bakalkin. 2001. p53 Latency. C-terminal domain prevents binding of p53 core to target but not to nonspecific DNA sequences. *J. Biol. Chem.* 276: 15650–15658.

Yum, S. W. & A. I. Faden. 1990. Comparison of the neuroprotective effects of the N-methyl-D-aspartate antagonist MK-801 and the opiate-receptor antagonist nalmefene in experimental spinal cord ischemia. *Arch. Neurol.* 47: 277–281.

Zamir, N., M. Palkovits & M. J. Brownstein. 1984. Distribution of immunoreactive dynorphin A1-8 in discrete nuclei of the rat brain: comparison with dynorphin A. *Brain Res.* 307: 61–68.

Zhang, W. Q., W. R. Mundy, L. Thai, P. M. Hudson, M. Gallagher, H. A. Tilson & J. S. Hong. 1991. Decreased glutamate release correlates with elevated dynorphin content in the hippocampus of aged rats with spatial learning deficits. *Hippocampus.* 1: 391–397.

Zhang, L., R. W. Peoples, M. Oz, J. Harvey-White, F. F. Weight & U. Brauneis. 1997. Potentiation of NMDA receptor-mediated responses by dynorphin at low extracellular glycine concentrations. *J. Neurophysiol.* 78: 582–590.

Ziak, D., A. Chvatal & E. Sykova. 1998. Glutamate-, kainate- and NMDA-evoked membrane currents in identified glial cells in rat spinal cord slice. *Physiol. Res.* 47: 365–375.

Zoli, M., A. Jansson, E. Sykova, L. F. Agnati & K. Fuxe. 1999. Volume transmission in the CNS and its relevance for neuropsychopharmacology. *Trends Pharmacol. Sci.* 20: 142–150.

7 Postnatal Stress Procedures Induce Long-Term Endocrine and Metabolic Alterations Involving Different Proopiomelanocortin-Derived Peptides

Alberto Loizzo, Gabriele Campana,
Stefano Loizzo, and Santi Spampinato

CONTENTS

7.1   INTRODUCTION

7.1.1   Definition of Stress

The term *stress* still has a sort of *shadowy anonymity*. According to Hans Selye (1976), "stress is the nonspecific response of the body to any demand." A stressor is an agent that produces stress at any time. The general adaptation syndrome represents the chronological development of the response to stressors when their action is prolonged. It consists of three phases: the alarm reaction, the stage of resistance, and the stage of exhaustion. The manifestations of the alarm reaction included the secretion of adrenocorticotropin hormone (ACTH), corticoids and catecholamines, thymicolymphatic involution, eosinopenia, and peptic ulceration (Selye 1976). The definition of the diseases of adaptation or stress-induced maladies is still a cause for debate. The diseases of adaptation depend primarily on an excessive or inappropriate response to indirect pathogens. Included are all "psychosomatic" diseases, allergies, and other immunologic responses, as well as excessive inflammatory reactions. However, at least two topics still need further definition—first, connections between the type of stress and the type of disease and, second, qualitative and quantitative definition of stress.

7.1.2   Stress Hormones—Which Ones?

The importance of the enhanced level or activity of the stress hormones, ACTH and cortisol in human beings or corticosterone in rodents, above all, in the pathogenesis of several metabolic disturbances in humans and models of metabolic diseases in animals are widely acknowledged (Hermanowski-Vosatka and Thieringer 2005; Walker 2006; Berthiaume et al. 2007; Drake, Tang, and Nyirenda 2007; Lee et al. 2008; Anagnostis et al. 2009; Cooper and Stewart 2009). Recently, several studies have suggested that stress hormones are not the sole pathogenetic mechanisms for triggering metabolic conditions in men or for dysmetabolic models in animals (Coccurello, D'Amato, and Moles 2009; D'Argenio et al. 2009). Several physiological bases can support this hypothesis.

7.1.3 STRESS MECHANISMS

Stress mechanisms can be summarized as follows: stress induces release of corticotropin-releasing hormone (CRH) from the hypothalamus, which, in turn, stimulates the synthesis and release of proopiomelanocortin (POMC) from the pituitary. In turn, POMC produces ACTH and the endogenous opioid β-endorphin, together with other peptides, by cleavage mechanisms. Finally, ACTH stimulates cortisol/corticosterone release from the adrenal gland. Thus, the importance of the ACTH–adrenal axis in the stress mechanism is well known. However, an unsolved question concerns the role of β-endorphin and of the endogenous opioid system as important stress chain promoters (Loizzo, Vella, et al. 2010). Endogenous opioid system, in fact, could participate in stress-induced dysregulation and, thus, it could concur in triggering some (in not all) signs of a major metabolic disease or even of a diabetes syndrome in animal models.

7.1.4 BRAIN RECEPTORS FUNCTION AND INTERCONNECTIONS

It is extremely difficult to investigate and prove such an assumption in humans for ethical and technical reasons. Crucial information has been provided by studies performed in developing animals; that is, postnatal stressful procedures in rodents gave important information on the critical period of development of brain structures and of some stress mechanisms and, above all, on the hypothalamus–pituitary–adrenal (HPA) connections. During the first 3 weeks of a rodent's postnatal life, several brain receptors gain their complete maturation (Lauder and Krebs 1986; McDowell and Kitchen 1987; Rosenfeld et al. 1993). Among these, opioid receptors and glucocorticoid receptors can be considered of importance because even mild stressful procedures applied daily to newborn rodents for 3 weeks up to weaning are able to produce enduring behavioral alterations, lasting up to adult age (Francis et al. 1999; Moles, Kieffer, and D'Amato 2004). The great majority of these studies was performed with the "classical" handling procedure, which is based on a mild psychological stress, that is, newborn pups are mother deprived for a few (10–15) minutes daily for a few days, usually up to weaning (21 days of age). In the adulthood, neonatally exposed animals present hormonal alterations (chiefly, reduced pituitary ACTH and adrenal corticosterone responses to stress compared with nonhandled animals) and also show behavioral alterations. However, following this handling model, no systematic studies on possible metabolic alterations were performed.

7.1.5 COMPLEX STRESS MODEL AND THE PHARMACOLOGICAL APPROACH

The addition of a physical distress to the psychological stress, however, induces radical changes in the stress consequences: mice receiving brief mother deprivation, daily (10 minutes), plus mild pain (subcutaneous, sc, injection of saline) showed a completely different enduring metabolic pattern, which resembles a mild type-2 diabetes model or a mild metabolic syndrome, in adulthood (Loizzo et al. 2006). These alterations occur in adult mice and are accompanied by several dysmetabolic signs: fasting hyperglycemia, mild overweight and abdominal overweight, blood lipid abnormality, and others. Moreover, other signs closely related to diabetes are evidenced, such as functional vascular alterations, disruption of hormonal feedback equilibrium in the HPA system, alterations of the endogenous opioid system, alterations of immune system, alterations of the glucocorticoid synthesis, and other behavioral and pathophysiological changes. Another novelty introduced by these studies was the adoption of a pharmacological approach. The theoretical basis of these studies consisted in the pharmacological blockade of μ- and δ-opioid receptors, which is performed through the daily administration of naloxone, 1 mg/kg sc, instead of saline, to stressed pups. Naloxone was administered with the aim of preventing, at least partially, any stress consequence possibly triggered by β-endorphin or other endogenous opioids released by the stressful procedure. Furthermore, the block of the hormonal chain induced by stress was attained by inhibiting POMC production. This was done with the administration of an antisense oligonucleotide versus POMC (AS), which was shown to block POMC mRNA transcription and β-endorphin and ACTH production *in vitro* (Spampinato et al. 1994), and was effective in blocking *in vivo* both arms of hormone overproduction, that is, both β-endorphin and ACTH-corticosterone, following sc administration (Loizzo et al. 2003). Therefore, postnatally stressed mice treated with AS are expected to show no or negligible diabetes signs as adults. No information is still available in the literature whether specific metabolic signs can be attributed to the opioid arm of the endogenous POMC overproduction, following postnatal stress. Therefore, one working hypothesis was devoted to study whether naloxone would be able to prevent some signs in postnatally stressed mice, whereas other signs could be attributed to the ACTH-corticosterone hyperactivity after stress.

7.2 EXPERIMENTAL MODEL

7.2.1 ANIMALS

Different series of multiparous pregnant mice from the CD-1 outbred strain, received on the 14th day of pregnancy, were set in single cages. Animals were housed in light- (12 hours on and 12 hours off), temperature- (20°C ± 2°C), and humidity- (55% ± 5%) controlled environments, with food and water available *ad libitum*. Food contained 3.95 kcal/g equivalent to assimilable 2.7 kcal/g. Pregnant females were checked for litters twice daily. About 12 hours after birth, male pups of similar weight were selected, and six pups per litter were randomly put together and cross-fostered, and randomly assigned to one of the groups. All experiments were performed in winter as the opioid receptor sensitivity is maximal in this period because of their circannual variations (De Ceballos and De Felipe 1985). Pregnant females were exposed to a certain stress travel on their 14th day. The possible interference due to prenatal travel stress was minimized by its extension to all experimental groups: control mice and handled, but not injected, mice (Wsp mice) did not develop any metabolic, hormonal, and behavioral alterations described for pre- and postnatal stress in the literature (Loizzo, Campana, et al. 2010). Experiments were carried out in accordance with the guidelines of the Council of European Communities 86/609/

EEC and the protocols were approved by the Bioethical Committee of the Istituto Superiore di Sanità (Roma, Italy).

7.2.2 Treatments

The model basically included the following groups: (1) nonneonatal-handled mice (controls—C): the pups were left undisturbed with their mothers, except for cage cleaning twice a week; (2) vehicle-treated handled mice (W): for 21 days, the pups were removed daily from the home cage (10 minutes mother deprived) and grouped in a container with fresh bedding material, then they were weighed and injected sc with sterile saline (1 mL/kg body weight per day); (3) naloxone-treated handled mice (Na): these pups were handled as in the vehicle groups and received (–)naloxone hydrochloride, 1 mg/kg body weight per day, as weight of the base; (4) As-treated handled mice (AS): pups were handled as in the vehicle groups but were treated with As-POMC (0.1 nmol/g body weight) per day; (5) handled but not injected mice (Wsp): the pups were removed daily (10 minutes mother deprived) from the home cage and grouped in a container with fresh bedding material, then returned to the home cage with the mother, according to the "classical" handling model. Other experiments were performed, including "positive" controls (i.e., stressed mice not treated with drugs or oligonucleotides). After weaning, the animals were rehoused, three per cage, and left undisturbed until the 35th day, in which they underwent pain threshold test. At adult age (usually at the 90th day, in some experiments at the 120th day or later), animals underwent other experiments before being sacrificed.

Two compounds were important for managing the neonatal stress model: (–)naloxone and the AS oligonucleotide. Naloxone, a μ- and δ-opioid receptor antagonist, was administered to pups in order to produce a brief-lasting block of the endogenous opioid system, including β-endorphin. In preliminary experiments, several doses of naloxone were used. Figure 7.1 reports a dose–response relationship for naloxone, which was able to prevent a typical effect induced by the model, that is, the enhancement of the pain threshold in postnatally stressed mice, at 35 days of age.

Therefore, the 1 mg/kg dose was found suitable for the following investigations, as it was the lowest dose, which was able to prevent completely the enhancement of pain threshold over control in stressed mice. Conversely, the (+)naloxone enantiomer did not induce any biological effects, as expected (Loizzo et al. 2002).

The other compound was AS, an antisense oligonucleotide phosphorothioate, which was first described by Spampinato et al. (1994). The use of antisense oligonucleotides is a common tool to downregulate the expression of a given gene; they have already been proposed to treat cancer and neurological diseases (Jansen and Zangemeister-Wittke 2002; Forte et al. 2005). However, their effective intracellular delivery remains an important issue for their clinical application, but phosphorothioate oligonucleotides cross the blood–brain barrier (Banks et al. 2001). In addition, we have previously shown that parenteral administration of the phosphorothioate antisense versus POMC in neonates dose-dependently reduces the anterior pituitary immunoreactive-ACTH, as well as plasma corticosterone in newborn mice, whereas its mismatch analog (MM-ACTH) is inactive (Loizzo et al. 2003).

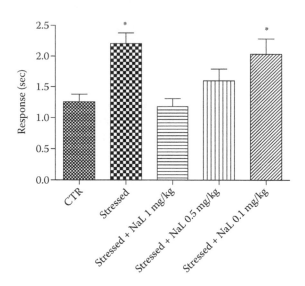

FIGURE 7.1 Dose–response effect of naloxone on tail-flick latency in 35-day-old mice (18 mice per group). Mice were subcutaneously treated, as described in Section 7.2.2, after birth for 21 days with saline (stressed) or naloxone (stressed + Nal). Tail-flick test was carried out 14 days after the last stress section. Abscissa: treatment (stress is for brief mother separation plus saline injection; Nal is for naloxone, dose in mg/kg, daily up to weaning). Ordinate: response latency in seconds, mean ± SEM. *$p < .05$ versus nonstressed mice (CTR). For methods, see Loizzo, A., Loizzo, S., Lopez, L. et al. 2002. Naloxone prevents cell-mediated immune alterations in adult mice following repeated mild stress in the neonatal period. *Br. J. Pharmacol.* 135: 1219–26.

The 21-base sequence of As-POMC and MM-POMC are:

5'-TCT GGC TCT TCT CGG AGG TCA-3' and 5'-T**GT** G**CC** TCT **TTC** CGG **T**GG AC**A**-3' (the mutated bases are in bold), respectively.

It has been previously reported that the antisense oligonucleotide reduces the synthesis of POMC-derived hormones both *in vitro* and *in vivo* (Spampinato et al. 1994; Loizzo et al. 2003). To avoid the potential confounding factor of a huge decrease in the activity of HPA axis, we selected doses that were able to reduce the activity increase due to handling, but left the basal level of POMC-derived molecules unaltered (Galietta et al. 2006).

7.3   RESULTS

7.3.1   Behavioral, Hormonal, and Metabolic Alterations Induced by the Postnatal Stress Procedure: Effect of Antisense Oligonucleotide and Naloxone

Postnatal stress procedure represents the starting mechanism for triggering all metabolic alterations observed in the adult age. Conversely, at weaning, no or scarce metabolic alterations are still evident. Figure 7.2 describes the effects induced by the complex stress model on HPA hormones in mice at weaning, on

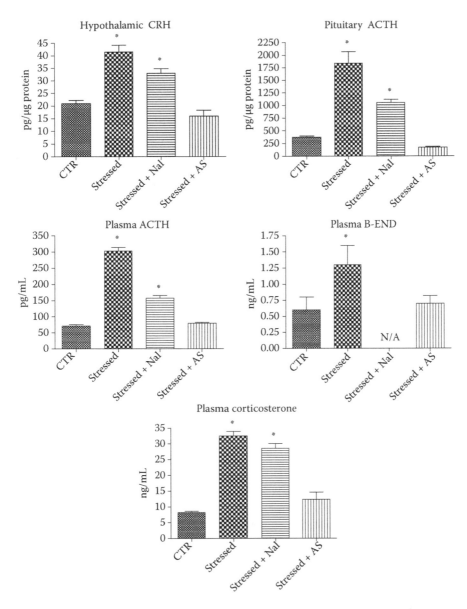

FIGURE 7.2 Stress hormone levels evaluated in mice neonatally exposed to the adopted stressful procedure (see section 7.2). Stress hormones were evaluated in the hypothalamus, pituitary, and plasma. Data were gathered in 21-day-old male mice, 30 minutes after the last stress section. Mice were subcutaneously treated, as described in section 7.2.2, with saline (stressed), naloxone (stressed + Nal), or an antisense oligonucleotide against proopiomelanocortin (stressed + AS). Values are mean ± SEM of at least six animals per group. *$p < .01$ versus nonstressed mice (CTR). For methods, see Galietta, G., Loizzo, A., Loizzo, S. et al. 2006. Administration of antisense oligonucleotide against pro-opiomelanocortin prevents enduring hormonal alterations induced by neonatal handling in male mice. *Eur. J. Pharmacol.* 550: 180–5.

the last day of treatment (21st day), and the blocking effect produced by AS, whereas naloxone induces only a partial antagonism, if any. Figure 7.3 shows the effects induced by the stressful procedure on the HPA hormones in adult mice, at the 90th day of life. Plasma corticosterone and ACTH levels, as well as pituitary ACTH levels, are strongly enhanced in neonatally stressed mice. On the contrary, hypothalamic ACTH and CRH levels are very low. This indicates a disruption of negative feedback mechanisms at the pituitary level and an enduring overproduction and release of ACTH and corticosterone. Conversely, at the adult age, plasma β-endorphin did not show any variations among the groups, although other investigators have reported endogenous opioid alterations in the brain of adult rodents, which had received postnatal manipulation (Ploj, Roman, and Nylander 2003).

The following paragraphs will focus on the most important signs produced by the adopted complex stress model that resemble those observed in the human metabolic syndrome (Alberti et al. 2005), that is, (1) overweight and abdominal overweight; (2) glucide abnormalities; (3) lipid abnormalities; and (4) cardiovascular abnormalities. Thereafter, other signs, such as behavioral, immune, and neurophysiological and neurometabolic changes, will be considered. Finally, single signs shall be put in relationship to the two arms of the previously described stress mechanism, that is, signs which are prevented by naloxone treatment and signs which are prevented by AS treatment.

7.3.2 DEFINITION OF THE MILD METABOLIC SYNDROME MODEL: MILD OVERWEIGHT, ABDOMINAL OVERWEIGHT, AND FOOD CALORIC EFFICIENCY

Stressed mice show a body weight similar to controls up to the 60th day of life. However, starting from this age, the weight increment curve begins to differentiate from controls and at 90 days, and thereafter, the body weight of stressed mice is consistently higher (+7.5%) versus controls. Postnatal naloxone or AS treatment prevents this overweight (Figure 7.4a). Abdominal overweight, measured as the weight of periepididymal fat pads, is also significantly heavier than controls. Moreover, a significant hypertrophic evolution of epididymal adipocytes with increased volume in at least 60% of cells was confirmed through morphometric image analysis. Naloxone-treated animals show fat pads weight quite analogous to controls; furthermore, AS-treated mice have epididymal fat pad similar to controls (Figure 7.4b) (Loizzo et al. 2006; Loizzo, Campana, et al., 2010).

Food caloric efficiency is measured as the body weight gain in grams, divided by food ingested in grams in a certain period of time. Previous investigations showed that the caloric efficiency computed over the whole period (23–118 days of life of mice employed in our experiments) is quite analogous for control and stressed animals (Loizzo et al. 2006). A detailed analysis has revealed that the caloric efficiency was quite identical in the pre-puberal period, but it increased consistently after puberty in stressed animals, and this contributed in creating the body weight difference at the adult age (Loizzo et al. 2006). It is interesting to underline that

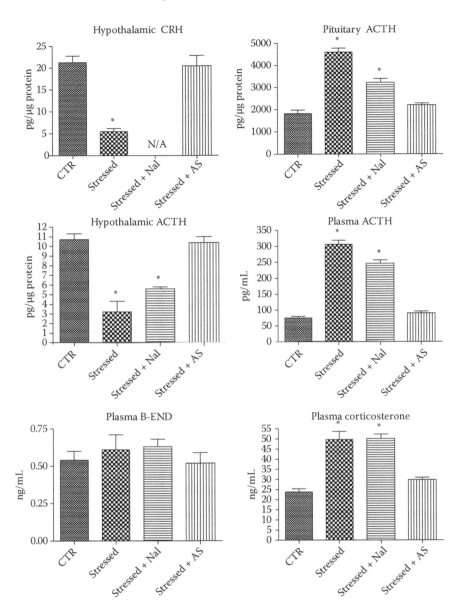

FIGURE 7.3   Stress hormone levels at 90 days of life in mice exposed for 21 days after birth to the adopted stressful procedure (see section 7.2). Stress hormones levels in the hypothalamus, pituitary, and plasma. Mice were subcutaneously treated, as described in section 7.2.2, with saline (stressed), naloxone (stressed + Nal), or an antisense oligonucleotide against proopiomelanocortin (stressed + AS). Values are mean ± SEM of at least six animals per group. Data were gathered in adult, 90-day-old mice (at least six animals per group) in normal (nonfasting) conditions. *$p > .01$ versus nonstressed mice (CTR). N/A, not determined. For methods, see Galietta, G., Loizzo, A., Loizzo, S. et al. 2006. Administration of antisense oligonucleotide against pro-opiomelanocortin prevents enduring hormonal alterations induced by neonatal handling in male mice. *Eur. J. Pharmacol.* 550: 180–5.

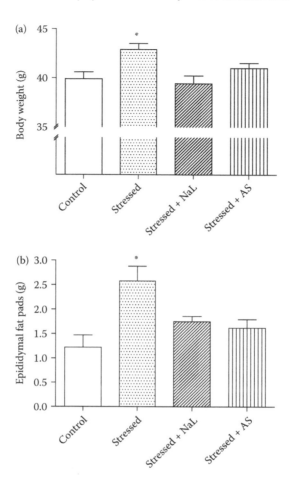

FIGURE 7.4 Body weight and epididymal fat pads in mice at 90 days of life. Mice were stressed and subcutaneously treated for 21 days after birth, as described in section 7.2.2, with saline (stressed), naloxone (stressed + Nal), or an antisense oligonucleotide against proopiomelanocortin (stressed + AS). (a) The weight of stressed mice is consistently heavier than the weight of controls. Both naloxone and AS, neonatally administered, were able to prevent weight increase. (b) Epididymal fat pads. The weight of fat pads is consistently heavier than the weight of controls; naloxone or AS were able to prevent this increase. *$p < .05$ versus nonstressed mice (CTR).

naloxone-treated animals show a caloric efficiency similar to controls (Table 7.1). Moreover, the spontaneous locomotor activity did not show consistent differences, but a (nonsignificant) reduction of 4.8% in stressed mice versus controls and versus naloxone-treated mice during the light phase (not shown). Also, food consumption was not consistently increased in the stressed group as a whole, but the food intake increase for the total period of observation was about 5 g per stressed mouse, which was enough to explain a part of the whole overweight in stressed mice, about 20% of the difference versus controls.

TABLE 7.1

Caloric Efficiency Was Measured as Body Weight Gain in Grams, Divided by Food Ingested in Grams during the Same Period of Time

Groups	Caloric Efficiency, 23–58 Days	Caloric Efficiency, 78–118 Days
Control	0.096 ± 0.005	0.012 ± 0.001
Stressed	0.101 ± 0.004	0.025 ± 0.003*
Stressed, Na treated	0.109 ± 0.006	0.014 ± 0.002

Two different periods were considered: pre- and peripubertal (23–58 days) and after the onset of overweight found in stressed mice (78–118 days). Nine mice per group were considered (three cages, three mice per cage).

*$p < .01$ versus other groups. (for methods, see Loizzo, A., Loizzo, S., Galietta, G. et al. 2006. Overweight and metabolic and hormonal parameter disruption are induced in adult male mice by manipulations during lactation period. *Pediatr. Res.* 59: 111–15.)

7.3.3 DEFINITION OF THE MILD METABOLIC SYNDROME MODEL: GLUCIDE METABOLISM

The International Diabetes Federation (Alberti et al. 2005) recommends a raised fasting plasma glucose, ≥ 100 mg/dL (5.6 mmol/L), for a diagnosis of metabolic syndrome in humans. Postnatally stressed mice present increased fasting plasma glycemia and naloxone-treated mice showed hyperglycemia as well. Conversely, AS treatment was able to prevent hyperglycemia (Figure 7.5).

However, after glucose loading, incremental area under the glucose curve (iAUC) values were not different in stressed versus control mice, and so was the iAUC after insulin loading (data not shown) (for methods and results, see Loizzo, Vella, et al. 2010; Loizzo, Campana, et al. 2010). Furthermore, basal insulin plasma levels were increased in stressed mice. Fasted control mice showed insulin level of 0.42 ± 0.11 µg/L, versus 0.93 ± 0.30 µg/L in stressed mice. Also in this case, as for glycemia levels, stressed and naloxone-treated mice did not show decreased levels of insulin, but they showed a consistent increase of fasting insulinemia (2.74 ± 0.58 µg/L, 11 mice per group, $p < .01$) (unpublished data, methods in Loizzo et al. 2006).

7.3.4 DEFINITION OF THE MILD METABOLIC SYNDROME MODEL: LIPID METABOLISM

At 90 days of age, triglycerides showed consistent increase by about 21% ($p < .05$), and total cholesterol by about 7%, in stressed mice versus controls. At 120 days of life, stressed mice showed an increase in triglycerides by 33% and total cholesterol by 13% (69.9 ± 2.4 versus 93.0 ± 3.9 and 97.9 ± 3.1 versus 110.2 ± 3.4 mg/dL, respectively, $p < .05$). Lipids did not show consistent variations in stressed and

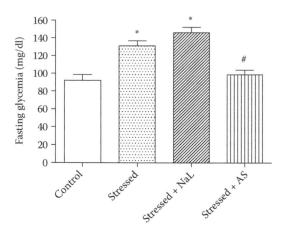

FIGURE 7.5   Fasting glycemia in mice at 90 days of life. Mice were stressed and subcutaneously treated for 21 days after birth, as described in section 7.2.2, with saline (stressed), naloxone (stressed + Nal), or an antisense oligonucleotide against proopiomelanocortin (stressed + AS). Glycemia in stressed mice was consistently higher than in controls. Naloxone treatment did not induce any consistent change, whereas AS treatment drew glycemia levels to those of nonstressed controls. Glycemia does not appear to be influenced by the opioid system, whereas it seems to be controlled by the ACTH-corticosterone component. $*p < .01$ versus nonstressed mice (CTR); $^{\#}p < .01$ versus stressed or versus stressed + Nal mice.

naloxone-treated mice. Plasma phospholipids did not show consistent changes in stressed mice. Erythrocyte membrane fatty acids were also analyzed to have a steady-state pattern of lipid composition but no consistent variations were found in stressed male mice as well (Loizzo et al. unpublished data; Loizzo et al. 2006; Loizzo, Vella, et al. 2010).

7.3.5   Definition of the Mild Metabolic Syndrome Model: Cardiovascular Alterations

No direct measures of cardiac parameters were produced in the stressed mice, but only indirect pathophysiological data on the reactivity of isolated aorta to drugs. Effects induced by postnatal stress on the vascular reactivity of adult animals were studied observing contractile (noradrenaline, NA) and relaxing (acetylcholine, ACh) responses of isolated aorta from mice aged 90 days. The effect of the NOS inhibitor, L-NNA, was also tested (for methods, see Cordellini and Vassilief 1998, modified). Concentration–response curve of NA-induced aortic ring contractions from the postnatally stressed group was significantly shifted rightward, thus indicating a lower effect in stressed mice versus controls. Conversely, the relaxing responses of aorta to ACh and the dose–relaxation curve were significantly shifted to the left, thus indicating a stronger effect exerted by ACh in stressed mice versus controls. Neonatal treatment with the AS prevented the rightward shift of the contractile–concentration response curve of aorta to NA and also reduced the ACh maximal relaxing effect observed in aorta from stressed animals; whereas, in the naloxone group, the dose–response curves to NA and ACh overlapped those obtained in the stressed group (Loizzo et al. unpublished data).

7.3.6 BEHAVIORAL RESULTS

With the help of different animal models, a little data in this field is presented in the literature (Moles et al. 2004). In the following paragraph, two main behavioral effects are described, that is, effects of postnatal stress on pain sensitivity and on passive avoidance test.

1. *Pain threshold enhancement.* Stressed mice showed consistent increase of pain threshold sensitivity at prepubertal ages, that is, at 25 to 45 days of age (Pieretti, d'Amore, and Loizzo 1991). This effect was evidenced using both tail-flick and hot-plate tests; it was completely prevented in stressed and naloxone-treated mice, in stressed and AS-treated mice, whereas mother-deprived not injected mice did not show any difference versus controls (Table 7.2).
2. *Passive avoidance test.* Mice were tested at 54 days of age at the one-trial avoidance test. Data showed that stressed mice remained in the starting box for a longer time than controls and naloxone-treated mice ($p < .001$). A significant difference was also evident in the entrance latency during the retention session, with stressed mice showing latencies longer than control and naloxone groups ($p < .001$) and, therefore, showing better performances (Table 7.3) (for methods, see Qin, Kang, and Smith 2002). These data can be put in relationship to brain neurophysiological and metabolic results.

7.3.7 NEUROPHYSIOLOGICAL–NEUROMETABOLIC DATA: BIPHASIC FLUORESCENCE TRANSIENTS EVOKED BY A SINGLE TETANUS TO CORTICAL PATHWAY INPUTS IN AN *EX VIVO* MODEL

Light pulses (360 nm, 100 ms) were used for excitation, whereas the emission (>410 nm) was collected from isolated slices in the primary visual cortex. The electrical stimulus (100 pulses, 70 μs, 100 Hz, 1 mA) is followed by a sudden decrease in fluorescence. This transient (initial component) after decrease is followed by a more sustained fluorescence increase (overshoot). Brain slices gathered from stressed mice showed a reaction in the initial component similar to controls, whereas the reaction of the overshoot component was consistently greater and appeared with a lower latency versus controls. Both naloxone and AS groups showed an overshoot curve similar to controls (Loizzo, Pieri, et al. 2010).

7.3.8 IMMUNE SYSTEM ALTERATIONS

The complex stress model also induces long-term (up to 110 days of life) splenocyte modifications, consisting of increased release of the Th-1 type cytokines; decreased release of some Th-2 type cytokines; enhanced natural killer-cell activity; enhanced proliferative splenocyte properties in resting conditions; and phytohemoagglutinin

TABLE 7.2

Nociceptive Threshold (in s ± SEM) Evaluated by the Tail-Flick and Hot-Plate Assays in Mice Exposed for 21 Days after Birth to the Stressful Procedures Described in Section 7.2.2

Groups	C	W	Na	AS	Wsp
Tail-Flick in s (mean ± SEM)	1.5 ± 0.2	6.3 ± 0.9**	1.8 ± 0.4	2.3 ± 0.6	2.0 ± 0.2
Hot-plate in s (mean ± SEM)	14.9 ± 1.5	33.4 ± 4.6**	13.5 ± 2.2	20.1 ± 2.5	N/A

**$p < .01$ versus other groups.

N/A, not performed. Newborn mice were randomly assigned to one of the following groups: controls (C); 10 minutes mother deprived and saline treated (W); 10 minutes mother deprived and naloxone treated (Na); 10 minutes mother deprived and antisense treated (AS); 10 minutes mother deprived and nonstressed (Wsp).

For methods, see Pieretti, S., d'Amore, A., Loizzo, A. 1991. Long-term changes induced by developmental handling on pain threshold: effects of morphine and naloxone. *Behav. Neurosi.* 105: 215–18. Values in seconds, mean ± SEM of 12 animals per group.

TABLE 7.3

Passive Avoidance Test

Groups	First Day Latency (s) Median ± SEM	Second Day Latency (s) Median ± SEM
Control	11 ± 0.6	105 ± 22
Stressed	16 ± 1*	248 ± 32*
Stressed and Na treated	12 ± 0.5	137 ± 9

At 54 days of age, that is, 33 days after the end of stressful procedures, animals underwent a passive avoidance test by a step-through to darkness paradigm, with one training trial and one retention trial. Mice were placed in a lighted platform (8 × 8 cm) and the timer run until the mouse completely entered an obscure compartment (20 × 20 × 20 cm), thus determining the training entrance latency. Then, the mouse received an electric footshock (0.35 mA for 6 seconds). Memory retention was tested 24 hours later without footshock. Entrance latency during the retention session was measured up to 300 seconds (cutoff time).

*$p < .01$ versus control or stressed and Na treated.

and concanavalin-A stimulation. Immunological changes are prevented by naloxone but not by the biologically inactive enantiomorph (+)naloxone (Loizzo et al. 2002).

7.3.9 PREVALENT OPIOID OR PREVALENT ACTH-CORTICOSTERONE INFLUENCES THE MILD METABOLIC SYNDROME MODEL

Table 7.4 summarizes the possible connections of mild metabolic syndrome/type-2 diabetes model signs to the endogenous opioid and ACTH-corticosteroid systems.

TABLE 7.4

Differential Effects Induced by Naloxone and an Antisense Oligonucleotide against Proopiomelanocortin (AS) on Postnatal Stress-Induced Residual Alterations to Adult Male Mice

Naloxone Prevention of Effects Due to Prevalent Opioid Influence	AS Prevention of Effects Due to Prevalent ACTH–Corticosteroid Influence
Transitory increase in pain threshold (up to 45 days of age) (Pieretti et al.1991)	Hyperglycemia (unpublished data)
Body weight increment curve (Loizzo et al. 2002)	Increased corticosterone and ACTH plasma levels (Galietta et al. 2006)
Abdominal overweight (Loizzo, Campana, et al. 2010)	Increased pituitary ACTH level (Galietta et al. 2006)
Fat cell hypertrophy (Loizzo et al. 2006)	Decreased hypothalamic ACTH and CRH levels (Galietta et al. 2006)
Better performance at the passive avoidance test (unpublished data)	Increased plasma ACTH and β-endorphin levels at 21 days of age (Galietta et al. 2006)
Anticipation of visual-evoked potentials latency (N1 peak) (unpublished data)	Increased amplitude and decreased latency of NAD(P)H fluorescence imaging overshoot (unpublished data)
Increased amplitude and decreased latency of NAD(P)H fluorescence imaging overshoot (in part) (unpublished data)	Alterations of isolated vas deferens sensitivity to δ1 selective opioid agonist (Loizzo et al. 2003)
Decreased hypothalamic ACTH level (in part) (Galietta et al. 2006)	Alterations of isolated aorta sensitivity to neurohumoral stimuli (NAd, ACh) (unpublished data)
Increased pituitary ACTH level (in part) (Galietta et al. 2006)	[Furthermore, AS prevents all effects prevented by naloxone, listed in the column at left]
Increased ACTH plasma level (in part) (Galietta et al. 2006)	
Increased immune response of the Th-1-type (Loizzo et al. 2002)	

All data were obtained at adult age, with the exception of some hormonal data, pain threshold level, and passive avoidance test.

7.4   DISCUSSION

Alteration of the HPA system is allegedly admitted as a cause or an important component of stress diseases, and plasma ACTH and cortisol levels in men (or corticosterone in rodents) are often suggested as a quantitative estimation of stress severity. However, from the data exposed in this review, we can suggest that different stress procedures in newborn mice induce activation of different hormones/neuropeptide mechanisms, which are responsible, together with ACTH-corticosterone, for the appearance of specific clusters of signs in the adult animal. The postnatal stressful procedures adopted in our investigations produced, at weaning, an enhancement of POMC-derived hormones and neuropeptides, that is, both opioids and ACTH. When mice became adults, in steady-state conditions, a feedback mechanism failure was evident. Plasma ACTH and corticosterone levels were abnormally high, and the increase was accompanied by low CRH and ACTH in the hypothalamus, as expected. However, ACTH in the pituitary was not sensitive to a negative feedback regulation and appeared abnormally high as well. Therefore, the enhanced ACTH appeared to be independent by lowered hypothalamic CRH. In the adult mice, these hormonal alterations were accompanied by several metabolic, immune, behavioral, and neurophysiological–neurometabolic alterations. Some alterations were prevented by naloxone treatment; others were not prevented by naloxone, but only by AS treatment. Therefore, we suggest that clusters of metabolic alterations depended mainly on the dysregulation of the endogenous opioid system, whereas others depended on the dysregulation of the ACTH–corticosterone system. The naloxone-sensitive endogenous opioid system is involved in all three components, which contribute to overweight and abdominal overweight, because stressed and naloxone-treated animals did not show any increase in food consumption and food caloric efficiency, neither did they show diminution in spontaneous activity. Conversely, naloxone did not counteract raised fasting plasma glucose and insulin observed in stressed animals and did not influence lipid parameters, neither did vascular physiology studied in the isolated aorta parameters.

Several results obtained in our hormonal–metabolic studies seem to introduce a certain novelty. Although several articles in the literature deal with stress (in humans and animals) and its relationship versus certain adult metabolic alterations (D'Argenio et al. 2009; Ravelli et al. 1999; Darnaudéry and Maccari 2008; Coccurello et al. 2009), we were not able to find out in the literature systematic investigations on the possible relationship between the type of stress, metabolic alterations, and the neurochemical/neurohumoral disruption mechanisms, which may represent a pathogenetic link for the three phenomena. This chapter may suggest future studies and define the term stress and stress-derived diseases. Some connections evidenced in postnatally stressed mice (e.g., between some immune alterations and the opioid system) have been widely studied in the literature; however, studies have been performed almost exclusively in adult subjects and, therefore, may reach different conclusions—different duration, partly different effects exerted by opioids (Sacerdote et al. 2000). This problem is further complicated by the numerous and complex physiological interconnections, which have been demonstrated between the endogenous opioid system and the endogenous ACTH-corticosteroid system in the central and peripheral nervous system of the adult animal (Pieretti et al. 1994; Capasso et al. 1996).

We are aware that our investigations were intended to help answering some questions, but new questions have surfaced. For example, what is the mechanism of pituitary feedback failure following postnatal stress? Our model apparently works in an outbred mouse strain, but we do not know whether it works in other (e.g., inbred) strains. According to some preliminary studies, it works in DBA/2J inbred strain, but it does not work in female mice of the same CD1 strain (Loizzo, Vella et al. 2010; Loizzo et al. unpublished data). Anyway, within our experimental model, some indications can be given. For instance, one interesting indication is that a mild but repeated pain (as a sham injection is) applied to developing organisms can bring about consequences, which are impossible to preview. Animal studies cannot be transferred to humans or to children, but we should remember that human newborns at risk may undergo long duration mother deprivation (they may be put in an incubator) and also painful maneuvers for medical reasons. We wonder whether these children are at increased risk for metabolic diseases as adults, and whether part of the risk can be avoided by reducing pain related to medical treatment.

ACKNOWLEDGMENTS

This chapter was realized in part with the contribution of Italian Istituto Superiore di Sanità and US-NIH Research Collaborative Project, "Gender difference in seizure sensitivity: role of steroids and neuroactive steroids" to Stefano Vella, Alberto Loizzo, and in part with the contribution of Italian Istituto Superiore di Sanità and US-NIH Research Collaborative Project, "Proposal for an integrated approach to rare diseases: a study between basic laboratory models and clinical epidemiology in amyotrophic lateral sclerosis (A.L.S.)" to S.L., Stefano Vella, and A.L.

REFERENCES

Alberti, K.M.M., Zimmet, P., Shaw, J. et al. 2005. The metabolic syndrome—a new worldwide definition. *Lancet.* 366: 1059–62.

Anagnostis, P., Athyros, V.G., Tziomalos, K., Karagiannis, A., Mikhailidis, D.P. 2009. Clinical review: the pathogenetic role of cortisol in the metabolic syndrome: a hypothesis. *J. Clin. Endocrinol. Metab.* 94: 2692–701.

Banks, W.A., Farr, S.A., Butt, W., Kumar, V.B., Franko, M.W., Morley, J.E. 2001. Delivery across the blood–brain barrier of antisense directed against amyloid beta: reversal of learning and memory deficits in mice overexpressing amyloid precursor protein. *J. Pharmacol. Exp. Ther.* 297: 1113–21.

Berthiaume, M., Laplante, M., Festuccia, W. et al. 2007. Depot-specific modulation of rat intraabdominal adipose tissue lipid metabolism by pharmacological inhibition of 11beta-hydroxysteroid dehydrogenase type 1. *Endocrinology.* 148: 2391–7.

Capasso, A., Di Giannuario, A., Loizzo, A., Pieretti, S., Sagratella, S., Sorrentino, L. 1996. Dexamethasone selective inhibition of acute opioid physical dependence in isolated tissues. *J. Pharmacol. Exp. Ther.* 276: 743–51.

Coccurello, R., D'Amato, F.R., Moles, A. 2009. Chronic social stress, hedonism and vulnerabilità to obesity: lessons from rodents. *Neurosci. Biobehav. Rev.* 33: 537–50.

Cooper, M.S., Stewart, P.M. 2009. 11Beta-hydroxysteroid dehydrogenase type 1 and its role in the hypothalamus-pituitary-adrenal axis, metabolic syndrome, and inflammation. *J. Clin. Endocrinol. Metab.* 94: 4645–54.

Cordellini, S., Vassilieff, V.S. 1998. Decreased endothelium-dependent vasoconstriction to noradrenaline in acute-stressed rats is potentiated by previous chronic stress: nitric oxide involvement. *Gen. Pharmac.* 30: 79–83.

D'Argenio, A., Mazzi, C., Pecchioli, L., Di Lorenzo, G., Siracusano, A., Troisi, A. 2009. Early trauma and adult obesity: is psychological dysfunction the mediating mechanism? *Physiol. Behav.* 98:543–6.

Darnaudéry, M., Maccari, S. 2008. Epigenetic programming of the stress response in male and female rats by prenatal restraint stress. *Brain Res. Rev.* 57: 571–85.

De Ceballos, M.L., De Felipe, C. 1985. Circannual variation in opioid receptor sensitivity in mouse vas deferens. *Eur. J. Pharmacol.* 106: 227–8.

Drake, A.J., Tang, J.I., Nyirenda, M.J. 2007. Mechanisms underlying the role of glucocorticoids in the early life programming of adult disease. *Clin. Sci.* 113: 219–32.

Forte, A., Cipollaro, M., Cascino, A., Galderisi, U. 2005. Small interfering RNAs and antisense oligonucleotides for treatment of neurological diseases. *Curr. Drug. Targets.* 6: 21–9.

Francis, D.D., Champagne, F.A., Liu, D., Meaney, M.J. 1999. Maternal care, gene expression, and the development of individual differences in stress reactivity. *Ann. N.Y. Acad. Sci.* 896: 66–84.

Galietta, G., Loizzo, A., Loizzo, S. et al. 2006. Administration of antisense oligonucleotide against pro-opiomelanocortin prevents enduring hormonal alterations induced by neonatal handling in male mice. *Eur. J. Pharmacol.* 550: 180–5.

Hermanowski-Vosatka, A., Thieringer, R. 2005. Inhibition of 11beta-HSD1 as a novel treatment for the metabolic syndrome: do glucocorticoids play a role? *Expert. Rev. Cardiovasc. Ther.* 3: 911–24.

Jansen, B., Zangemeister-Wittke, U. 2002. Antisense therapy for cancer—the time of truth. *Lancet. Oncol.* 3: 672–83.

Lauder, J.M., Krebs, H. 1986. Do neurotransmitters, neurohumors and hormones specify critical periods? In *Developmental Neuropsychobiology*, ed. W.T. Greenough, and J.M. Juraska, NY: Academic Press.

Lee, M.J., Fried, S.K., Mundt S.S. et al. 2008. Depot-specific regulation of the conversion of cortisone to cortisol in human adipose tissue. *Obesity.* 16: 1178–85.

Loizzo, S., Campana, G., Vella, S. et al. 2010. Post-natal stress-induced endocrine and metabolic alterations in mice at adulthood involve different pro-opiomelanocortin-derived peptides. *Peptides.* 31: 2123–9.

Loizzo, A., Capasso, A., Galietta, G., Severini, C., Campana, G., Spampinato, S. 2003. Vas deferens response to selective opioid receptor agonists in adult mice is impaired following postnatal repeated mild stress. *Eur. J. Pharmacol.* 458: 201–5.

Loizzo, A., Loizzo, S., Galietta, G. et al. 2006. Overweight and metabolic and hormonal parameter disruption are induced in adult male mice by manipulations during lactation period. *Pediatr. Res.* 59: 111–15.

Loizzo, A., Loizzo, S., Lopez, L. et al. 2002. Naloxone prevents cell-mediated immune alterations in adult mice following repeated mild stress in the neonatal period. *Br. J. Pharmacol.* 135: 1219–26.

Loizzo, S., Pieri, M., Ferri, A. et al. 2010. Dynamic NAD(P)H post-synaptic autofluorescence signals for the assessment of mitochondrial function in a neurodegenerative disease: monitoring the primary motor cortex of G93A mice, an amyotrophic lateral sclerosis model. *Mitochondrion.* 10: 108–14.

Loizzo, S., Vella, S., Loizzo, A. et al. 2010. Sexual dimorphic evolution of metabolic programming in non-genetic non-alimentary mild metabolic syndrome model in mice depends on feed-back mechanisms integrity for pro-opiomelanocortin-derived endogenous substances. *Peptides.* 31: 1598–605.

McDowell, J., Kitchen, I. 1987. Development of opioid systems: peptides, receptors and pharmacology. *Brain Res. Rev.* 12: 397–421.

Moles, A., Kieffer, B.L., D'Amato, F.R. 2004. Deficit in attachment behavior in mice lacking the mu-opioid receptor gene. *Science.* 304: 1983–6.

Pieretti, S., d'Amore, A., Loizzo, A. 1991. Long-term changes induced by developmental handling on pain threshold: effects of morphine and naloxone. *Behav. Neurosi.* 105: 215–18.

Pieretti, S., Di Giannuario, A., Domenici, M.R. et al. 1994. Dexamethasone-induced selective inhibition of the central mu opioid receptor: functional in vivo and in vitro evidence in rodents. *Br. J. Pharmacol.* 113: 1416–22.

Ploj, K., Roman, E., Nylander, I. 2003. Long-term effects of short and long periods of maternal separation on brain opioid peptide levels in male Wistar rats. *Neuropeptides.* 37: 149–56.

Qin, M., Kang, J., Smith, C.B. 2002. Increased rates of cerebral glucose metabolism in a mouse model of fragile X mental retardation. *Proc. Natl. Acad. Sci. U. S. A.* 99: 15758–63.

Ravelli, A.C., van Der Meulen, J.H., Osmond, C., Barker, D.J., Bleker, O.P. 1999. Obesity at the age of 50 y in men and women exposed to famine prenatally. *Am. J. Clin. Nutr.* 70: 811–16.

Rosenfeld, P., van Eekelen, J.A., Levine, S., de Kloet, E.R. 1993. Ontogeny of corticosteroid receptors in the brain. *Cell Mol. Neurobiol.* 13: 295–319.

Sacerdote, P., Manfredi, B., Gaspani, L., Panerai, A.E. 2000. The opioid antagonist naloxone induces a shift from type 2 to type 1 cytokine pattern in BALB/cJ mice. *Blood.* 96: 2031–6.

Selye, H. 1976. Forty years of stress research: principal remaining problems and misconceptions. *Can. Med. Ass. J.* 115: 53–6.

Spampinato, S., Canossa, M., Carboni, L., Campana, G., Leanza, G., Ferri, S. 1994. Inhibition of proopiomelanocortin expression by an oligodeoxynucleotide complementary to β-endorphin mRNA. *Proc. Natl. Acad. Sci. U. S. A.* 91: 8072–6.

Walker, B.R. 2006. Cortisol—cause and cure for metabolic syndrome? *Diabet. Med.* 23: 1281–8.

8 Prothymosin α—A Novel Endogenous Neuroprotective Polypeptide against Ischemic Damages

Hiroshi Ueda, Hayato Matsunaga, and Sebok K. Halder

CONTENTS

8.1 INTRODUCTION

Stroke is a common medical emergency, which can cause permanent neurological damage with severe complications, leading to a bedridden life, and even death (Feigin 2005; Ueda et al. 2007). Stroke is the third leading cause of death in developed countries and the leading cause of major disability in adults. The consequences of a stroke include the loss of functions, such as dysfunctions of motor skills, memory, and sensory perception, which are caused by various kinds of brain ischemia, leading to neuronal death (White et al. 2000; Ueda 2009; Sims and Muyderman 2010). In most cases with brain ischemia, necrotic cell death occurs first in the ischemic core, followed by apoptosis several days later in the region surrounding the ischemic core, referred to as the penumbra (Dirnagl, Iadecola, and Moskowitz 1999; Lipton 1999; Ueda and Fujita 2004; Hossmann 2006). Neuronal necrosis in the ischemic core is caused by deprivation of oxygen, glucose, and some neurotrophic factors by stopping blood flow, and results in the release of cytotoxic substances, which, in turn, generate reactive oxygen species (ROS) (Szeto 2006; Niizuma et al. 2010) and cause further damage to the surrounding neurons (Ueda 2009). These cytotoxic substances also activate nonneuronal cells, astrocytes, and microglia and further release other types of cytotoxic substances, such as cytokines and nitric oxide, which cause further damage to surrounding neurons to increase the necrotic expansion and, in turn, cause neuronal apoptosis surrounding the ischemic core (Danton and Dietrich 2003; Nakajima and Kohsaka 2004; Swanson, Ying, and Kauppinen 2004; Lai and Todd 2006; Zhao and Rempe 2010). Although apoptosis has a nature of being a converging type of cell death, it is interesting to hypothesize that the expansion of neuronal death by necrosis is terminated by late apoptosis (Ueda 2009). This hypothesis seems to be partly contradicted with previous findings that apoptotic inhibitors or antiapoptotic neurotrophins have significant, but limited, protective effects against ischemia-induced brain damage (Cheng et al. 1998; Brines et al. 2000; Gilgun-Sherki et al. 2002; Gladstone, Black, and Hakim 2002; Lin et al. 2006). The limited effects may be attributed to the fact that neuronal apoptosis is observed at the early stage, as well as at the late stage of stroke (Lipton 1999; Ueda and Fujita 2004). Therefore, the inhibition of necrosis is more preferable for the rapid treatment of stroke, as seen in the case with the fact that the quick treatment is currently emphasized by the efficacy of thrombolytic tissue plasminogen activator (tPA) treatments (The National Institute of Neurological Disorders and Stroke rt-PA Stroke Study Group 1995; Gladstone et al. 2002; Borsello et al. 2003). It is also evident that only 3% of all stroke patients receive tPA because of very narrow therapeutic time window (Liu et al. 2009). However, further investigations of novel molecules that can completely inhibit rapid cell death by necrosis are necessary for the rapid treatment of stroke.

8.2 APOPTOSIS AND NECROSIS

Among representative cell death modes, apoptosis is a programmed and noninflammatory cell death, characterized by nuclear fragmentation, chromatin condensation, membrane blebbing, and loss of cell membrane asymmetry (Fujita and Ueda 2003a, 2003b; Danial and Korsmeyer 2004). Apoptosis is triggered by various stimuli, such as binding of extracellular death signals, deprivation of growth factors, or treatments with cytotoxic drugs (Sprick and Walczak 2004). Activation of initiator caspases, such as caspase-8 and caspase-9, will proteolytically activate procaspase-3, and activated caspase-3 is responsible for the cleavage of cellular components and subsequent execution of DNA fragmentation and cell death. The Bcl-2 family and proapoptotic (Bax, Bak) proteins decide whether a cell lives or dies. In contrast to apoptosis, necrosis is characterized to be bioenergetically catastrophic (Edinger and Thompson 2004), and to cause mitochondrial swelling, loss of electron density, and early plasma membrane rupture. However, the molecular bases of mechanisms remain elusive. In addition to these cell death modes, autophagy, a unique and self-destructive cell death mode through a lysosomal pathway should be also considered (Levine and Yuan 2005). Unlike apoptosis, autophagy plays a cell-adaptive role for survival or repair under stress conditions. However, a massive autophagic process can kill damaged cells (Gozuacik and Kimchi 2004, 2007). Most recently, the Nomenclature Committee on Cell Death reported updated guidelines and understanding of the various cell death modes (Kroemer et al. 2009).

8.2.1 *In Vitro* Apoptosis and Necrosis

In apoptosis mechanisms (Figure 8.1), three major pathways share caspase-3 activation and nuclear fragmentation as the final step (Baker and Reddy 1998; Ferri and Kroemer 2001; Chen and Goeddel 2002; Danial and Korsmeyer 2004). In the first mechanism effected through mitochondrial pathways, the release of apoptosis-initiation factors from mitochondria plays key roles in apoptosome formation and caspase-3 activation (Kroemer 1997; Ferri and Kroemer 2001; Martinou and Green 2001; Zamzami and Kroemer 2001). These factors are released through mitochondrial permeability transition pores (mPTPs) composed of adenine nucleotide translocase and voltage-dependent anion channels, Apaf1 and adenosine triphosphate (ATP), and generate activated caspase-9, which, in turn, further activates caspase-3, followed by DNA fragmentation through activation of caspase-activated DNase (Enari et al. 1998; Sakahira, Enari, and Nagata 1998). The opening of mPTPs is positively regulated by proapoptotic Bax and Bak and negatively regulated by antiapoptotic Bcl-2 and Bcl-xL (Tsujimoto 2002; Cory, Huang, and Adams 2003). Growth factors or neurotrophic factors are known to inhibit apoptosis by activating Akt, which phosphorylates Bad and abolishes its interactions with Bcl-2 and Bcl-xL (Zha et al. 1996; Shimizu et al. 2000; Belzacq et al. 2003). The second mechanism is mediated by death receptors, namely, Fas and tumor necrosis factor (TNF)-αl receptors, which are activated by Fas ligand and TNF-α, respectively. These receptors mediate the activation of caspase-8 and caspase-3 through Fas-associated death domain (FADD), Fas-TNF-α-related receptor interacting protein (RIP), and TNF-α receptor-associated protein (TRADD) (Ashkenazi and Dixit 1998; Walczak and Krammer 2000). In the third mechanism, accumulation of abnormal proteins in the endoplasmic reticulum (ER) initiates so-called ER stress, which leads to cleavage and activation of IRE1α/β and caspase-12 (Nakagawa et al. 2000; Diaz-Horta et al. 2002), and to activation of TRAF-2 and caspase-3, respectively (Urano et al. 2000; Yoneda et al. 2001; Nishitoh et al. 2002).

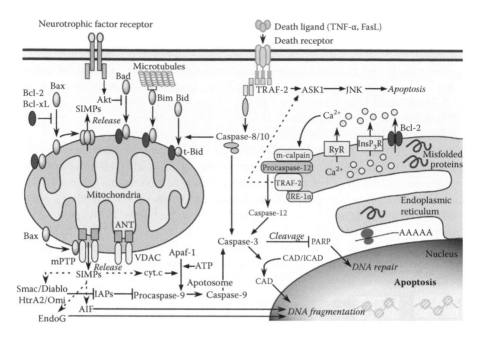

FIGURE 8.1   The three major pathways of apoptosis. Mitochondrial pathways are closely related to the expression of members of the Bcl-2 family of proteins. Proapoptotic Bak and Bax open mitochondrial permeability transition pores (mPTPs) to release soluble intermembrane proteins (SIMPs), including cytochrome c (cyt.c), apoptosis-inducing factor (AIF), Smac/DIABLO, EndoG and HtrA2/Omi (Li, L. Y., Luo, X., and Wang, X., *Nature*, 412:95–9, 2001). Among these, cyt.c plays a major role in inducing apoptosis through activation of caspase-3 and caspase-activated DNase (CAD). Bcl-2 and Bcl-xL are major antiapoptotic proteins that inhibit the functions of these proapoptotic proteins. The other two pathways through death receptors [Fas and tumor necrosis factor-α (TNF-α) receptors] or endoplasmic reticulum stress also use caspase-3 activation as the common execution pathway. Other details are described in the text.

In the mechanisms of necrosis, energy failure or a drastic decrease in the cellular ATP levels is the most likely accepted mechanism (Eguchi, Shimizu, and Tsujimoto 1997; Leist et al. 1997). Three major parameters, namely, supply, synthesis, and consumption, are known to determine the cellular ATP levels. Although insulin receptors mainly determine glucose uptake by recruiting the glucose transporters, GLUT1 and GLUT4 (GLUT1/4), to the plasma membrane through phosphorylation in peripheral cells (Ducluzeau et al. 2002), glucose supply in neurons is mediated by transporters that are constitutively expressed in the membranes (Burkhalter et al. 2003). A characteristic model of neuronal necrosis using low-density and serum-free culture of cortical neurons has been established (Fujita and Ueda 2003a, 2003b), in which rapid decreases in [^{3}H]-2-deoxyglucose (DG) uptake and cellular ATP levels under stress, possibly due to endocytosis of GLUT1/4 under such starvation conditions (Ueda et al. 2007), are observed. There is a report that Bcl$_2$/adenovirus E1B 19-kDa protein-interacting protein (BNIP$_3$) opens mPTPs to allow mitochondrial influx of H$_2$O and Ca^{2+} (Vande Velde et al. 2000), which, in turn, causes activation

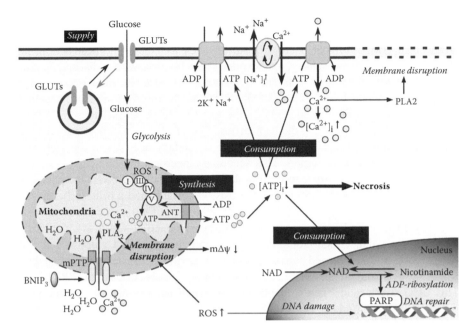

FIGURE 8.2　Roles of ATP metabolism in neuronal necrosis. Glucose transporters (GLUTs) are involved in the supply of cellular glucose (depicted as *supply*), a substrate for ATP production through glycolysis and oxidative phosphorylation (depicted as *synthesis*) in mitochondria. Some species of GLUT are constitutively localized, whereas others are translocated to the membrane upon cell stimulation by extracellular signals. Poly(ADP-ribose) polymerase (PARP) restores the DNA damage caused by cellular stress, by using abundant cellular ATP molecules (depicted as *consumption*). A rapid decrease in the cellular ATP levels leads to necrosis.

of phospholipase A2 (PLA2) and mitochondrial membrane destruction (Malis and Bonventre 1988). Finally, nicotinamide adenine dinucleotide (NAD)/ATP consumption may be caused by activation of poly(ADP-ribose) polymerase (PARP), which restores the damage to nuclear DNA caused by ROS (Kim and Koh 2002). It is also interesting that caspase-3 catalyzes PARP and thereby inhibits ATP depletion for the repair of DNA damage (Figure 8.2). In other words, apoptosis mechanisms also play a role in inhibiting necrosis.

8.2.2　CHARACTERISTIC NECROSIS MODEL

Neurons also require various signals from target tissues, glial cells, or other neurons via synapses. Indeed, we found that cortical neurons under serum-free conditions died by necrosis in low-density cultures, but interestingly they died by apoptosis in high-density cultures (Fujita et al. 2001). These findings suggest that some soluble factors in the high-density cultures may switch the cell death mode from necrosis to apoptosis. To study this unique cell death mode switch mechanism, we developed a successful method using serum-free and low-density culture of embryonic rat cortical neurons without any supplements (Fujita et al. 2001; Fujita and Ueda 2003a, 2003b). In the system, scanning electron microscopy analysis of cortical neurons showed many pores at 6 hours on their surfaces and cell surface membranes retaining the nuclei at 12 hours were largely destroyed (Fujita and Ueda 2003a, 2003b; Ueda et al. 2007). Typical necrotic features, such as membrane destruction, loss of cytoplasmic electron density, and swollen mitochondria with a disrupted cristae structure, were observed at 6 hours by transmission electron microscopy (TEM) analysis (Fujita and Ueda 2003a, 2003b). Necrotic features were also detected by staining with propidium iodide (PI). In the focal brain ischemia, neurons adjacent to the ischemic region rapidly die by necrosis. Therefore, a clinically strategic target for the prevention of disability would be the secondary necrosis in the vicinity of the initial ischemic core region. This secondary necrosis may be caused by mild ischemia or various cytotoxic molecules derived from the dead cells. To specify the molecules that induce the secondary necrosis, we established a mild ischemia (and reperfusion) model under low-oxygen and low-glucose (LOG) stress as an alternative protocol. In this model, neurons cultured in the presence of serum were subjected to culture in balanced salt solution medium containing 1 mM glucose and 0.4% oxygen for 2 hours, followed by reperfusion with normal serum-containing medium, which showed marked neuronal necrosis without apoptosis (Ueda et al. 2004, 2007). This LOG stress model is also applicable to retinal neuron–neuroblastoma hybrid N18-RE105 cells (Ueda et al. 2004, 2007).

A marked decrease in glucose transport was also observed due to excess internalization of the glucose transporter, GLUT1/4, in the cortical neurons under LOG stress conditions (Ueda et al. 2007). Although the detailed mechanisms of the GLUT1/4 internalization under starvation or ischemic conditions remain elusive, the underlying mechanisms seem to be related to the ischemia-induced deficiency of growth factors, as GLUT membrane trafficking is regulated by growth factor-related signaling pathways through Akt, PKC, or Grb2-dynamin complexes (Ando et al. 1994; Watson and Pessin 2001, 2006). Taken together, the LOG stress model combined with measurement of GLUT translocation may facilitate the use of drug treatments to study the signal transduction of necrosis and its inhibition by endogenous and exogenous compounds.

8.3 IDENTIFICATION OF PROTHYMOSIN α AS AN ANTINECROTIC FACTOR

The search for antinecrotic factors was initiated by density-dependent survival of cortical neurons in serum-free culture. After various approaches, we finally discovered an efficient way to obtain significant amounts of active materials through molecular weight cutoff ultrafiltration, ion-exchange filtration, and sodium dodecyl sulfate and polyacrylamide gel electrophoresis (SDS-PAGE) separation. After SDS-PAGE, the survival activity was recovered by a protein band at 20 kDa. This protein was analyzed by matrix-assisted laser desorption/ionization-time of flight mass spectrometry (MALDI-TOF MS), and a subsequent search of the nonredundant NCBI protein database for matching peptide mass fingerprints revealed 17 peptides that were unique to rat prothymosin α (ProTα) (Ueda et al. 2007). Moreover, tandem MS analysis confirmed that the N-terminal of purified ProTα was an acetylated serine (129.612 m/z versus Ser 87.343 m/z, see Figure 8.3), in agreement with a

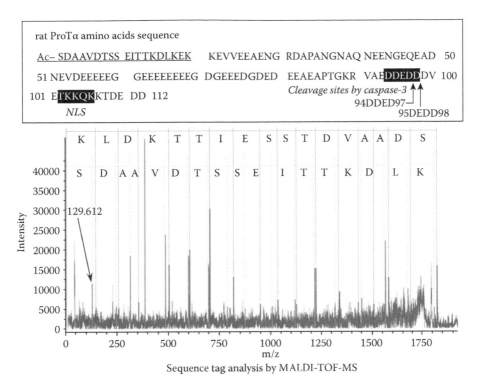

FIGURE 8.3 (See color insert) Identification of ProTα by MALDI-TOF-MS. Amino acid sequence of rat ProTα (upper panel). Underlined sequence is predicted amino acids. White-colored sequences indicate the cleavage sites by activated-caspase-3 and nuclear localization signal (NLS). Peptide sequence search by MS/MS sequence tag analysis (lower panel). Arrow indicates acetylated serine.

previous report (Pineiro, Cordero, and Nogueira 2000). The structure of ProTα has several unique characteristics in that it is highly hydrophilic and acidic (pI = 3.55) owing to the abundance of glutamic and aspartic acids (50% of the total residues) in the middle part of the protein. The cluster of acidic amino acids in this region seems to resemble a putative histone-binding domain. This region does not contain any histidine, sulfated, or aromatic amino acid residues. A small stretch of basic residues, corresponding to the thymic hormone thymosin α1 (28 amino acids), is found at the N-terminal, whereas another stretch of basic residues at the C-terminal includes a nuclear localization signal (NLS; TKKQKK). The fact is that ProTα, a monomeric protein without any regular secondary structures under physiological conditions (Gast et al. 1995), may explain its poor immunogenicity and a favorable property in terms of its clinical use. Other characteristics of ProTα have been well described (Segade and Gomez-Marquez 1999). In the experiments using anti-ProTα IgG-conjugated beads, the eluates obtained from acid-treated beads produced a single band corresponding to recombinant ProTα (rProTα). A large proportion of ProTα and the survival activity in the conditioned medium were recovered

Induction
    E. coli. trasformed with ProTα/pET20b
    ▼ Added IPTG (0.1 mM for 3 h)
    E. coli. expressing ProTα
Isolation of ProTα by acid phenol method
    E. coli. (expressing ProTα)
      ▼ Lysed with SDS containing buffer
      ▼ Added phenol soln. (pH 4.5)
      ▼ Centrifugation
    Supernatant
      ▼ Phase separation with $CHCl_3$
    Upper aqueous phase
      Phase separation with Phenol soln. (pH 4.5)
    ▼   &
                3 M NaOAc (pH 5.2)
    Lower phenol phase
      ▼ Precipitation with ethanol
    Pellet
      ▼ Dissolved in water

    ▼ Extraction of SDS with n-butanol
Lower aqueous phase
    ▼ Added 3 M NaOAc (pH 5.2)
    ▼ Precipitation with Ethanol
Pellet (ProTα)
    ▼ Dissolved in Tris buffer
Purification of ProTα by column chromatography
1) VIVAPURE Q MINI (ion-exchange spin column)
    ▼ Elution with step gradient
    ▼ Dialysis with HEPES buffer
2) Mono Q HR (ion-exchange column chromatography)
    ▼ Elution with linear gradient
3) EndoTrap (affinity column chromatography)
    ▼ Collect flow-through fraction
    ▼ Dialysis with PBS
High-purity ProTα

FIGURE 8.4   Flow chart of purification of recombinant ProTα.

in the eluates. ProTα mutants lacking the *N*-terminal region (Δ1-29), including thymosin-α_1 (Pineiro, Cordero, and Nogueira 2000), or the *C*-terminal region (Δ102-112), including the NLS, retained the original activity of ProTα. Detection of ProTα by SDS-PAGE analysis is often difficult. Blotting onto a membrane and detection by silver staining are extremely ineffective because of highly acidic properties of ProTα (pI = 3.75). Therefore, we often follow the highly sensitive Gelcode™ Blue Stain method without a blotting procedure. One-step purification via acid phenol extraction of most preparations is sufficient to concentrate ProTα to a detectable level for the Gelcode Blue Stain method. Recently, we established the method to purify endotoxin-free ProTα from *Escherichia coli* expressing rProTα (Figure 8.4). The endotoxin level of purified rProTα was below 0.5 EU/μg.

8.4   ProTα AND CELL DEATH MODE SWITCH IN CULTURED NEURONS

Based on the original findings that cortical neurons show cell density-dependent survival under serum-free starvation conditions, ProTα was identified as a necrosis-inhibitory molecule. However, as neurons in high-density culture die by apoptosis, we needed to characterize the mechanisms underlying the more complicated cell death mode switch by using ProTα.

8.4.1   Inhibition of Neuronal Necrosis by ProTα

The addition of rProTα to the low-density and serum-free culture of primary cortical neurons abolished the typical necrosis features, such as disrupted plasma membranes and swollen mitochondria in TEM analysis at 6 hours, but induced apoptosis at 12 hours. When the cell death mode was evaluated by double staining with PI (necrosis)/

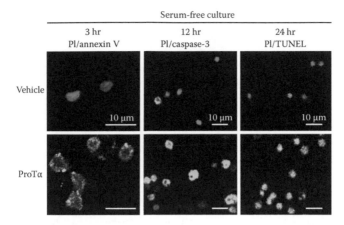

FIGURE 8.5 (See color insert) ProTα-induced cell death mode switch in neuronal cultures. Double staining of low-density cultures with necrosis and apotosis makers. Cell death mode was evaluated by double staining with PI (necrosis)/annexin V (apoptosis) at 3 hours, PI/ antiactivated-caspase-3 IgG (apoptosis) at 12 hours, and PI/TUNEL (apoptosis) at 24 hours. Under serum-free stress, 69% (12 hours), 86% (24 hours), and 92% (48 hours) of neurons died by necrosis, whereas only 15%, 22%, and 5% of neurons died by apoptosis. Addition of ProTα totally switched the cell death mode form necrosis to apoptosis.

annexin V (apoptosis) at 3 hours, PI/antiactivated-caspase-3 IgG (apoptosis) at 12 hours, and PI/terminal deoxynucleotidyl transferase biotin-dUTP nick end labeling (TUNEL) (apoptosis) at 24 hours, most of the neurons died, but the addition of ProTα totally switched the cell death mode from necrosis to apoptosis (Figure 8.5). As stated earlier, the cellular ATP level and [^{3}H]-2-DG uptake of cortical neurons were rapidly decreased immediately under serum-free culture, but these changes were markedly inhibited by the addition of ProTα (Ueda et al. 2007). As previously reported (Fujita and Ueda 2003a, 2003b), the rapid decrease and its reversal by ProTα seem to be parallel to the activity of glucose transport, as the [^{3}H]-2-DG uptake was markedly decreased by serum-free culture, but reversed by ProTα. Similar results were also observed with ischemic model using LOG stress, which caused GLUT1/4 endocytosis. Pharmacological studies revealed that the signaling of ProTα-induced membrane translocation of internalized GLUT1/4 was mediated through activation of putative $G_{i/o}$-coupled receptors, PLC and $PKC\beta_{II}$ (Figure 8.6).

8.4.2 Induction of Apoptosis by ProTα

Caspase-3 is primarily believed to be the final execution molecule for apoptotic cell death linked to DNA breakdown and nuclear fragmentation (Ferri and Kroemer 2001; Danial and Korsmeyer 2004). As ProTα activates caspase-3 as well as caspase-9 in serum-free and permanent ischemia models, but not caspase-8 or caspase-12 (Ueda et al. 2007), it is suggested that ProTα causes apoptosis through the mitochondrial pathway. Indeed, ProTα increased the expression level of proapoptotic Bax and Bim, and slightly decreased the expression of antiapoptotic Bcl-2 and Bcl-xL, which regulate mitochondrial apoptotic signaling. As the pretreatments with antisense

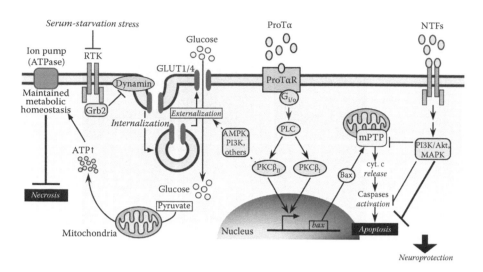

FIGURE 8.6 Mechanism of ProTα-induced cell death mode switch. Serum-free or starvation stress leads to endocytosis of the glucose transporters GLUT1/4, which in turn causes bioenergetic catastrophe-mediated necrosis through a rapid loss of glucose supply. Addition of ProTα to ischemia-treated neurons causes translocation of GLUT1/4 to the membrane to allow sufficient glucose supply through activation of $G_{i/o}$, PLC, and $PKCβ_{II}$. ProTα-induced apoptosis occurs later at 12 hours after the start of serum-free stress. The machinery is mediated by upregulation of Bax, which in turn causes mitochondrial cyt. c release and subsequent apoptosis. Bax upregulation is also mediated by activation of $G_{i/o}$, PLC, and PKC, similar to the case of necrosis. However, both $PKCβ_I$ and $PKCβ_{II}$ upregulations mediate this apoptotic mechanism. Since caspase-3-mediated PARP degradation minimizes the ATP consumption, the apoptosis induction may have a crucial role in inhibiting the rapid necrosis. In addition, as pyruvate, a substrate for ATP production in mitochondria, inhibits necrosis but does not cause apoptosis, the apoptosis machinery seems to be independent of the necrosis inhibition. Neurotrophins, such as BDNF or EPO, which are expected in the ischemic brain and retina, can inhibit the apoptosis machinery at a later stage.

oligodeoxynucleotide (AS-ODN) for $PKCβ_I$ and $PKCβ_{II}$ reversed the ProTα-induced proapoptotic Bax expression, it is suggested that ProTα-induced apoptosis induction is mediated by the activation of $G_{i/o}$-coupled receptors, PLC and $PKCβ_I/β_{II}$.

8.4.3 INHIBITION OF ProTα-INDUCED APOPTOSIS BY NEUROTROPHIC FACTORS

Although the addition of ProTα delayed the cell death of cortical neurons in serum-free culture, most of the neurons completely died by apoptosis after 24 hours. However, when neurons were treated with ProTα under conditions of ischemia and subsequent reperfusion with serum-containing medium, no significant cell death was observed for at least 48 hours (Ueda et al. 2007). These findings indicate that some factors existing in the serum prevented ProTα-induced apoptosis. Indeed, further addition of nerve growth factor (NGF), brain-derived growth factor (BDNF), basic fibroblast growth factor (bFGF), or interleukin (IL)-6, comprising representative apoptosis inhibitors (Kaplan and Miller 2000; Ay, Sugimori, and Finklestein 2001; Huang and

Reichardt 2001; Patapoutian and Reichardt 2001; Sofroniew, Howe, and Mobley 2001; Yamashita et al. 2005), ensured the survival of the cell in serum-free culture for 48 hours, whereas these factors alone had no effects on the survival of the cell (Ueda et al. 2007). Similarly, BIP-V5, which blocks the translocation of Bax to mitochondria (Yoshida et al. 2004), selectively inhibited the ProTα-induced apoptosis.

8.4.4 Evidence for Apoptosis-Induced Inhibition of Rapidly Occurring Necrosis

It should be noted that concomitant addition of N-benzyloxycarbonyl-Val-Ala-Asp (OMe)-fluoromethylketone (zVAD-fmk), a pan-type caspase inhibitor, with ProTα did not lead to long-lasting survival, but caused marked cell death by necrosis at a later stage (Figure 8.7). Alternatively, the blockade of caspase activity may allow a large PARP-mediated depletion of NAD/ATP, as stated above (Figures 8.1 and 8.2). The fact that zVAD-fmk inhibits apoptosis, but causes necrosis, is consistent with previous studies' findings that decreases in intracellular ATP levels changed the type of cell death from apoptosis to necrosis, after stimulation of cells by extracellular apoptosis signals (Eguchi et al. 1997; Leist et al. 1997). The addition of PARP inhibitor 3-aminobenzamide reversed the rapid decrease in the intracellular ATP levels and inhibited necrosis in low-density and serum-free culture, whereas zVAD-fmk significantly reversed the increased ATP levels and inhibited necrosis by ProTα. In other words, apoptosis-induction in the early stage by ProTα after ischemia may play a defensive role in inhibiting rapid cell death by necrosis (Figure 8.6). However, as the late phase of apoptosis is also induced by many other cytotoxic cytokines and nitric oxide through different pathways, beneficial roles of ProTα-induced apoptosis

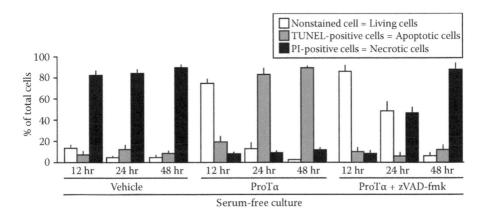

FIGURE 8.7 Apoptosis-induced inhibition of rapidly occurring necrosis. Demonstration of the time-dependent changes in cell death mode status in the presence of ProTα and zVAD-fmk. The addition of ProTα alone inhibited necrosis throughout 48 hours and increased the number of living cells more prominently at the early stage (12 hours), but not at the later stage (24 or 48 hours). On the contrary, the number of cells showing apoptosis time-dependently increase in the presence of ProTα. Further addition of zVAD-fmk, a pan-type caspase inhibitor, caused a marked cell death by necrosis at a later stage.

in terms of antinecrosis would be limited. Furthermore, it remains to be determined how potent the apoptosis-induced antinecrosis effect is, particularly in the *in vivo* status. This issue, however, is not important, as the ProTα-induced apoptosis is effectively blocked by several kinds of neurotrophins.

8.4.5 Multiple Functions of ProTα in Cell Death Regulation

ProTα is a highly acidic nuclear protein of the α-thymosin family and is found in the nuclei of virtually all mammalian cells (Haritos et al. 1985; Clinton et al. 1991). ProTα is generally thought to be an oncoprotein that is correlated with cell proliferation by sequestering the anticoactivator factor, a repressor of estrogen receptor activity, in various cells (Martini et al. 2000; Bianco and Montano 2002). With regard to cell death regulation, intracellular ProTα was reported to play a cytoprotective role by inhibiting apoptosome formation in HeLa cells subjected to apoptotic stress (Jiang et al. 2003). On the other hand, ProTα has been reported to act as an extracellular signaling molecule, as observed in the activation of macrophages, natural killer cells, lymphokine-activated killer cells, and in the production of IL-2 and TNF-α (Pineiro, Cordero, and Nogueira 2000). Considering that ProTα converts uncontrollable neuronal necrosis to growth factor-reversible apoptosis, ProTα has multiple functions inside and outside the cell, particularly in terms of cell survival and proliferation (Gomez-Marquez 2007).

Recently, there was a report that exogenous full-length ProTα and endogenous ProTα released by CD8[+] T-cells may act as a signaling ligand for toll-like receptor-4 (TLR4) and trigger TRIF-mediated IFN-β induction and MyD88-mediated induction of proinflammatory cytokines, such as TNF-α, to suppress HIV-1 after the entry into macrophages (Mosoian et al. 2010). It has also been reported that there are some unidentified ProTα-binding proteins (Pineiro et al. 2001; Salgado et al. 2005). However, further studies to examine how these candidate molecules play such multiple functions in terms of cell death regulation would be a fantastic subject.

8.5 *IN VIVO* NEUROPROTECTIVE ACTIONS OF ProTα IN ISCHEMIA

8.5.1 Middle Cerebral Artery Occlusion Model

In the focal cerebral ischemia model, middle cerebral artery occlusion (MCAO) in rats is often performed by using monofilament nylon surgical sutures (3–0 in size), which is removed at 60 minutes after the occlusion to restore the blood flow. After 1-hour of MCAO, marked loss of triphenyltetrazolium chloride staining, which detects the presence of mitochondrial enzymes, was specifically observed in the ipsilateral regions of the cerebral cortex, hippocampus, and striatum at 24 hours after reperfusion. Systemic administration of recombinant rat ProTα (rrProTα; 100 μg/kg, intraperitoneally) at 30 minutes and 3 hours after reperfusion significantly reversed the brain damage and suppressed ischemia-induced motor dysfunction and lethality (Fujita and Ueda 2007). A similar degree of ProTα-induced prevention of ischemic damage was observed when rrProTα was administered as a single injection at 30 minutes or 3 hours after reperfusion or in a pair of later injections at 3 hours and 6 hours after reperfusion. Since Myc-tagged rProTα administered intraperitoneally was detected in the cortex at 3 hours after MCAO stress (Fujita and Ueda 2007), the neuroprotective actions of ProTα administered through systemic routes are probably because of transient disruption of the blood–brain barrier (BBB) in the ischemic brain (Paul et al. 2001).

For the thrombolytic therapies for acute ischemic stroke, tPA was first approved by the Food and Drug Administration in 1996. Although tPA therapy has significant benefits (The National Institute of Neurological Disorders and Stroke rt-PA Stroke Study Group 1995), it has a restrictive time window of 3 hours, which allows only 1–2% of patients with acute ischemic stroke to receive tPA therapy. Furthermore, thrombolytic therapies have frequent risks of cerebral hemorrhage, which restrict their use in certain patients. Therefore, there is a requirement for the development of neuroprotective therapies as sole regimens or in combination with tPA or thrombolytic therapies. Most of ~30 neurotrophins have been shown to exhibit neuroprotective effects in brain ischemia, injury, or neurodegenerative diseases. In animal studies, NGF, BDNF, and bFGF were reported to show significant neuroprotective actions (Hefti 1986; Williams et al. 1986; Henderson et al. 1993; Tsukahara et al. 1994; Ay et al. 1999). However, clinical trials with these neurotrophins have failed, possibly due to their poor BBB permeability or unexpected side effects (Olson et al. 1992; The BDNF Study Group (Phase III) 1999; Turner, Parton, and Leigh 2001; Bogousslavsky et al. 2002). Chimeric peptide approaches for targeting transfer receptors on the BBB are now being evaluated as a new type of approach (Wu 2005). In light of the present situation, ProTα, which penetrates the BBB and shows potent neuroprotection in the ischemic brain even after delayed and systemic administration, would be a strong candidate as a sole regimen, as well as in combination with tPA and thrombolytic therapies for stroke.

8.5.2 RETINAL ISCHEMIA MODEL

In the retinal ischemic model, mouse eyes were subjected to hydrostatic pressure (130 mmHg) for 45 minutes followed by reperfusion (Adachi et al. 1998; Ueda et al. 2004). Following retinal ischemia, decrease in the thickness of retinal cell layers, namely, the ganglion cell layer (GCL), inner nuclear layer (INL), and outer nuclear layer (ONL), were observed in a time-dependent manner. The maximal decrease of retinal thickness was observed on day 7, but no further damage was observed on day 14, after ischemia and reperfusion. Intravitreous application of recombinant mouse ProTα (rmProTα) 30 minutes prior to ischemia or at 3 or 24 hours after ischemia prevented the decrease in the thickness of the layers (Fujita et al. 2009; Ueda et al. 2010). Systemic administration of rmProTα for 3 hours after ischemia also showed complete protection. Similar ProTα-induced retinal protection was observed when PI staining, caspase-3/TUNEL staining, and TEM analysis were performed. Similar to the cerebral ischemia model, ischemia-specific retinal transport of biotinylated ProTα administered through a systemic route (intravenously) was also observed. Recently, we established the protocol to purify rProTα using acid–phenol extraction, FPLC column chromatography, and endotoxin-free column (Figure 8.4). Throughout this protocol, it was found that the endotoxin level was as low as 0.5 endotoxin unit. At 24 hours after retinal ischemia, intravitreous injection of this purified rProTα (0.1 pmol/μL/eye) completely prevented the decrease in retinal thickness and cell damage compared to control animals at day 7 after ischemia, as evaluated by electroretinogram (ERG) and hematoxylin-eosin (HE) staining, respectively. Further analysis of ERG and HE staining showed that the neuroprotective activity of purified rProTα against retinal ischemia-induced damages was completely abolished by the digestion of rProTα with V8, a protease that breaks peptide bonds on the carboxyl side of aspartic and glutamic acid residues, whereas V8 alone has no effect on the retinal cells, as shown in Figure 8.8. Ischemia-induced functional damage was also evaluated by ERG. In this method, the amplitudes called a-waves represent the function of photoreceptor cells, while b-waves represent the functions of bipolar and Muller cells. The a- and b-wave amplitudes were markedly decreased to approximately 25% and 15%, respectively, relative to those in control animals at day 7 after ischemia. Purified rProTα also completely prevented the damage in ERG activity.

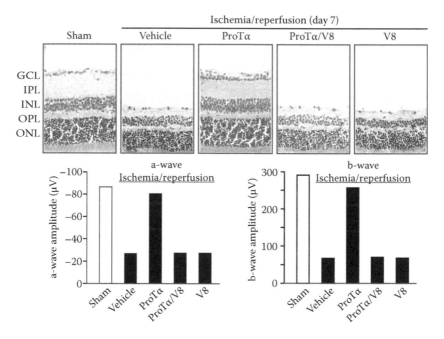

FIGURE 8.8 ProTα prevents retinal ischemia-induced cell death. HE staining of retinal sections at 7 days after ischemia and reperfusion (upper panel). ProTα was administrated at 24 hours after retinal ischemic stress. ProTα completely prevented the decrease in retinal thickness and cell damages. Digested ProTα by V8 protease loses neuroprotective activity. Prevention of retinal ischemic damage evaluated by a- and b-wave ERG analysis (lower panel). The results were evaluated on the basis of a- and b-wave amplitudes. The a- and b-wave amplitudes were markedly decreased to approximately 25% and 15%, respectively, relative to those in control animals at day 7 after ischemia. ProTα also completely prevented the damage in ERG activity.

8.6   *IN VIVO* PᵣₒTα-INDUCED CELL DEATH MODE SWITCH

Both necrosis and apoptosis were observed in damaged brain regions at 24 hours after MCAO by evaluating PI and activated-caspase-3 staining, respectively. Systemic ProTα injection markedly inhibited both necrotic and apoptotic cell death (Fujita and Ueda 2007). When anti-BDNF or antierythropoietin (EPO) IgG (1 µg each) was introduced into the subarachnoid space through a parietal bone at 30 minutes prior to MCAO stress, the level of MCAO-induced apoptosis, but not the necrosis level, was further deteriorated by ProTα (Fujita and Ueda 2007). These findings suggest that the apoptosis caused by ProTα may also be inhibited by brain neurotrophins. Thus, it is evident that ProTα is a unique cell death regulatory molecule, in that it converts irretrievable necrotic cell death into controllable apoptosis. As this apoptosis can be inhibited by growth factors secreted upon ischemic stress or ischemia plus ProTα, it is expected that ProTα may have an overall neuroprotective role in the treatment of stroke.

By analogy with cerebral ischemia, *in vivo* neurotrophins are also expected to inhibit ProTα-induced apoptosis in the retinal ischemia model (Fujita et al. 2009). In immunohistochemical studies, weak BDNF immunoreactivities were observed throughout the retina, in the INL, inner plexiform layer, ONL, and outer segment. Ischemic stress selectively elevated the BDNF level in the GCL at 24 hours after the stress, whereas administration of ProTα (100 µg/kg, intravenously) further enhanced the BDNF expression throughout the retinal layers. As ProTα treatment alone without ischemia had no effect, the combination of ischemia and ProTα seems to potentiate BDNF expression through unknown mechanisms. On the other hand, EPO expression was specifically elevated in the GCL after ischemic stress. ProTα administration did not change the EPO levels in the GCL or INL, but increased the level in the outer segment. Western blot analysis confirmed that ProTα plus ischemic stress caused significant elevations of BDNF and EPO, but not NGF or bFGF, in the retina. When anti-BDNF or anti-EPO IgG (1 µg/eye) was intravitreously administered at 30 minutes prior to ischemic stress, ProTα-induced TUNEL staining reappeared more intensely, compared with the vehicle control, but no significant change in the intensity of PI staining was observed (Fujita et al. 2009; Ueda et al. 2010). However, these IgG treatments alone had no effect on the levels of either necrosis or apoptosis. These findings strongly suggest that ProTα inhibits the necrosis of retinal cells, but causes apoptosis. Furthermore, the apoptosis caused is inhibited by neurotrophins, such as BDNF or EPO, which are upregulated by ProTα under ischemic conditions (Figure 8.9).

8.7   STRESS-INDUCED NONVESICULAR RELEASE UNDERLYING *IN VIVO* ROLE OF PᵣₒTα

To examine the *in vivo* role of ProTα, intravitreous pretreatment with an AS-ODN against ProTα was carried out. This AS-ODN, but not a mismatched or scrambled oligodeoxynucleotide (MS-ODN), significantly worsened the histological damage at day 4 after ischemia. Similar results were observed for intravitreous pretreatment (1 µg/eye 30 minutes prior to ischemic stress) of anti-ProTα IgG, which absorbs

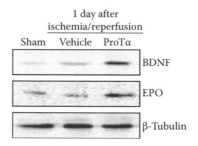

FIGURE 8.9 Upregulation of various neurotrophic factors after retinal ischemia stress with ProTα. ProTα was administrated 3 hours after retinal ischemic stress. Western blot analysis of BDNF and EPO in retinas was performed at 24 hours after ischemia–reperfusion.

ProTα (Fujita et al. 2009). Functional damage was also deteriorated by this antibody treatment, as evaluated by ERG. As ProTα-like immunoreactivities completely disappeared without exception at 3 hours after the stress, it is evident that ProTα released upon ischemic stress plays *in vivo* neuroprotective roles.

To further study the molecular basis of ischemia-induced ProTα release, we used C6 glioma cells, because of their robustness against ischemia stress. On the analogy of the nonclassical release of FGF-1 (Matsunaga and Ueda 2006a, 2006b), which lacks signal peptide sequence, the ischemia stress caused an exhausting extracellular release of ProTα, which also lacks signal peptide sequence. Our recent study revealed that the mechanisms underlying the nonclassical ProTα release from C6 glioma cells are mediated by the loss of ATP and elevated Ca^{2+} (Matsunaga and Ueda 2010). In this mechanism, the first step is the release of ProTα from the nucleus due to ATP loss, followed by extracellular release through a Ca^{2+}-dependent interaction with Ca^{2+}-binding protein, S100A13. It should be of interest to note that ProTα release did not occur when the cells were treated with apoptogenic reagents, although it is released from the nucleus. Detailed studies revealed that caspase-3 activated by apoptogenic reagent treatments cleaved the *C*-terminal of ProTα, which contains NLS, a key domain responsible for the interaction with S100A13. Thus, it is suggested that the released ProTα may play necrosis-inhibitory roles at the early phase of ischemic stress, contrasting the view that neurotrophins play antiapoptotic roles at a later phase (Figure 8.10).

8.8   PREVENTION OF ISCHEMIA-INDUCED MEMORY AND LEARNING DYSFUNCTION

In a global ischemic model using bilateral common carotid artery occlusion for 30 minutes, there were marked loss of neurons in the pyramidal cell layers and dentate gyri of the surviving mice at 28 days after occlusion (Fujita and Ueda 2007). A single systemic (intraperitoneal) injection of rmProTα (100 μg/kg) at 24 hours after occlusion completely prevented brain damage, learning, and memory deficits in a step-through passive avoidance task and lethality.

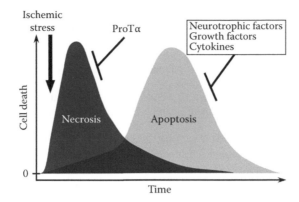

FIGURE 8.10 Schematic illustration of neuroprotective role of endogenous factors against ischemic stress.

8.9 CONCLUSIONS

The hypothesis is that ProTα acts as a mediator in cell death mode switch from uncontrollable necrosis to neurotrophin-reversible apoptosis and may provide a novel strategy for the prevention of serious brain damage in stroke. Although TLR-4 has been identified as a receptor candidate for ProTα in the immunoprotection, it remains elusive whether this receptor plays a key role in the potent neuroprotective actions of ProTα in cerebral and retinal ischemia. Regarding to clinical issues, it is evident that rProTα, by itself, has unique therapeutic potentials against acute ischemic stroke.

ACKNOWLEDGMENTS

Parts of this study were supported by Grants-in-Aid for Scientific Research (to H.U., B: 13470490 and B: 15390028) on Priority Areas—Research on Pathomechanisms of Brain Disorders (to H.U., 17025031) from the Ministry of Education, Culture, Sports, Science, and Technology (MEXT) and by Health and Labour Sciences Research Grants on Biological Resources and Animal Models for Drug Development.

REFERENCES

Adachi, K., Fujita, Y., Morizane, C., Akaike, A., Ueda, M., Satoh, M., et al. 1998. Inhibition of NMDA receptors and nitric oxide synthase reduces ischemic injury of the retina. *Eur J Pharmacol* 350:53–7.

Ando, A., Yonezawa, K., Gout, I., Nakata, T., Ueda, H., Hara, K., et al. 1994. A complex of GRB2-dynamin binds to tyrosine-phosphorylated insulin receptor substrate-1 after insulin treatment. *EMBO J* 13:3033–8.

Ashkenazi, A., and Dixit, V. M. 1998. Death receptors: signaling and modulation. *Science* 281:1305–8.

Ay, H., Ay, I., Koroshetz, W. J., and Finklestein, S. P. 1999. Potential usefulness of basic fibroblast growth factor as a treatment for stroke. *Cerebrovasc Dis* 9:131–5.

Ay, I., Sugimori, H., and Finklestein, S. P. 2001. Intravenous basic fibroblast growth factor (bFGF) decreases DNA fragmentation and prevents downregulation of Bcl-2 expression in the ischemic brain following middle cerebral artery occlusion in rats. *Brain Res Mol Brain Res* 87:71–80.

Baker, S. J., and Reddy, E. P. 1998. Modulation of life and death by the TNF receptor superfamily. *Oncogene* 17:3261–70.

Belzacq, A. S., Vieira, H. L., Verrier, F., Vandecasteele, G., Cohen, I., Prevost, M. C., et al. 2003. Bcl-2 and Bax modulate adenine nucleotide translocase activity. *Cancer Res* 63:541–6.

Bianco, N. R., and Montano, M. M. 2002. Regulation of prothymosin alpha by estrogen receptor alpha: molecular mechanisms and relevance in estrogen-mediated breast cell growth. *Oncogene* 21:5233–44.

Bogousslavsky, J., Victor, S. J., Salinas, E. O., Pallay, A., Donnan, G. A., Fieschi, C., et al. 2002. Fiblast (trafermin) in acute stroke: results of the European-Australian phase II/III safety and efficacy trial. *Cerebrovasc Dis* 14:239–51.

Borsello, T., Clarke, P. G., Hirt, L., Vercelli, A., Repici, M., Schorderet, D. F., et al. 2003. A peptide inhibitor of c-Jun N-terminal kinase protects against excitotoxicity and cerebral ischemia. *Nat Med* 9:1180–6.

Brines, M. L., Ghezzi, P., Keenan, S., Agnello, D., de Lanerolle, N. C., Cerami, C., et al. 2000. Erythropoietin crosses the blood-brain barrier to protect against experimental brain injury. *Proc Natl Acad Sci U S A* 97:10526–31.

Burkhalter, J., Fiumelli, H., Allaman, I., Chatton, J. Y., and Martin, J. L. 2003. Brain-derived neurotrophic factor stimulates energy metabolism in developing cortical neurons. *J Neurosci* 23:8212–20.

Chen, G., and Goeddel, D. V. 2002. TNF-R1 signaling: a beautiful pathway. *Science* 296:1634–5.

Cheng, Y., Deshmukh, M., D'Costa, A., Demaro, J. A., Gidday, J. M., Shah, A., et al. 1998. Caspase inhibitor affords neuroprotection with delayed administration in a rat model of neonatal hypoxic-ischemic brain injury. *J Clin Invest* 101:1992–9.

Clinton, M., Graeve, L., el-Dorry, H., Rodriguez-Boulan, E., and Horecker, B. L. 1991. Evidence for nuclear targeting of prothymosin and parathymosin synthesized in situ. *Proc Natl Acad Sci U S A* 88:6608–12.

Cory, S., Huang, D. C., and Adams, J. M. 2003. The Bcl-2 family: roles in cell survival and oncogenesis. *Oncogene* 22:8590–607.

Danial, N. N., and Korsmeyer, S. J. 2004. Cell death: critical control points. *Cell* 116:205–19.

Danton, G. H., and Dietrich, W. D. 2003. Inflammatory mechanisms after ischemia and stroke. *J Neuropathol Exp Neurol* 62:127–36.

Diaz-Horta, O., Kamagate, A., Herchuelz, A., and Van Eylen, F. 2002. Na/Ca exchanger overexpression induces endoplasmic reticulum-related apoptosis and caspase-12 activation in insulin-releasing BRIN-BD11 cells. *Diabetes* 51:1815–24.

Dirnagl, U., Iadecola, C., and Moskowitz, M. A. 1999. Pathobiology of ischaemic stroke: an integrated view. *Trends Neurosci* 22:391–7.

Ducluzeau, P. H., Fletcher, L. M., Vidal, H., Laville, M., and Tavare, J. M. 2002. Molecular mechanisms of insulin-stimulated glucose uptake in adipocytes. *Diabetes Metab* 28:85–92.

Edinger, A. L., and Thompson, C. B. 2004. Death by design: apoptosis, necrosis and autophagy. *Curr Opin Cell Biol* 16:663–9.

Eguchi, Y., Shimizu, S., and Tsujimoto, Y. 1997. Intracellular ATP levels determine cell death fate by apoptosis or necrosis. *Cancer Res* 57:1835–40.

Enari, M., Sakahira, H., Yokoyama, H., Okawa, K., Iwamatsu, A., and Nagata, S. 1998. A caspase-activated DNase that degrades DNA during apoptosis, and its inhibitor ICAD. *Nature* 391:43–50.

Feigin, V. L. 2005. Stroke epidemiology in the depeloping world. *Lancet* 365:2160–1.

Ferri, K. F., and Kroemer, G. 2001. Organelle-specific initiation of cell death pathways. *Nat Cell Biol* 3:E255–63.

Fujita, R., and Ueda, H. 2003a. Protein kinase C-mediated cell death mode switch induced by high glucose. *Cell Death Differ* 10:1336–47.

Fujita, R., and Ueda, H. 2003b. Protein kinase C-mediated necrosis-apoptosis switch of cortical neurons by conditioned medium factors secreted under the serum-free stress. *Cell Death Differ* 10:782–90.

Fujita, R., and Ueda, H. 2007. Prothymosin-alpha1 prevents necrosis and apoptosis following stroke. *Cell Death Differ* 14:1839–42.

Fujita, R., Ueda, M., Fujiwara, K., and Ueda, H. 2009. Prothymosin-alpha plays a defensive role in retinal ischemia through necrosis and apoptosis inhibition. *Cell Death Differ* 16:349–58.

Fujita, R., Yoshida, A., Mizuno, K., and Ueda, H. 2001. Cell density-dependent death mode switch of cultured cortical neurons under serum-free starvation stress. *Cell Mol Neurobiol* 21:317–24.

Gast, K., Damaschun, H., Eckert, K., Schulze-Forster, K., Maurer, H. R., Muller-Frohne, M., et al. 1995. Prothymosin alpha: a biologically active protein with random coil conformation. *Biochemistry* 34:13211–18.

Gilgun-Sherki, Y., Rosenbaum, Z., Melamed, E., and Offen, D. 2002. Antioxidant therapy in acute central nervous system injury: current state. *Pharmacol Rev* 54:271–84.

Gladstone, D. J., Black, S. E., and Hakim, A. M. 2002. Toward wisdom from failure: lessons from neuroprotective stroke trials and new therapeutic directions. *Stroke* 33:2123–36.

Gomez-Marquez, J. 2007. Function of prothymosin alpha in chromatin decondensation and expression of thymosin beta-4 linked to angiogenesis and synaptic plasticity. *Ann N Y Acad Sci* 1112:201–9.

Gozuacik, D., and Kimchi, A. 2004. Autophagy as a cell death and tumor suppressor mechanism. *Oncogene* 23:2891–906.

Gozuacik, D., and Kimchi, A. 2007. Autophagy and cell death. *Curr Top Dev Biol* 78:217–45.

Haritos, A. A., Blacher, R., Stein, S., Caldarella, J., and Horecker, B. L. 1985. Primary structure of rat thymus prothymosin alpha. *Proc Natl Acad Sci U S A* 82:343–6.

Hefti, F. 1986. Nerve growth factor promotes survival of septal cholinergic neurons after fimbrial transections. *J Neurosci* 6:2155–62.

Henderson, C. E., Camu, W., Mettling, C., Gouin, A., Poulsen, K., Karihaloo, M., et al. 1993. Neurotrophins promote motor neuron survival and are present in embryonic limb bud. *Nature* 363:266–70.

Hossmann, K. A. 2006. Pathophysiology and therapy of experimental stroke. *Cell Mol Neurobiol* 26:1055–81.

Huang, E. J., and Reichardt, L. F. 2001. Neurotrophins: roles in neuronal development and function. *Annu Rev Neurosci* 24:677–736.

Jiang, X., Kim, H. E., Shu, H., Zhao, Y., Zhang, H., Kofron, J., et al. 2003. Distinctive roles of PHAP proteins and prothymosin-alpha in a death regulatory pathway. *Science* 299:223–6.

Kaplan, D. R., and Miller, F. D. 2000. Neurotrophin signal transduction in the nervous system. *Curr Opin Neurobiol* 10:381–91.

Kim, Y. H., and Koh, J. Y. 2002. The role of NADPH oxidase and neuronal nitric oxide synthase in zinc-induced poly(ADP-ribose) polymerase activation and cell death in cortical culture. *Exp Neurol* 177:407–18.

Kroemer, G. 1997. The proto-oncogene Bcl-2 and its role in regulating apoptosis. *Nat Med* 3:614–20.

Kroemer, G., Galluzzi, L., Vandenabeele, P., Abrams, J., Alnemri, E. S., Baehrecke, E. H., et al. 2009. Classification of cell death: recommendations of the Nomenclature Committee on Cell Death. *Cell Death Differ* 16:3–11.

Lai, A. Y., and Todd, K. G. 2006. Microglia in cerebral ischemia: molecular actions and interactions. *Can J Physiol Pharmacol* 84:49–59.

Leist, M., Single, B., Castoldi, A. F., Kuhnle, S., and Nicotera, P. 1997. Intracellular adenosine triphosphate (ATP) concentration: a switch in the decision between apoptosis and necrosis. *J Exp Med* 185:1481–6.

Levine, B., and Yuan, J. 2005. Autophagy in cell death: an innocent convict? *J Clin Invest* 115:2679–88.

Li, L. Y., Luo, X., and Wang, X. 2001. Endonuclease G is an apoptotic DNase when released from mitochondria. *Nature* 412:95–9.

Lin, C. H., Cheng, F. C., Lu, Y. Z., Chu, L. F., Wang, C. H., and Hsueh, C. M. 2006. Protection of ischemic brain cells is dependent on astrocyte-derived growth factors and their receptors. *Exp Neurol* 201:225–33.

Lipton, P. 1999. Ischemic cell death in brain neurons. *Physiol Rev* 79:1431–568.

Liu, W., Hendren, J., Qin, X. J., and Liu, K. J. 2009. Normobaric hyperoxia reduces the neurovascular complications associated with delayed tissue plasminogen activator treatment in a rat model of focal cerebral ischemia. *Stroke* 40:2526–31.

Malis, C. D., and Bonventre, J. V. 1988. Susceptibility of mitochondrial membranes to calcium and reactive oxygen species: implications for ischemic and toxic tissue damage. *Prog Clin Biol Res* 282:235–59.

Martini, P. G., Delage-Mourroux, R., Kraichely, D. M., and Katzenellenbogen, B. S. 2000. Prothymosin alpha selectively enhances estrogen receptor transcriptional activity by interacting with a repressor of estrogen receptor activity. *Mol Cell Biol* 20:6224–32.

Martinou, J. C., and Green, D. R. 2001. Breaking the mitochondrial barrier. *Nat Rev Mol Cell Biol* 2:63–7.

Matsunaga, H., and Ueda, H. 2006a. Evidence for serum-deprivation-induced co-release of FGF-1 and S100A13 from astrocytes. *Neurochem Int* 49:294–303.

Matsunaga, H., and Ueda, H. 2006b. Voltage-dependent N-type Ca^{2+} channel activity regulates the interaction between FGF-1 and S100A13 for stress-induced non-vesicular release. *Cell Mol Neurobiol* 26:237:46.

Matsunaga, H., and Ueda, H. 2010. Stress-induced non-vesicular extracellular release of prothymosin-α initiated by an interaction with S100A13, and its blockade by caspase-3 cleavage. *Cell Death Differ* 17:1760–72.

Mosoian, A., Teixeira, A., Burns, C. S., Sander, L. E., Gusella, L. G., He, C., et al. 2010. Prothymosin-α inhibits HIV-1 via Toll-like receptor 4-mediated type I interferon induction. *Proc Natl Acad Sci U S A* 107:10178–83.

Nakagawa, T., Zhu, H., Morishima, N., Li, E., Xu, J., Yankner, B. A., et al. 2000. Caspase-12 mediates endoplasmic-reticulum-specific apoptosis and cytotoxicity by amyloid-beta. *Nature* 403:98–103.

Nakajima, K., and Kohsaka, S. 2004. Microglia: neuroprotective and neurotrophic cells in the central nervous system. *Curr Drug Targets Cardiovasc Haematol Disord* 4:65–84.

Niizuma, K., Yoshioka, K., Chen, H., Kim, G. S., Jung, J. E. Katsu, et al. 2010. Mitochondrial and apoptotic neuronal death signaling pathways in cerebral ischemia. *Biochim Biophys Acta* 1802:92–9.

Nishitoh, H., Matsuzawa, A., Tobiume, K., Saegusa, K., Takeda, K., Inoue, K., et al. 2002. ASK1 is essential for endoplasmic reticulum stress-induced neuronal cell death triggered by expanded polyglutamine repeats. *Genes Dev* 16:1345–55.

Olson, L., Nordberg, A., von Holst, H., Backman, L., Ebendal, T., Alafuzoff, I., et al. 1992. Nerve growth factor affects 11C-nicotine binding, blood flow, EEG, and verbal episodic memory in an Alzheimer patient (case report). *J Neural Transm Park Dis Dement Sect* 4:79–95.

Patapoutian, A., and Reichardt, L. F. 2001. Trk receptors: mediators of neurotrophin action. *Curr Opin Neurobiol* 11:272–80.

Paul, R., Zhang, Z. G., Eliceiri, B. P., Jiang, Q., Boccia, A. D., Zhang, R. L., et al. 2001. Src deficiency or blockade of Src activity in mice provides cerebral protection following stroke. *Nat Med* 7:222–7.

Pineiro, A., Begona Bugia, M., Pilar Arias, M., Cordero, O. J., and Nogueira, M. 2001. Identification of receptors for prothymosin alpha on human lymphocytes. *Biol Chem* 382:1473–82.

Pineiro, A., Cordero, O. J., and Nogueira, M. 2000. Fifteen years of prothymosin alpha: contradictory past and new horizons. *Peptides* 21:1433–46.

Sakahira, H., Enari, M., and Nagata, S. 1998. Cleavage of CAD inhibitor in CAD activation and DNA degradation during apoptosis. *Nature* 391:96–9.

Salgado, F. J., Pineiro, A., Canda-Sanchez, A., Lojo, J., and Nogueira, M. 2005. Prothymosin alpha-receptor associates with lipid rafts in PHA-stimulated lymphocytes. *Mol Membr Biol* 22:163–76.

Segade, F., and Gomez-Marquez, J. 1999. Prothymosin alpha. *Int J Biochem Cell Biol* 31:1243–8.

Shimizu, S., Konishi, A., Kodama, T., and Tsujimoto, Y. 2000. BH4 domain of antiapoptotic Bcl-2 family members closes voltage-dependent anion channel and inhibits apoptotic mitochondrial changes and cell death. *Proc Natl Acad Sci U S A* 97:3100–5.

Sims, N. R., and Muyderman, H. 2010. Mitochondria, oxidative metabolism and cell death in stroke. *Biochim Biophys Acta* 1802:80–91.

Sofroniew, M. V., Howe, C. L., and Mobley, W. C. 2001. Nerve growth factor signaling, neuroprotection, and neural repair. *Annu Rev Neurosci* 24:1217–81.

Sprick, M. R., and Walczak, H. 2004. The interplay between the Bcl-2 family and death receptor-mediated apoptosis. *Biochim Biophys Acta* 1644:125–32.

Swanson, R. A., Ying, W., and Kauppinen, T. M. 2004. Astrocyte influences on ischemic neuronal death. *Curr Mol Med* 4:193–205.

Szeto, H. H. 2006. Mitochondria-targeted peptide antioxidants: novel neuroprotective agents. *AAPS J* 8:E521–31.

The BDNF Study Group (Phase III). 1999. A controlled trial of recombinant methionyl human BDNF in ALS. *Neurology* 52:1427–33.

The National Institute of Neurological Disorders and Stroke rt-PA Stroke Study Group. 1995. Tissue plasminogen activator for acute ischemic stroke. *N Engl J Med* 333:1581–7.

Tsujimoto, Y. 2002. Bcl-2 family of proteins: life-or-death switch in mitochondria. *Biosci Rep* 22:47–58.

Tsukahara, T., Yonekawa, Y., Tanaka, K., Ohara, O., Wantanabe, S., Kimura, T., et al. 1994. The role of brain-derived neurotrophic factor in transient forebrain ischemia in the rat brain. *Neurosurgery* 34:323–31.

Turner, M. R., Parton, M. J., and Leigh, P. N. 2001. Clinical trials in ALS: an overview. *Semin Neurol* 21:167–75.

Ueda, H. 2009. Prothymosin alpha and cell death mode switch, a novel target for the prevention of cerebral ischemia-induced damage. *Pharmacol Ther* 123:323–33.

Ueda, H., and Fujita, R. 2004. Cell death mode switch from necrosis to apoptosis in brain. *Biol Pharm Bull* 27:950–5.

Ueda, M., Fujita, R., Koji, T., and Ueda, H. 2004. The cognition-enhancer nefiracetam inhibits both necrosis and apoptosis in retinal ischemic models *in vitro* and *in vivo*. *J Pharmacol Exp Ther* 309:200–7.

Ueda, H., Fujita, R., Yoshida, A., Matsunaga, H., and Ueda, M. 2007. Identification of prothymosin-alpha1, the necrosis-apoptosis switch molecule in cortical neuronal cultures. *J Cell Biol* 176:853–62.

Ueda, H., Matsunaga, H., Uchida, H., and Ueda, M. 2010. Prothymosin α as robustness molecule against ischemic stress to brain and retina. *Ann NY Acad Sci* 1194:20–6.

Urano, F., Wang, X., Bertolotti, A., Zhang, Y., Chung, P., Harding, H. P., et al. 2000. Coupling of stress in the ER to activation of JNK protein kinases by transmembrane protein kinase IRE1. *Science* 287:664–6.

Vande Velde, C., Cizeau, J., Dubik, D., Alimonti, J., Brown, T., Israels, S., et al. 2000. BNIP3 and genetic control of necrosis-like cell death through the mitochondrial permeability transition pore. *Mol Cell Biol* 20:5454–68.

Walczak, H., and Krammer, P. H. 2000. The CD95 (APO-1/Fas) and the TRAIL (APO-2L) apoptosis systems. *Exp Cell Res* 256:58–66.

Watson, R. T., and Pessin, J. E. 2001. Intracellular organization of insulin signaling and GLUT4 translocation. *Recent Prog Horm Res* 56:175–93.

Watson, R. T., and Pessin, J. E. 2006. Bridging the GAP between insulin signaling and GLUT4 translocation. *Trends Biochem Sci* 31:215–22.

White, B. C., Sullivan, J. M., DeGracia, D. J., O'Neil, B. J., Neumar, R. W., Grossman, et al. 2000. Brain ischemia and reperfusion: molecular mechanisms of neuronal injury. *J Neurol Sci* 179:1–33.

Williams, L. R., Varon, S., Peterson, G. M., Wictorin, K., Fischer, W., Bjorklund, A., et al. 1986. Continuous infusion of nerve growth factor prevents basal forebrain neuronal death after fimbria fornix transection. *Proc Natl Acad Sci U S A* 83:9231–5.

Wu, D. 2005. Neuroprotection in experimental stroke with targeted neurotrophins. *NeuroRx* 2:120–8.

Yamashita, T., Sawamoto, K., Suzuki, S., Suzuki, N., Adachi, K., Kawase, T., et al. 2005. Blockade of interleukin-6 signaling aggravates ischemic cerebral damage in mice: possible involvement of Stat3 activation in the protection of neurons. *J Neurochem* 94:459–68.

Yoneda, T., Imaizumi, K., Oono, K., Yui, D., Gomi, F., Katayama, T., et al. 2001. Activation of caspase-12, an endoplastic reticulum (ER) resident caspase, through tumor necrosis factor receptor-associated factor 2-dependent mechanism in response to the ER stress. *J Biol Chem* 276:13935–40.

Yoshida, T., Tomioka, I., Nagahara, T., Holyst, T., Sawada, M., Hayes, P., et al. 2004. Bax-inhibiting peptide derived from mouse and rat Ku70. *Biochem Biophys Res Commun* 321:961–6.

Zamzami, N., and Kroemer, G. 2001. The mitochondrion in apoptosis: how Pandora's box opens. *Nat Rev Mol Cell Biol* 2:67–71.

Zha, J., Harada, H., Yang, E., Jockel, J., and Korsmeyer, S. J. 1996. Serine phosphorylation of death agonist BAD in response to survival factor results in binding to 14-3-3 not BCL-X(L). *Cell* 87:619–28.

Zhao, Y., and Rempe, D. A. 2010. Targeting astrocytes for stroke therapy. *Neurotherapeutics* 7:439–51.

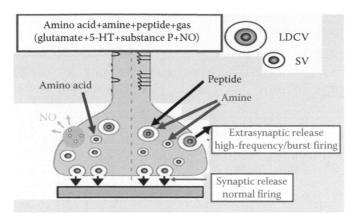

FIGURE 1.1 The one neuron multiple transmitters concept. This nerve ending stores four different types of messenger molecules, partly in different compartments: Peptides in large dense core vesicles (LDCVs, diameter ~1,000Å), amines in LDCVs and synaptic vesicles (SVs, diameter ~500Å), and glutamate in SVs. Nitric oxide (NO) is not stored but generated "upon demand" by NO synthase and diffuses through the membrane. Neuropeptides are released extrasynaptically in response to burst firing/high firing frequency, amino acids mainly into the synapse, and amines presumably in both ways. (From Hökfelt 1991; Lundberg 1996; Merighi 2002.)

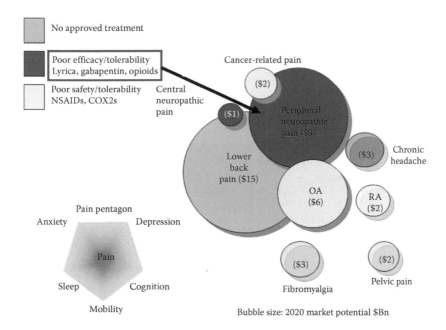

FIGURE 1.2 Schematic presentation of various pain states and their estimated values in the market. Dark grey indicates that available drugs have poor efficacy, as is the case for those presently used to treat peripheral neuropathic pain. The market for neuropathic pain drugs is large, here estimated to be 9 billion USD in 2020. NSAID, nonsteroidal anti-inflammatory drugs; COX, cyclooxygenase; OA, osteoarthritis; RA, rheumatoid arthritis. (Courtesy of Dr Andy, Dray, AstraZeneca, Montreal.)

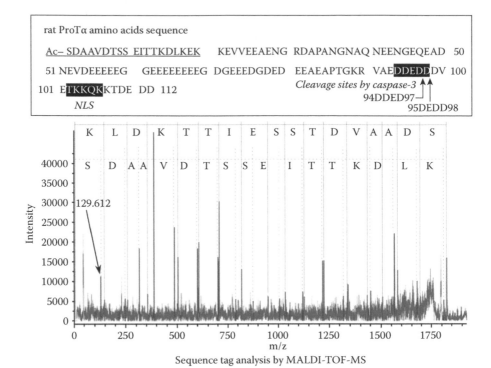

FIGURE 8.3 Identification of ProTα by MALDI-TOF-MS. Amino acid sequence of rat ProTα (upper panel). Underlined sequence is predicted amino acids. White-colored sequences indicate the cleavage sites by activated-caspase-3 and nuclear localization signal (NLS). Peptide sequence search by MS/MS sequence tag analysis (lower panel). Arrow indicates acetylated serine.

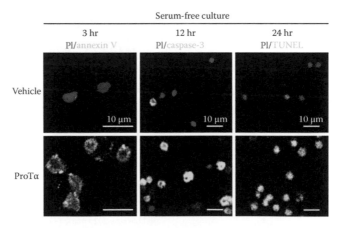

FIGURE 8.5 ProTα-induced cell death mode switch in neuronal cultures. Double staining of low-density cultures with necrosis and apoptosis makers. Cell death mode was evaluated by double staining with PI (necrosis)/annexin V (apoptosis) at 3 hours, PI/antiactivated-caspase-3 IgG (apoptosis) at 12 hours, and PI/TUNEL (apoptosis) at 24 hours. Under serum-free stress, 69% (12 hours), 86% (24 hours), and 92% (48 hours) of neurons died by necrosis, whereas only 15%, 22%, and 5% of neurons died by apoptosis. Addition of ProTα totally switched the cell death mode form necrosis to apoptosis.

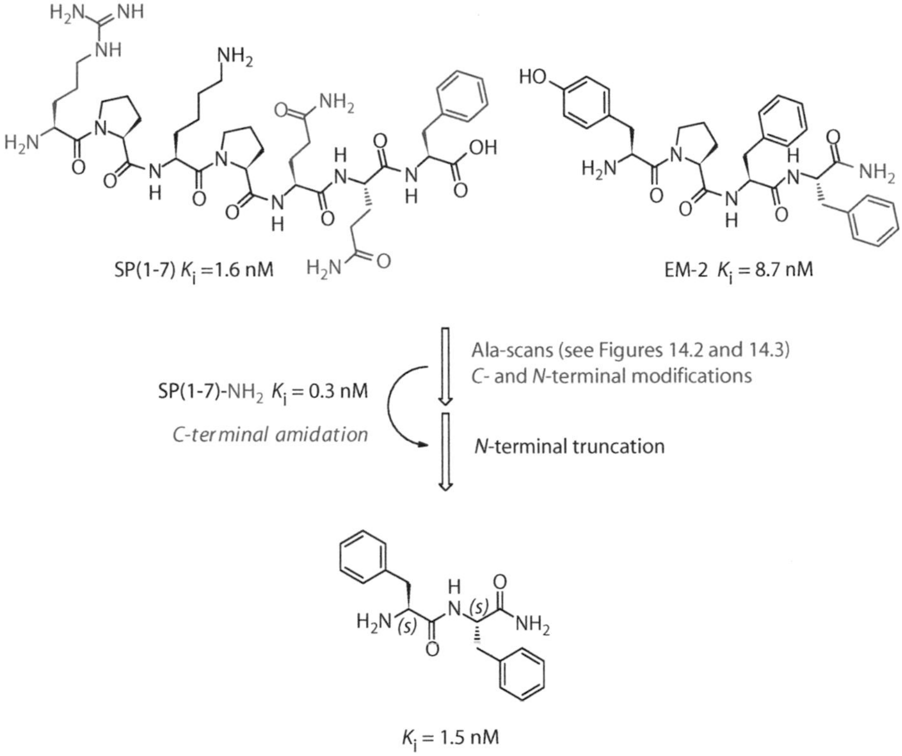

SP(1-7) $K_i = 1.6$ nM

EM-2 $K_i = 8.7$ nM

SP(1-7)-NH$_2$ $K_i = 0.3$ nM

C-terminal amidation

Ala-scans (see Figures 14.2 and 14.3)
C- and N-terminal modifications

N-terminal truncation

$K_i = 1.5$ nM

FIGURE 14.4 Summary of the optimization process starting from the mother peptides SP(1-7) and EM-2 and leading to a dipeptide with equal binding affinity as SP(1-7) itself.

9 Influence of Anabolic Androgenic Steroids on Dynorphinergic Pathways in Rat's Brain

Kristina Magnusson

CONTENTS

9.1 INTRODUCTION

Anabolic androgenic steroids (AAS) abuse is becoming increasingly common and may result in a range of physical as well as psychiatric effects, such as altered behavior in terms of increased aggression, cognitive dysfunction, and addictive behavior. AAS comprise testosterone and its derivatives, of which nandrolone is very common. Studies have shown the nandrolone-induced effects in male rats at various neuropeptide levels in several areas of the central nervous system (CNS). Effects of nandrolone on the dynorphinergic systems may be linked to some of the reported behavior alterations. This chapter is aimed to review mechanisms underlying alterations in the dynorphin system and also to describe effects attributed to nandrolone administration on its selective receptor, the κ-opioid peptide (KOP) receptor. The results display significant effects on the enzymatic

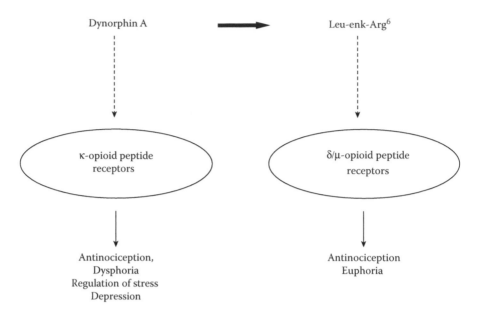

FIGURE 9.1 Conversion of dynorphin to Leu-enkephalin-Arg[6] includes an alteration in receptor activation profile, leading to different biological effects.

conversion of dynorpin A into its bioactive metabolite, Leu-enkephalin-Arg[6], as a result of nandrolone treatment (Figure 9.1). More profound investigations on the dynorphinergic system displayed effects on the KOP receptor density in various brain regions. Studies have also confirmed a significant increase in the expression of the gene transcript of the dynorphinergic precursor, prodynorphin (*PDYN*), in the hippocampus—the region of the brain associated with cognitive processes. In addition, in studies of male rats, impaired spatial learning and memory as recorded in the Morris water maze (MWM) task following nandrolone administration has also been encountered. Thus, reviewed data strengthen the role of the dynorphin system in mechanisms underlying behavioral effects seen in individuals taking AAS.

9.2 ANABOLIC ANDROGENIC STEROIDS

9.2.1 Definition

AAS is the designation of the endogenous male sex hormone testosterone and synthetic derivatives of the same. Common effects of all AAS are enhanced tissue building, that is, anabolic effects, and the development and maintenance of male sexual characteristics, that is, androgenic effects (Marshall 1988).

9.2.2 Use in Society

The general belief in present society is that the use of AAS is a problem mainly connected to athletes and bodybuilders. However, there are reports, which indicate that the abuse of AAS is spreading among adolescents and young adults not connected to sports (Kindlundh et al. 1999; Sjoqvist et al. 2008), and media constantly report cases of AAS use, for example, in connection to violent acts, where the perpetrator has been under the influence of AAS.

The prevalence of AAS use in different countries has been investigated with a number of surveys. Among nonathletes, the estimated lifetime prevalence typically ranges between 1% and 5% (Kindlundh et al. 1998; Nilsson et al. 2001; Wanjek et al. 2007; Tahtamouni et al. 2008). The prevalence is somewhat higher among athletes (Stilger and Yesalis 1999).

The intake of AAS to enhance physical performance is however not a recent phenomenon; testosterone, as well as testosterone-derived substances, has been used for this purpose since the isolation and characterization of the male hormone. As often is the case with substances of abuse, the original intentions were to find pharmaceutical substances to treat different diseases and improve health. There is still a clinical use of AAS in the treatment of hypogonadism, impotence, osteoporosis, and anemia. But, the clinical use of AAS is negligible compared with the illegal use.

9.2.3 Administration Patterns

Bodybuilders and athletes usually administer AAS according to strict regimes. These regimes normally comprise two to three cycles of administration per year, where each cycle lasts for 6–12 weeks. A resting period is usually included between the cycles, with the intention of minimizing adverse effects. In addition, doses are gradually increased, then decreased over the period of the cycle, called *pyramiding*, to further reduce adverse behavioral effects caused by sudden withdrawal of the AAS. Furthermore, several different AAS are frequently administered within each cycle, also called *stacking* (Pagonis et al. 2006). The theory behind stacking is that receptor downregulation is less likely to occur when several steroids are used instead of a single steriod. In addition, some steroids work synergistically, when combined. However, this has not been proven scientifically. Doses used within these regimes are often found to be up to 100 times greater than the therapeutic levels (Rogol and Yesalis 1992; Pagonis et al. 2006). So, it is not surprising that a number of physiological, as well as psychological, adverse effects usually develop after some time of abuse. It is, therefore, not unusual for a person under AAS abuse to administer a wide range of other drugs at the same time to reduce AAS-induced adverse effects or to enhance the anabolic effects. For instance, aromatase inhibitors are used to decrease the conversion to estrogens and reduce feminizing effects. In addition, as endogenous testosterone declines when exogenous androgens are administered, substances, such as human chorionic gonadotropin, are used at the end of a cycle to increase the production of endogenous testosterone. The list of additional substances can be made long, and AAS users usually possess great knowledge of the drugs to be used to avoid or minimize unwanted effects as well as to reach the desired effects.

The use of AAS is also associated with other forms of drug abuse, such as excess alcohol consumption and the use of illicit drugs (Kindlundh et al. 1999; Arvary and Pope 2000; Skarberg et al. 2009). This multidrug abuse has become apparent among adolescents, even outside athletic and bodybuilding contexts, who take AAS to boost their self-esteem or become intoxicated (Kindlundh et al. 1999).

9.2.4 Psychological Aspects

There are a number of reports pointing at psychological effects as a result of AAS abuse. However, there is extreme variability in physiological symptoms caused by AAS abuse, because of differences between the AAS, dosage, duration of use, personality of the abuser, as well as previous or present use of other drugs. Furthermore, the biochemical mechanisms behind these effects are less understood than those behind physical effects. A cocktail of steroids are also commonly used, making it even harder to distinguish and correlate psychological effects of a specific steroid.

Some of the more prominent psychological effects are manic-like states defined by irritability, aggressiveness, euphoria, grandiose beliefs, hyperactivity, and reckless or dangerous behavior (Pope and Katz 1994; Pope et al. 2000; Daly et al. 2003; Papazisis et al. 2007). It has been shown that AAS abusers are involved in more fights, are more verbally aggressive, and more violent toward their significant others when using AAS than when not using it. The abuse of AAS has unfortunately added a new term to dictionaries, *roid rage*, where individuals suddenly commit terrible acts of violence (Pope et al. 2000; Pagonis et al. 2006). In fact, increased aggression is the most consistent behavioral effect of high-dose AAS exposure in surveys and prospective studies. The aggressive behavior as a result of chronic treatment with AAS is also reflected in animal models in rats and hamsters (Harrison, Connor, et al. 2000; Johansson, Lindqvist, et al. 2000).

A number of studies report of AAS dependence, where, in all cases, supraphysiological doses of AAS have been chronically administered (Brower et al. 1991; Gruber and Pope 2000; Skarberg et al. 2009). As an example, results from a study conducted on 100 Australian competitive and recreational AAS users showed that 23% met the *Diagnostic and Statistical Manual of Mental Disorders*—Fourth Edition (*DSM IV*) criteria for substance dependence and 25% qualified for substance abuse (Copeland et al. 2000). Also, a recent study in Sweden reported of AAS abuse and dependence according to the *DSM IV* criteria among 32 patients who were attending an addiction center (Skarberg et al. 2009). Furthermore, individuals who have been under supraphysiological doses of AAS for a long time may display withdrawal-like effects, such as depressive symptoms, impaired concentration, fatigue, and suicidal thoughts, upon cessation of the abuse (Pope and Katz 1994; Malone et al. 1995; Papazisis et al. 2007). In addition, reinforcing properties of androgens were also shown in conditioned place preference models in animals, such as mice (Arnedo et al. 2002) and rats (Alexander et al. 1994). Other studies demonstrate that rats and hamsters self-administer testosterone when given orally as well as intravenously (Tincello et al. 1997; Johnson and Wood 2001; Wood et al. 2004), thus further confirming the rewarding properties of androgens. It should, however, be noted that AAS are not addictive in the same manner as compounds such as amphetamine or heroine. Nevertheless, this does not imply that the steroids are safe and AAS dependence is undoubtedly also reliant on individual susceptibility.

Other psychological features associated with administration of AAS are the development of depression as well as cognitive dysfunctions, such as forgetfulness, distractibility, and confusion or delirious states (Su et al. 1993; Daly et al. 2003). However, the literature describing the effects of AAS on cognition is limited. Only a few controlled studies regarding cognitive effects of AAS in humans and animals exist. These will be further reviewed in the following section.

9.2.5 Pharmacological Aspects

Testosterone and other AAS are thought to exert their actions in the body in a similar manner. Different mechanisms have been proposed and include direct interaction with the androgenic receptor, interference with the glucocorticoid receptor, as well as nongenomic pathways mainly found in the CNS, as described more thoroughly in a review by Kicman (2008). It is likely that a range of pathways rather than a single constant mechanism conduct the cellular actions of steroids.

The most investigated mechanism of action for androgens involves the activation of androgenic receptors (Mangelsdorf et al. 1995). These intracellular receptors belong to the nuclear receptor superfamily and possess a ligand-binding domain, as well as at least two transcriptional activation domains; this enables interactions with the DNA and modulation of transcription. In the absence of a steroid ligand, the androgenic receptor is found inactive in the cytoplasm of the cell, associated to a chaperone complex. Steroids are small, fatty molecules that are able to diffuse passively into the cell. Subsequent to passing the cell membrane, the steroid ligand binds to the receptor, which conveys conformational changes that result in activation of the receptor. The activated receptor is dissociated from the chaperone complex and translocated into the nucleus of the cell, where it forms homodimers with another receptor through cooperative interaction with the DNA (Evans 1988). This activates coregulators, that is, coactivators or corepressors, which form a transcription complex that affects transcription, which will further affect protein translation and alteration in cell function, growth, or differentiation.

Furthermore, AAS are clinically used as anticatabolic agents at catabolic conditions, such as severe burns (Demling and Orgill 2000). It has been suggested that AAS exert this anticatabolic effect by antagonizing the glucocorticoid receptor (Hickson et al. 1990). Indeed, a case report showed that a patient gave some response to testosterone treatment despite the occurrence of an amino acid mutation in the androgen receptor DNA-binding domain, thus suggesting this effect to be exerted by mechanisms other than the androgenic receptor (Tincello et al. 1997). However, AAS generally show a low binding affinity to the glucocorticoid receptor, which indicates a more complex explanation.

The effects of steroids conducted via the classical genomic pathway outlined earlier, require comparatively long time, usually hours or even days. Interestingly, it has become obvious that steroids are able to activate more rapid nongenomic pathways, as effects of steroids have been seen after a much shorter lag time. As early as in 1963, a study by Klein and Henk showed acute cardiovascular effects of aldosterone in men after only 5 minutes. Further on, effects of aldosterone were seen on sodium ion exchange in mammalian erythrocytes, known to be lacking a nucleus (Spach and Streeten 1964). However, much of the nongenomic actions of androgens are yet to be elucidated, particularly with regard to effects in the CNS, which can be linked to behavioral mechanisms. Studies on AAS-induced effects in the CNS that are independent of nuclear receptor signaling have been made. Chronic treatment with AAS, including nandrolone, has been shown to allosterically modulate the function of the $GABA_A$ receptor, when given acutely (Jorge-Rivera et al. 2000; Yang et al. 2002). Furthermore, AAS have been suggested to affect the serotonergic system (Kindlundh et al. 2003), the dopaminergic system (Birgner et al. 2007, 2008), as well as the glutamate system (Le Greves et al. 2002).

In addition, alterations within neuropeptidergic systems, such as the endogenous opioid system (Johansson et al. 1997; Johansson, Hallberg, et al. 2000; Johansson, Lindqvist et al. 2000; Guarino and Spampinato 2008) and the tachykinin system (Hallberg et al. 2000, 2005), have been displayed. Previous studies have in fact shown altered levels of the tachykinin substance P, as well as of dynorphinergic endogenous opioids, as a result of nandrolone treatment in rats (Hallberg et al. 2000; Johansson, Hallberg, et al. 2000). This may give rise to effects, such as altered receptor activation. Hence, the mechanisms generating altered peptide levels are of great interest in a behavioral context. The substance P system and the dynorphinergic system are of relevance in AAS abuse due to their association with regulation of behaviors, such as aggression, reward, cognition, and depression; these behaviors are also reported to be affected during AAS abuse.

9.3   THE DYNORPHINERGIC SYSTEM

Neuropeptides are derived by enzymatic cleavage of larger prepropeptides in the ribosoms, to yield propeptides. The propeptides will undergo enzymatic cleavage, which results in bioactive neuropeptides that can be released into the synapses. In contrast to classical neurotransmitters, which are often inactivated by reuptake processes, neuropeptides are inactivated or converted by proteases. Since no recycling occurs, newly synthesized neuropeptides are constantly delivered to the nerve terminal from the cell body. The neuropeptides will after the release be degraded by specific enzymes to either inactive or active peptide fragments, which will thus terminate, retain, or convert the activity of the released peptide (Figure 9.2).

Dynorphin is the designation of opioid peptides that derive from the *PDYN* precursor. Early research suggested them to be one single peptide and it was named dynorphin (*Dyn-* from the Greek *dynamis* = power), and dynorphin's extraordinary potency is showed to be 700 times that of Leu-enkephalin in guinea pig ileum bioassays (Cox et al. 1975; Goldstein et al. 1979, 1981). The *PDYN* gene contains four exons (1–4) in humans and in rodents, whereas exons 3 and 4 contain the entire coding sequence (Horikawa et al. 1983). After transcription and translation, the large inactive *PDYN* precursor is further processed into bioactive peptides by the prohormone convertases (PC), PC1 and PC2, and carboxypeptidase E (Day et al. 1998). Moderate to high *PDYN* mRNA levels have been found in amygdala, dentate gyrus, nucleus accumbens, hypothalamus, caudate putamen, and hippocampus. Major naturally occurring dynorphins that have been identified are, dynorphin A(1-17), dynorphin A(1-8), dynorphin B(1-13), dynorphin B(1-29), big dynorphin, α-neoendorphin, and β-neoendorphin, displayed in Table 9.1. Dynorphins exhibit binding affinity toward all three opioid peptide receptors, that is, the KOP receptor, the μ-opioid peptide (MOP) receptor, and the δ-opioid peptide (DOP) receptor.

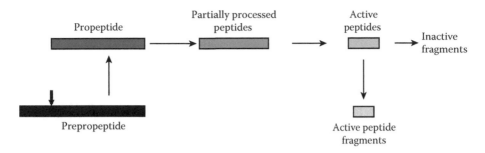

FIGURE 9.2   Pathways for neuropeptide processing and inactivation.

TABLE 9.1
Sequences of Major Naturally Occurring Dynorphins

Peptide	Amino Acid Sequence
Dynorphin A(1-17)	Tyr-Gly-Gly-Phe-Leu-Arg-Arg-Ile-Arg-Pro-Lys-Leu-Lys-Trp-Asp-Asn-Gln
Dynorphin A(1-8)	Tyr-Gly-Gly-Phe-Leu-Arg-Arg-Ile
Dynorphin B(1-13)	Tyr-Gly-Gly-Phe-Leu-Arg-Arg-Gln-Phe-Lys-Val-Val-Thr
Dynorphin B(1-29)	Tyr-Gly-Gly-Phe-Leu-Arg-Arg-Gln-Phe-Lys-Val-Val-Thr-Arg-Ser-Gln-Glu-Asp-Pro-Asn-Ala-Tyr-Ser-Gly-Glu-Leu-Phe-Asp-Ala
α-Neoendorphin	Tyr-Gly-Gly-Phe-Leu-Arg-Lys-Tyr-Pro-Lys
β-Neoendorphin	Tyr-Gly-Gly-Phe-Leu-Arg-Lys-Tyr-Pro
Big dynorphin	Tyr-Gly-Gly-Phe-Leu-Arg-Arg-Ile-Arg-Pro-Lys-Leu-Lys-Trp-Asp-Asn-Gln-Lys-Arg- Tyr-Gly-Gly-Phe-Leu-Arg-Arg-Gln-Phe-Lys-Val-Val-Thr

However, dynorphins show a clear preference for the KOP receptor (Goldstein et al. 1981; Chavkin et al. 1982; Corbett et al. 1982). The selectivity toward the KOP receptor is mediated by the *C*-terminal of the dynorphin peptides with Arg[7] and Lys[11] making the greatest contribution, whereas the *N*-terminal, containing the pentapeptide, Leu-enkephalin, contributes to the opioid activity (Chavkin and Goldstein 1981).

Dynorphins are involved in a number of physiological systems, such as neuroendocrine regulation, pain regulation, motor activity, cardiovascular function, respiration, temperature regulation, feeding behavior, and stress responsivity (Fallon and Leslie 1986). Furthermore, dynorphins are suggested to mediate negative emotional states, such as depression and dysphoria, and be involved in cognitive functions (Pfeiffer et al. 1986; Jiang et al. 1989; Bals-Kubik et al. 1993; Nguyen et al. 2005). Activation of the KOP receptor can produce actions similar to MOP and DOP receptor activation. However, KOP receptor activation also produces effects that oppose the actions of MOP and DOP receptor activation. This can be seen in addiction-relevant brain areas, where MOP and DOP receptor activation mediate conditioned place preference, whereas KOP receptor activation mediate conditioned place aversion in rats (Bals-Kubik et al. 1993).

9.4 EFFECTS OF AAS ON THE DYNORPHINERGIC SYSTEM AND COGNITIVE FUNCTIONS

Nandrolone-induced effects have been observed on the density of the KOP receptor measured by the specific binding of [^{3}H]Cl-977 (Magnusson, Birgner, et al. 2009). These effects could be the result of the direct effects exerted by nandrolone or may be attributed to compensatory mechanisms. Thus, nandrolone administration may result in cellular adaptations, such as altered KOP receptor density, and affinity for ligands in order to maintain synaptic homeostasis. KOP receptors are found in a vast number of brain regions, with a dense distribution in the amygdala, caudate putamen, endopiriform nucleus, hypothalamus, and the stria terminalis. These receptors have been suggested to be involved in the modulation of drug-seeking behavior, stress response, mood alterations, as well as cognitive functions (Zhang et al. 2004; Carlezon et al. 2006; McLaughlin et al. 2006).

Studies have shown connections between high-dose AAS administration and behaviors known to be associated with the dynorphinergic system, such as anxiety and cognition (Naghdi et al. 2003; Rocha et al. 2007; Wittmann et al. 2009). KOP receptor density in amygdala was affected following nandrolone administration (Magnusson, Birgner, et al. 2009). This region is suggested to influence neural activity and memory storage in the limbic system, as activation of amygdala improved caudate- and hippocampal-dependent learning tasks (McGaugh and Cahill 1997). Furthermore, injections of testosterone enanthate in this region in rats resulted in impaired spatial memory and learning in the MWM task (Naghdi et al. 2003).

Administration of nandrolone decanoate every third day for 2 weeks prior to the MWM test has been shown to prolong the latencies before the rats found the hidden platform (Magnusson, Hanell, et al. 2009). In addition, the probe trial, which was performed 3 days after the last trial day, displayed that nandrolone-treated animals spent significantly less time in the quadrant where the platform was previously positioned compared with the controls. There were no differences regarding swimming speed between or within the groups on different days. Thus, no motoric effects were observed as a result of nandrolone treatment.

Hormones such as testosterone are shown to be associated with different forms of cognitive processes (Galea et al. 1995). For example, a study on aging mice with a loss of testosterone exhibited spatial learning deficits, which were reversed by administration of therapeutical levels of testosterone (Flood et al. 1995). Furthermore, endogenous testosterone concentrations have been suggested to be related to gender differences in spatial cognitive ability in humans, as well (O'Connor et al. 2001). Nevertheless, investigations of the effects of endogenous testosterone concentrations on spatial cognitive ability have produced conflicting results, with researchers reporting either a positive correlation between increased testosterone and improved spatial ability, the opposite pattern, or no effect at all (Clark et al. 1995; Silverman, et al. 1999; O'Connor et al. 2001; Liben et al. 2002). Varying study designs and different steroids may cause discrepancies in the results. It should also be noted that the cognitive effects of supraphysiological levels of androgens might differ from the effects of endogenous levels.

A number of studies have examined the effects of exogenous testosterone or other AAS on the performance of male rats in reference memory tasks, such as the MWM task. For example, one study showed impairments in probe trial performance of young and middle-aged rats in the MWM task, following long-term treatment with testosterone (Goudsmit et al. 1990). In addition, microinjections of testosterone directly into the CA1 region of the hippocampus of male rats also showed impaired performance in the MWM task (Naghdi et al. 2001, 2005). Similar effects of testosterone were also shown, following injections directed at the basolateral amygdala (Naghdi et al. 2003). A resent study indicates negative effects of high-dose nandrolone on social memory in the olfactory social memory test. The authors suggest it to be effects via activation of central androgen receptors, as coadministration of the androgen antagonist, flutamide, eliminated the effects (Kouvelas et al. 2008).

The mechanisms behind this impairment are not fully understood. However, one might speculate through some explanations: Apart from effects in the CNS via the intracellular androgenic receptor, steroid hormones have been shown to modulate memory processes by the regulation of acetylcholine transferase and acetylcholinesterase activities, affecting the levels of acetylcholine (McEwen et al. 1982). Injections of nandrolone have also been shown to alter glutaminergic transmission, which is reported to affect recognition in visual and olfactory memory tasks (Hlinak et al. 2005; Rossbach et al. 2007). Moreover, endogenous levels of testosterone decline, when high doses of exogenous androgens are administered due to a negative feedback mechanism. Low levels of endogenous testosterone give rise to cognitive impairments and could be a contributing factor in this matter.

In addition, elevated *PDYN* mRNA levels in the male rat hippocampus following treatment with nandrolone decanoate have been shown (Magnusson, Hanell, et al. 2009). The nandrolone-treated rats displayed impairments in MWM performance, indicating effects on learning and memory. Provided that the observed increase in *PDYN* mRNA expression reflects an increase in *PDYN* protein translation and biologically active dynorphinergic peptides, increased dynorphin levels could be one of the mechanisms behind the effects on cognition observed in the AAS-treated animals.

Research regarding the contribution of the dynorphinergic system to learning and memory performance has been performed earlier. Intracerebroventricular administration of dynorphin A(1-17) or dynorphin B produced dose-dependent increases in step-through latency in a passive avoidance test (Kuzmin et al. 2006). The authors suggested that dynorphins might facilitate learning and memory based on aversive stimulation, often of importance for survival. In contrast, based on positive reinforcement studies, dynorphins were reported to impair some forms of memory (Colombo et al. 1992; Velazquez and Alter 2004). Dynorphins have been suggested to play a modulatory role on cognitive acquisition and appear to enhance memory retention in negative reinforcement-based tasks (Wall and Messier 2000; Kuzmin et al. 2006). On the contrary, they have an inhibitory effect on memory and learning in tasks based on positive or neutral stimuli. For example, studies on mice have revealed that aged *PDYN* knockout mice performed better in the MWM task than similarly aged wild-type mice (Jiang et al. 1989; Nguyen et al. 2005). Studies on rats showed that learning-deficient rats display an increased abundance of *PDYN* mRNA, as well as higher hippocampal dynorphin A(1-8) levels than control animals

(Jiang et al. 1989). Furthermore, microinjections of a KOP receptor agonist into the CA3 region of the rat hippocampus suppressed learning and memory performance in the MWM task (Daumas et al. 2007). A recent study showed that prolonged stress exposure impairs learning and memory performance. Since the impairment of novel object recognition was abolished by antagonism of the KOP receptor or *PDYN* gene disruption, the authors suggested this effect to be mediated by the activation of KOP receptors (Carey et al. 2009). Association between the dynorphinergic system and decreased memory function has not only been observed in rodents but also in humans, and a recent study suggests a role of *PDYN* gene polymorphism in episodic memory and verbal fluency in healthy elderly humans (Kolsch et al. 2009). In congruence with this, and of particular relevance to human disease, patients with Alzheimer's disease (AD) displayed markedly elevated Dyn A levels in the frontal cortex and a correlation with neuritic plaque density was also observed (Yakovleva et al. 2007). Moreover, increased KOP receptor levels have been reported in brains of AD patients (Hiller et al. 1987), and cognitive performance in AD patients can be improved by administration of the nonselective opioid antagonist, naloxone (Reisberg et al. 1983).

It should be noted that the cognitive effects of exposure to supraphysiological levels of androgens may differ from the modulatory effects of endogenous levels of hormones. For example, low physiological levels of estradiol in ovariectomized female rats improved performance on a delayed alternation T-maze task, whereas high pharmacological levels of estradiol impaired performance (Wide et al. 2004).

Even though one should be careful with transferring results from animal studies to effects in humans, the fact still remains that nandrolone decanoate administration significantly impaired spatial learning and memory in rats and indications of similar effects exist in human high-dose AAS abusers (Su et al. 1993; Porcerelli and Sandler 1998).

ACKNOWLEDGMENTS

This work was supported by the Swedish Research Council (Grant 9459).

REFERENCES

Alexander, G. M., M. G. Packard, et al. (1994). "Testosterone has rewarding affective properties in male rats: implications for the biological basis of sexual motivation." *Behav Neurosci* **108**(2): 424–8.

Arnedo, M. T., A. Salvador, et al. (2002). "Similar rewarding effects of testosterone in mice rated as short and long attack latency individuals." *Addict Biol* **7**(4): 373–9.

Arvary, D. and H. G. Pope, Jr. (2000). "Anabolic-androgenic steroids as a gateway to opioid dependence." *N Engl J Med* **342**(20): 1532.

Bals-Kubik, R., A. Ableitner, et al. (1993). "Neuroanatomical sites mediating the motivational effects of opioids as mapped by the conditioned place preference paradigm in rats." *J Pharmacol Exp Ther* **264**(1): 489–95.

Birgner, C., A. M. Kindlundh-Hogberg, et al. (2007). "Altered extracellular levels of DOPAC and HVA in the rat nucleus accumbens shell in response to sub-chronic nandrolone administration and a subsequent amphetamine challenge." *Neurosci Lett* **412**(2): 168–72.

Birgner, C., A. M. Kindlundh-Hogberg, et al. (2008). "The anabolic androgenic steroid nandrolone decanoate affects mRNA expression of dopaminergic but not serotonergic receptors." *Brain Res* **1240**: 221–8.

Brower, K. J., F. C. Blow, et al. (1991). "Symptoms and correlates of anabolic-androgenic steroid dependence." *Br J Addict* **86**(6): 759–68.

Carey, A. N., A. M. Lyons, et al. (2009). "Endogenous kappa opioid activation mediates stress-induced deficits in learning and memory." *J Neurosci* **29**(13): 4293–300.

Carlezon, W. A., Jr., C. Beguin, et al. (2006). "Depressive-like effects of the kappa-opioid receptor agonist salvinorin A on behavior and neurochemistry in rats." *J Pharmacol Exp Ther* **316**(1): 440–7.

Chavkin, C. and A. Goldstein (1981). "Specific receptor for the opioid peptide dynorphin: structure–activity relationships." *Proc Natl Acad Sci U S A* **78**(10): 6543–7.

Chavkin, C., I. F. James, et al. (1982). "Dynorphin is a specific endogenous ligand of the kappa opioid receptor." *Science* **215**(4531): 413–5.

Clark, A. S., M. C. Mitre, et al. (1995). "Anabolic-androgenic steroid and adrenal steroid effects on hippocampal plasticity." *Brain Res* **679**(1): 64–71.

Colombo, P. J., J. L. Martinez, Jr., et al. (1992). "Kappa opioid receptor activity modulates memory for peck-avoidance training in the 2-day-old chick." *Psychopharmacology (Berl)* **108**(1–2): 235–40.

Copeland, J., R. Peters, et al. (2000). "Anabolic-androgenic steroid use disorders among a sample of Australian competitive and recreational users." *Drug Alcohol Depend* **60**(1): 91–6.

Corbett, A. D., S. J. Paterson, et al. (1982). "Dynorphin and dynorphin are ligands for the kappa-subtype of opiate receptor." *Nature* **299**(5878): 79–81.

Cox, B. M., K. E. Opheim, et al. (1975). "A peptide-like substance from pituitary that acts like morphine. 2. Purification and properties." *Life Sci* **16**(12): 1777–82.

Daly, R. C., T. P. Su, et al. (2003). "Neuroendocrine and behavioral effects of high-dose anabolic steroid administration in male normal volunteers." *Psychoneuroendocrinology* **28**(3): 317–31.

Daumas, S., A. Betourne, et al. (2007). "Transient activation of the CA3 Kappa opioid system in the dorsal hippocampus modulates complex memory processing in mice." *Neurobiol Learn Mem* **88**(1): 94–103.

Day, R., C. Lazure, et al. (1998). "Prodynorphin processing by proprotein convertase 2. Cleavage at single basic residues and enhanced processing in the presence of carboxypeptidase activity." *J Biol Chem* **273**(2): 829–36.

Demling, R. H. and D. P. Orgill (2000). "The anticatabolic and wound healing effects of the testosterone analog oxandrolone after severe burn injury." *J Crit Care* **15**(1): 12–17.

Evans, R. M. (1988). "The steroid and thyroid hormone receptor superfamily." *Science* **240**(4854): 889–95.

Fallon, J. H. and F. M. Leslie (1986). "Distribution of dynorphin and enkephalin peptides in the rat brain." *J Comp Neurol* **249**(3): 293–336.

Flood, J. F., S. A. Farr, et al. (1995). "Age-related impairment in learning but not memory in SAMP8 female mice." *Pharmacol Biochem Behav* **50**(4): 661–4.

Galea, L. A., M. Kavaliers, et al. (1995). "Gonadal hormone levels and spatial learning performance in the Morris water maze in male and female meadow voles, Microtus pennsylvanicus." *Horm Behav* **29**(1): 106–25.

Goldstein, A., W. Fischli, et al. (1981). "Porcine pituitary dynorphin: complete amino acid sequence of the biologically active heptadecapeptide." *Proc Natl Acad Sci U S A* **78**(11): 7219–23.

Goldstein, A., S. Tachibana, et al. (1979). "Dynorphin-(1-13), an extraordinarily potent opioid peptide." *Proc Natl Acad Sci U S A* **76**(12): 6666–70.

Goudsmit, E., N. E. Van de Poll, et al. (1990). "Testosterone fails to reverse spatial memory decline in aged rats and impairs retention in young and middle-aged animals." *Behav Neural Biol* **53**(1): 6–20.

Gruber, A. J. and H. G. Pope, Jr. (2000). "Psychiatric and medical effects of anabolic-androgenic steroid use in women." *Psychother Psychosom* **69**(1): 19–26.

Guarino, G. and S. Spampinato (2008). "Nandrolone decreases mu opioid receptor expression in SH-SY5Y human neuroblastoma cells." *Neuroreport* **19**(11): 1131–5.

Hallberg, M., P. Johansson, et al. (2000). "Anabolic-androgenic steroids affect the content of substance P and substance P(1-7) in the rat brain." *Peptides* **21**(6): 845–52.

Hallberg, M., A. Kindlundh, et al. (2005). "The impact of chronic nandrolone decanoate administration on the NK1 receptor density in rat brain as determined by autoradiography." *Peptides* **26**(7): 1228–34.

Harrison, R. J., D. F. Connor, et al. (2000). "Chronic anabolic-androgenic steroid treatment during adolescence increases anterior hypothalamic vasopressin and aggression in intact hamsters." *Psychoneuroendocrinology* **25**(4): 317–38.

Hickson, R. C., S. M. Czerwinski, et al. (1990). "Glucocorticoid antagonism by exercise and androgenic-anabolic steroids." *Med Sci Sports Exerc* **22**(3): 331–40.

Hiller, J. M., Y. Itzhak, et al. (1987). "Selective changes in mu, delta and kappa opioid receptor binding in certain limbic regions of the brain in Alzheimer's disease patients." *Brain Res* **406**(1–2): 17–23.

Hlinak, Z., D. Gandalovicova, et al. (2005). "Behavioral deficits in adult rats treated neonatally with glutamate." *Neurotoxicol Teratol* **27**(3): 465–73.

Horikawa, S., T. Takai, et al. (1983). "Isolation and structural organization of the human preproenkephalin B gene." *Nature* **306**(5943): 611–14.

Jiang, H. K., V. V. Owyang, et al. (1989). "Elevated dynorphin in the hippocampal formation of aged rats: relation to cognitive impairment on a spatial learning task." *Proc Natl Acad Sci U S A* **86**(8): 2948–51.

Johansson, P., M. Hallberg, et al. (2000). "The effect on opioid peptides in the rat brain, after chronic treatment with the anabolic androgenic steroid, nandrolone decanoate." *Brain Res Bull* **51**(5): 413–18.

Johansson, P., A. Lindqvist, et al. (2000). "Anabolic androgenic steroids affects alcohol intake, defensive behaviors and brain opioid peptides in the rat." *Pharmacol Biochem Behav* **67**(2): 271–9.

Johansson, P., A. Ray, et al. (1997). "Anabolic androgenic steroids increase beta-endorphin levels in the ventral tegmental area in the male rat brain." *Neurosci Res* **27**(2): 185–9.

Johnson, L. R. and R. I. Wood (2001). "Oral testosterone self-administration in male hamsters." *Neuroendocrinology* **73**(4): 285–92.

Jorge-Rivera, J. C., K. L. McIntyre, et al. (2000). "Anabolic steroids induce region- and subunit-specific rapid modulation of GABA(A) receptor-mediated currents in the rat forebrain." *J Neurophysiol* **83**(6): 3299–309.

Kicman, A. T. (2008). "Pharmacology of anabolic steroids." *Br J Pharmacol* **154**(3): 502–21.

Kindlundh, A. M., D. G. Isacson, et al. (1998). "Doping among high school students in Uppsala, Sweden: a presentation of the attitudes, distribution, side effects, and extent of use." *Scand J Soc Med* **26**(1): 71–4.

Kindlundh, A. M., D. G. Isacson, et al. (1999). "Factors associated with adolescent use of doping agents: anabolic-androgenic steroids." *Addiction* **94**(4): 543–53.

Kindlundh, A. M., J. Lindblom, et al. (2003). "The anabolic-androgenic steroid nandrolone induces alterations in the density of serotonergic 5HT1B and 5HT2 receptors in the male rat brain." *Neuroscience* **119**(1): 113–20.

Klein, K. and W. Henk (1963). "[Clinical experimental studies on the influence of aldosterone on hemodynamics and blood coagulation]." *Z Kreislaufforsch* **52**: 40–53.

Kolsch, H., M. Wagner, et al. (2009). "Gene polymorphisms in prodynorphin (*PDYN*) are associated with episodic memory in the elderly." *J Neural Transm* **116**(7): 897–903.

Kouvelas, D., C. Pourzitaki, et al. (2008). "Nandrolone abuse decreases anxiety and impairs memory in rats via central androgenic receptors." *Int J Neuropsychopharmacol* **11**(7): 925–34.

Kuzmin, A., N. Madjid, et al. (2006). "Big dynorphin, a prodynorphin-derived peptide produces NMDA receptor-mediated effects on memory, anxiolytic-like and locomotor behavior in mice." *Neuropsychopharmacology* **31**(9): 1928–37.

Le Greves, P., Q. Zhou, et al. (2002). "Effect of combined treatment with nandrolone and cocaine on the NMDA receptor gene expression in the rat nucleus accumbens and periaqueductal gray." *Acta Psychiatr Scand* (412): 129–32.

Liben, L. S., E. J. Susman, et al. (2002). "The effects of sex steroids on spatial performance: a review and an experimental clinical investigation." *Dev Psychol* **38**(2): 236–53.

Magnusson, K., C. Birgner, et al. (2009). "Nandrolone decanoate administration dose-dependently affects the density of kappa opioid peptide receptors in the rat brain determined by autoradiography." *Neuropeptides* **43**(2): 105–11.

Magnusson, K., A. Hanell, et al. (2009). "Nandrolone decanoate administration elevates hippocampal prodynorphin mRNA expression and impairs Morris water maze performance in male rats." *Neurosci Lett* **467**(3): 189–93.

Malone, D. A., Jr., R. J. Dimeff, et al. (1995). "Psychiatric effects and psychoactive substance use in anabolic-androgenic steroid users." *Clin J Sport Med* **5**(1): 25–31.

Mangelsdorf, D. J., C. Thummel, et al. (1995). "The nuclear receptor superfamily: the second decade." *Cell* **83**(6): 835–9.

Marshall, E. (1988). "The drug of champions." *Science* **242**(4876): 183–4.

McEwen, B. S., A. Biegon, et al. (1982). "Steroid hormones: humoral signals which alter brain cell properties and functions." *Recent Prog Horm Res* **38**: 41–92.

McGaugh, J. L. and L. Cahill (1997). "Interaction of neuromodulatory systems in modulating memory storage." *Behav Brain Res* **83**(1–2): 31–8.

McLaughlin, J. P., S. Li, et al. (2006). "Social defeat stress-induced behavioral responses are mediated by the endogenous kappa opioid system." *Neuropsychopharmacology* **31**(6): 1241–8.

Naghdi, N., N. Majlessi, et al. (2005). "The effect of intrahippocampal injection of testosterone enanthate (an androgen receptor agonist) and anisomycin (protein synthesis inhibitor) on spatial learning and memory in adult, male rats." *Behav Brain Res* **156**(2): 263–8.

Naghdi, N., N. Nafisy, et al. (2001). "The effects of intrahippocampal testosterone and flutamide on spatial localization in the Morris water maze." *Brain Res* **897**(1–2): 44–51.

Naghdi, N., S. Oryan, et al. (2003). "The study of spatial memory in adult male rats with injection of testosterone enanthate and flutamide into the basolateral nucleus of the amygdala in Morris water maze." *Brain Res* **972**(1–2): 1–8.

Nguyen, X. V., J. Masse, et al. (2005). "Prodynorphin knockout mice demonstrate diminished age-associated impairment in spatial water maze performance." *Behav Brain Res* **161**(2): 254–62.

Nilsson, S., A. Baigi, et al. (2001). "The prevalence of the use of androgenic anabolic steroids by adolescents in a county of Sweden." *Eur J Public Health* **11**(2): 195–7.

O'Connor, D. B., J. Archer, et al. (2001). "Activational effects of testosterone on cognitive function in men." *Neuropsychologia* **39**(13): 1385–94.

Pagonis, T. A., N. V. Angelopoulos, et al. (2006). "Psychiatric side effects induced by supraphysiological doses of combinations of anabolic steroids correlate to the severity of abuse." *Eur Psychiatry* **21**(8): 551–62.

Papazisis, G., D. Kouvelas, et al. (2007). "Anabolic androgenic steroid abuse and mood disorder: a case report." *Int J Neuropsychopharmacol* **10**(2): 291–3.

Pfeiffer, A., V. Brantl, et al. (1986). "Psychotomimesis mediated by kappa opiate receptors." *Science* **233**(4765): 774–6.

Pope, H. G., Jr. and D. L. Katz (1994). "Psychiatric and medical effects of anabolic-androgenic steroid use. A controlled study of 160 athletes." *Arch Gen Psychiatry* **51**(5): 375–82.

Pope, H. G., Jr., E. M. Kouri, et al. (2000). "Effects of supraphysiologic doses of testosterone on mood and aggression in normal men: a randomized controlled trial." *Arch Gen Psychiatry* **57**(2): 133–40.

Porcerelli, J. H. and B. A. Sandler (1998). "Anabolic-androgenic steroid abuse and psychopathology." *Psychiatr Clin North Am* **21**(4): 829–33.

Reisberg, B., S. H. Ferris, et al. (1983). "Effects of naloxone in senile dementia: a double-blind trial." *N Engl J Med* **308**(12): 721–2.

Rocha, V. M., C. M. Calil, et al. (2007). "Influence of anabolic steroid on anxiety levels in sedentary male rats." *Stress* **10**(4): 326–31.

Rogol, A. D. and C. E. Yesalis, 3rd. (1992). "Clinical review 31: anabolic-androgenic steroids and athletes: what are the issues?" *J Clin Endocrinol Metab* **74**(3): 465–9.

Rossbach, U. L., P. Steensland, et al. (2007). "Nandrolone-induced hippocampal phosphorylation of NMDA receptor subunits and ERKs." *Biochem Biophys Res Commun* **357**(4): 1028–33.

Silverman, I., D. Kastuk, et al. (1999). "Testosterone levels and spatial ability in men." *Psychoneuroendocrinology* **24**(8): 813–22.

Sjoqvist, F., M. Garle, et al. (2008). "Use of doping agents, particularly anabolic steroids, in sports and society." *Lancet* **371**(9627): 1872–82.

Skarberg, K., F. Nyberg, et al. (2009). "Multisubstance use as a feature of addiction to anabolic-androgenic steroids." *Eur Addict Res* **15**(2): 99–106.

Spach, C. and D. H. Streeten (1964). "Retardation of sodium exchange in dog erythrocytes by physiological concentrations of aldosterone, in vitro." *J Clin Invest* **43**: 217–27.

Stilger, V. G. and C. E. Yesalis (1999). "Anabolic-androgenic steroid use among high school football players." *J Community Health* **24**(2): 131–45.

Su, T. P., M. Pagliaro, et al. (1993). "Neuropsychiatric effects of anabolic steroids in male normal volunteers." *JAMA* **269**(21): 2760–4.

Tahtamouni, L. H., N. H. Mustafa, et al. (2008). "Prevalence and risk factors for anabolic-androgenic steroid abuse among Jordanian collegiate students and athletes." *Eur J Public Health* **18**(6): 661–5.

Tincello, D. G., P. T. Saunders, et al. (1997). "Correlation of clinical, endocrine and molecular abnormalities with in vivo responses to high-dose testosterone in patients with partial androgen insensitivity syndrome." *Clin Endocrinol (Oxf)* **46**(4): 497–506.

Wall, P. M. and C. Messier (2000). "Concurrent modulation of anxiety and memory." *Behav Brain Res* **109**(2): 229–41.

Wanjek, B., J. Rosendahl, et al. (2007). "Doping, drugs and drug abuse among adolescents in the State of Thuringia (Germany): prevalence, knowledge and attitudes." *Int J Sports Med* **28**(4): 346–53.

Velazquez, I. and B. P. Alter (2004). "Androgens and liver tumors: Fanconi's anemia and non-Fanconi's conditions." *Am J Hematol* **77**(3): 257–67.

Wide, J. K., K. Hanratty, et al. (2004). "High level estradiol impairs and low level estradiol facilitates non-spatial working memory." *Behav Brain Res* **155**(1): 45–53.

Wittmann, W., E. Schunk, et al. (2009). "Prodynorphin-derived peptides are critical modulators of anxiety and regulate neurochemistry and corticosterone." *Neuropsychopharmacology* **34**(3): 775–85.

Wood, R. I., L. R. Johnson, et al. (2004). "Testosterone reinforcement: intravenous and intracerebroventricular self-administration in male rats and hamsters." *Psychopharmacology (Berl)* **171**(3): 298–305.

Yakovleva, T., Z. Marinova, et al. (2007). "Dysregulation of dynorphins in Alzheimer disease." *Neurobiol Aging* **28**(11): 1700–8.

Yang, P., B. L. Jones, et al. (2002). "Mechanisms of anabolic androgenic steroid modulation of alpha(1)beta(3)gamma(2L) GABA(A) receptors." *Neuropharmacology* **43**(4): 619–33.

Zhang, Y., E. R. Butelman, et al. (2004). "Effect of the endogenous kappa opioid agonist dynorphin A(1-17) on cocaine-evoked increases in striatal dopamine levels and cocaine-induced place preference in C57BL/6J mice." *Psychopharmacology (Berl)* **172**(4): 422–9.

10 Angiotensin II Receptors and Neuroprotection

Robert E. Widdop, Ulrike Muscha Steckelings,
Claudia A. McCarthy, and Jennifer K. Callaway

CONTENTS

10.1 INTRODUCTION

It is well recognized that stroke and related cerebral ischemia is one of the most common causes of mortality in the developed world, not to mention the immense burden of morbidity and disability that afflicts stroke survivors. It is also well recognized that blood pressure (BP) control is critical in reducing the incidence of stroke (MacMahon et al. 1990). On this point, drugs that inhibit the rennin–angiotensin system (RAS) are increasingly used in preventative strategies against stroke or recurrent stroke (HOPE 2000; PROGRESS 2001; Dahlof et al. 2002). However, as will be discussed, data from animal studies investigating stroke also suggest that part of the neuroprotective effect of AT1 receptor antagonists is independent of BP. This fact also points to direct central interference of the brain RAS, which was originally described over 3 decades ago (Unger et al. 1988; Steckelings, Bottari, and Unger 1992).

Therefore, in this chapter, we will review the underlying molecular mechanisms that provide the protective feature of RAS inhibition in stroke. In doing so, it is important to discuss the interplay between the two main AT receptors, namely, the

AT1 receptors (AT1R) and AT2 receptors (AT2R), and the potential of brain AT2 receptors as new therapeutic targets for neuroprotection.

10.2 CENTRAL AT1 AND AT2 RECEPTORS

All components of the RAS have been identified in the mammalian brain (Allen et al. 1998; Unger et al. 1988; Steckelings, Bottari, and Unger 1992). Evidence from various sources indicates that brain RAS is regulated independently of the peripheral RAS, which can directly influence only those cerebral regions that lack the blood–brain barrier (McKinley et al. 1990). The identification of high-affinity-specific binding sites for angiotensin II (Ang II) in rat and bovine brain suggested a physiological role for angiotensin peptides in the central nervous system, and, indeed, interaction of Ang II with cognate brain receptors was found to induce changes, including osmoregulation, control of BP, modulation of the sympathetic nervous system, drinking behavior, and natriuresis (Steckelings et al. 1992; Unger et al. 1988; De Gasparo et al. 2000).

The discovery of the Ang II receptor subtypes revealed greatly differing distribution patterns, suggesting differences in the function of the two receptor subtypes in the brain (Chiu et al. 1989; Whitebread et al. 1989). The distribution of AT1R in the subfornical organ, paraventricular nucleus, dorsal vagal complex, and area postrema was consistent with its known functions of cardiovascular control, fluid balance, and hormone secretion. AT2R, on the other hand, are expressed in substantia nigra, putamen, caudate nucleus, thalamus, and cerebellum, with the ratio of AT1 to AT2 receptors varying greatly with brain location (Steckelings et al. 1992; Allen et al. 1998). It is also now well accepted that additional ATR subtypes, such as the Mas receptor and the AT4 receptor, also known as insulin-regulated aminopeptidase (IRAP), exist and exert unique functional effects (Allen et al. 1998; Albiston et al. 2007; Santos et al. 2008), although a discussion on these aspects is beyond the scope of the current chapter.

Although the majority of the classical effects of Ang II are unequivocally attributable to AT1R stimulation, the AT2R is thought to evoke counterregulatory effects that often oppose AT1R activation, including vasodilatation, antigrowth, and proapoptotic effects (Steckelings, Kaschina, and Unger 2005; Jones et al. 2008). Although such effects are well described, there is still some controversy mainly on growth-related aspects of AT2R function (Widdop et al. 2003; Porrello, Delbridge, and Thomas 2009; Steckelings et al. 2010). Much of the interest in AT2R function arises from the fact that this is a highly plastic receptor, as the AT2R is upregulated in various pathophysiological conditions, including left ventricular hypertrophy (Matsubara 1998), nerve damage (Gallinat et al. 1998), vascular injury (Nakajima et al. 1995), and myocardial infarction (Nio et al. 1995). Similarly, there is evidence to suggest that AT2R may protect the brain from adverse events.

Increased expression of the AT2R in brain tissue has been described in stroke (Zhu et al. 2000; Li et al. 2005) and in the neurodegenerative conditions of Alzheimer's disease (Ge and Barnes 1996; Wright and Harding 2010) and Huntington's disease (Ge and Barnes 1996). These findings suggest that brain AT2R may play a similar tissue remodeling role in the brain as they do in the periphery. Indeed, the potential role of ATR in neuroprotection in the brain has been studied most extensively in the context of cerebral ischemia, and these aspects will be studied in the remainder of this chapter.

10.3 PATHOPHYSIOLOGY OF CEREBRAL ISCHEMIA

In ischemic stroke, blood flow is impaired by an occlusion of a cerebral artery, either by an embolus or local thrombosis. A reduction in blood supply impairs glucose and oxygen delivery to the compromised tissue, leading to the depletion of ATP and a loss of membrane potential. The resulting depolarization causes the activation of presynaptic and somatodendritic voltage-dependent Ca^{2+} channels. Activation of Ca^{2+} channels stimulates the excessive release of excitatory amino acids, which in conjunction with the diminution of the cell's energy supplies, initiates a complicated cascade of biochemical events, activating a number of injury-related pathways. These pathways include inflammatory pathways, oxidative free radical production, and ultimately programmed cell death or necrosis (Dirnagl, Iadecola, and Moskowitz 1999). Following stroke, two distinct regions of tissue may be defined, the core and the surrounding penumbral (or periinfarct) regions. The core region is generally considered to be the area of tissue that is irreversibly damaged even if blood flow was to be reestablished (Kidwell and Warach 2003). The penumbral region can be functionally impaired but still contains tissue that is viable for up to 12 hours before the physiological features that define the core will spread into the penumbra (Paciaroni, Caso, and Agnelli 2009). It is this potentially salvageable tissue that is the main focus of therapy.

10.4 RAS INHIBITION AND NEUROPROTECTION IN ANIMAL MODELS OF STROKE

Animal studies have concluded that inhibition of the RAS protects against neuronal injury following stroke, often in a BP-independent manner (Ito et al. 2002; Groth et al. 2003). Ito and coworkers demonstrated that pretreatment with an AT1R antagonist or an angiotensin-converting enzyme (ACE) inhibitor resulted in a greater reduction in infarct volume, following an ischemic event when compared to controls or rats pretreated with a sodium channel blocker despite these drugs having similar effects on BP (Ito et al. 2002). Likewise, Groth et al. (2003) found that chronic but not acute AT1R blockade reduced infarct volume and improved neurological outcome, following middle cerebral artery occlusion (MCAO), despite the fact that both acute and chronic treatment reduced BP. Neuroprotection, independent of BP-lowering effects, has been demonstrated by studies showing that low-dose candesartan significantly reduced infarct volume poststroke, but the same was not true for a higher dose of candesartan, causing excessive BP-lowering effect (Brdon et al. 2007). Similarly, a nonhypotensive dose of candesartan reduced superoxide production and improved neurological outcome, following global ischemia (Sugawara et al. 2005), and neuroprotection occurred when AT1R blockade was given directly into cerebral ventricles of rats as a pretreatment (Dai et al. 1999; Lou et al. 2004; Li et al. 2005), or systemically using candesartan cilexetil, at doses that block central AT1R (Nishimura et al. 2000; Ito et al. 2002; Groth et al. 2003; Lu, Zhu, and Wong 2005). In addition, AT1R knockout mice exhibited smaller infarct size, following temporary MCAO (Walther et al. 2002). Underlying mechanisms may involve the inhibition of AT1R-mediated cerebrovascular pathological growth and inflammation, as well as improved cerebrovascular compliance and autoregulation (Nishimura et al. 2000; Saavedra, Benicky, and Zhou 2006). Thus, animal research has demonstrated that AT1R blockade can provide neuroprotection; often this protection extends beyond the control of BP.

10.4.1 AT2R in Stroke

Given the opposing roles of AT1R and AT2R in many settings, such as hypertension, heart failure, and vascular function (Widdop et al. 2003; Steckelings et al. 2005; Jones et al. 2008), it may be anticipated that AT2R may exert neuroprotective effects. AT2R is reported to be upregulated in stroke (Zhu et al. 2000; Li et al. 2005; Lu et al. 2005, Mogi et al. 2006), although this may be model specific (McCarthy et al. 2009). Knockout studies have suggested that the AT2R may be neuroprotective, as AT2R-deficient mice have a larger infarct area and a more severe neurological deficit, following stroke, when compared with wild type. In addition, the neuroprotective effect of valsartan was reduced in AT2R knockout animals (Iwai et al. 2004; Mogi et al. 2006), and either peripheral (Lu et al. 2005; Faure et al. 2008) or central (Li et al. 2005) administration of an AT2R antagonist reversed the neuroprotection evoked by AT1R blockade, suggesting opposing roles for AT1R and AT2R during stroke.

It is important to note that all the aforementioned studies have indirectly touched upon AT2R function, on the basis of worsened infarcts in AT2R deficient mice or on the ability of the AT2R antagonist to reverse neuroprotective effects of AT1R blockade. As such, these are not clinically relevant scenarios. Therefore, we addressed the importance of direct AT2R activation in a model of stroke, using a focal reperfusion model in conscious spontaneously hypertensive rats. To this end, we recently reported that an intracerebroventricular infusion of the AT2R agonist, CGP42112, commencing 5 days prior to stroke, caused marked neuroprotection when assessed 3 days after stroke (i.e., reduced infarct size, improved motor function, and increased neuronal survival) (McCarthy et al. 2009).

10.5 ANGIOTENSIN RECEPTORS AND NEUROPROTECTION

In the early days of AT2R research, cell lines, which only expressed the AT2R but not the AT1R, were popular. Such an expression pattern enabled the use of the natural ligand, Ang II, as a stimulus without being at risk to elicit AT1R-mediated effects. Two of the most frequently used cell lines of this kind were the PC12W and the NF-108 cells, both being of neuronal origin. Although PC12W cells are a substrain from a clonal isolation of a rat pheochromocytoma (Greene and Tischler 1976), NG108 cells are a hybridomal cell line of a rat neuroblastoma and a mouse glioma (Klee and Nirenberg 1974). Originally used to study AT2R signaling, researchers soon realized that these cells drastically changed morphology, when the AT2R was stimulated. It turned out that AT2R-stimulation promoted differentiation of these neuronal cells, which resulted in the outgrowth of neurites (Laflamme et al. 1996; Meffert et al. 1996).

Elongation of neurites and axons is a prerequisite for regenerative processes in the nervous system. Therefore, it was just consequent to evaluate whether AT2R stimulation has any regenerative potential in the central or peripheral nervous system. For this purpose, it was first studied whether AT2R expression undergoes any alteration in the case of neuronal injury. For both injuries, in peripheral nerves and within the brain, an upregulation of AT2R expression was reported (Gallinat et al. 1998; Zhu et al. 2000; Li et al. 2005). Interestingly, in the sciatic nerve, the peak in AT2R expression shifted with time from the center of injury to more distant parts of the nerve. AT2R upregulation occurred even in the respective dorsal root ganglion neurons (Gallinat et al. 1998).

Alterations in receptor expression certainly point to a functional role of the respective receptor, but are not proof enough. Therefore, the impact of AT2R stimulation on the repair of injured nerves was investigated in models of optic and sciatic nerve crush. In these models, the AT2R promoted axonal growth and remyelination, resulting in an increased number of axons bridging the site of injury (Lucius et al. 1998; Reinecke et al. 2003). The study on sciatic nerve injury additionally involved functional tests, which revealed a significant improvement of motor and sensory function after AT2R stimulation, thus demonstrating the neuroregenerative potential of the AT2R (Reinecke et al. 2003). The neuroregenerative potential of AT2R stimulation has not only been demonstrated for peripheral or central nerves but also for the brain in experimental stroke (see section 10.4.1).

In stroke, the AT1R clearly play a detrimental role and promote tissue damage. Consequently, pharmacological blockade of the AT1R has been shown in numerous studies to improve neurological outcome and reduce infarct size in various models of experimental stroke (Fournier et al. 2004; Saavedra et al. 2006; Thöne-Reineke, Steckelings, and Unger 2006). Regarding injury of peripheral or central nerves, much less is known about any therapeutic effectiveness of AT1R blockers (ARBs). A study by Iwasaki et al. (2002) showed that the ARB olmesartan is able to promote neurite outgrowth in a similar way as in AT2R stimulation. They further showed that olmesartan reduced spinal motor neuron death after postnatal sciatic nerve section in rats *in vivo*. However, the mechanism by which olmesartan exerted these actions remained unclear: Was it by inhibition of AT1R-mediated effects of Ang II (which is rather unlikely in the *in vitro* setting) or by pleiotropic, AT1R-independent effects? Olmesartan has in fact been described to possess pleiotropic effects, such as inhibition of oxidative stress (Kurita et al. 2008), which may well be a mechanism protecting from neuronal damage. However, the mechanism underlying neurite outgrowth remains unknown.

A very recent publication by Güler et al. (2010) reported improved neurological outcome and a reduction of tissue necrosis by olmesartan in a model of ischemia-reperfusion injury of spinal cord in rats. Again, attenuation of oxidative stress by olmesartan was described as an underlying mechanism of action—leaving the question, whether the antioxidant effect of olmesartan was because of blockade of the AT1R for Ang II actions or a pleiotropic mode of action, unanswered. Two recent studies examined the effectiveness of RAS blockade in experimental autoimmune encephalomyelitis, a mouse model of multiple sclerosis (Platten et al. 2009; Stegbauer et al. 2009). Blockade of AT1R-mediated Ang II actions by the ARBs, losartan or candesartan, the ACE inhibitors, enalapril or lisinopril, or the renin-inhibitor, aliskiren, led to the amelioration of neurological deficits accompanied by strong antiinflammatory effects, such as reduction of CD11b + or CD11c + antigen presenting cells, inhibition of synthesis of various chemokines (CCL2, CCL3, and CXCL10), or induction of CD4 + FoxP3 + regulatory T cells (Platten et al.. 2009; Stegbauer et al. 2009). The authors further show a significant activation of the local RAS within the inflamed spinal cord, making it most likely that the favorable therapeutic effects of AT1R blockade (and the other modes of RAS inhibition) were indeed due to the prevention of AT1R-mediated effects of Ang II (Stegbauer et al. 2009). Indirect stimulation of the AT2R (Widdop et al. 2003) may, of course, have contributed to the favorable outcome of the ARB-treated animals as well.

10.6 POTENTIAL MECHANISMS OF ACTION OF NEUROPROTECTION BY INTERFERENCE WITH THE RAS

The mechanism of action of AT1R blockade for neuroprotection is quite well understood as far as the inhibition of Ang II-induced effects is concerned. Ang II via its AT1R promotes tissue damage in neuronal injury, by enhancing inflammation, oxidative stress, and apoptosis (Thöne-Reineke et al. 2006). In this context, the AT1R is known to be coupled with classical proinflammatory signaling mechanisms, such as the JAK/STAT pathway or activation of NF-κB (De Gasparo et al. 2000; Das 2005). Furthermore, it was recently described to promote inflammation by increasing the activity of CYP4A/4F, resulting in enhanced 20-HETE synthesis (Rompe et al. 2010). Regarding oxidative stress, Ang II, via the AT1R, promotes the production of reactive oxygen species by direct activation of NADPH oxidase through stimulation of Rac1 or p47phox or by mitochondrial O_2-generation (Griendling et al. 1994; Chabrashvili et al. 2003; Yin et al. 2010).

The induction of apoptosis, via the AT1R, has been described to be either direct through activation of p53 DNA-binding activity, resulting in upregulation of the proapoptotic bax gene or secondary through generation of reactive oxygen species, which is a well-known direct inducer of apoptosis (Bonnet, Cao, and Cooper 2001). ARBs act by interrupting all these signaling cascades at the receptor level. In addition, some ARBs, such as telmisartan or irbesartan, have AT1R-independent, PPARγ (peroxisomeproliferator-activated-receptor-γ)-agonistic properties (Schupp et al. 2004). Tian et al. (2009) elegantly showed that there is a competition of the transcription factors, PPARγ and NF-kB, for certain cofactors, which are essential for initiation of transcription. In the event of PPARγ activation by ARBs, the transcription factor NF-kB runs short of cofactors, resulting in less transcription of proinflammatory mediators and, thus, an antiinflammatory effect. With respect to receptor-independent pleiotropic effects, some ARBs, such as olmesartan and candesartan, were reported to have direct antioxidant activity, which would counteract the proapoptotic effect of oxidative stress (Chen et al. 2008; Kurita et al. 2008).

In addition to the effects by AT1R blockade and pleiotropic effects, ARBs may also act via indirect stimulation of the AT2R. A study by Iwai et al. (2004) addressing the outcome of stroke in AT2R-deficient mice supports this assumption by showing that the antioxidant effect of the ARB valsartan is weaker in the absence of the AT2R. Furthermore, in a study from our group, blockade of central AT2R with PD123177 abolished the neuroprotective effects of central AT1R blockade with irbesartan on infarct size and neurological outcome in stroke in rats (Li et al.2005).

Not much is known about signaling of the AT2R in neuroprotection and neuroregeneration. The antiinflammatory effect of the AT2R certainly contributes to the neuroprotective effect and results from interference with proinflammatory, kinase-driven signaling pathways through AT2R-induced phosphatase activation. Such interference has been shown for the JAK/STAT pathway and for NF-kB (Horiuchi et al. 1999; Wu et al. 2004; Rompe et al. 2010). Furthermore, the AT2R has an antiinflammatory action by increasing synthesis of epoxyeicosatrienoic acid (EET), an arachidonic acid metabolite, with antiinflammatory properties (Rompe et al. 2010). PPARγ-activation seems an antiinflammatory mechanism of action that AT2R stimulation and some ARBs have in common. Moreover, PPARγ is an important transcriptional regulator of cell differentiation. Using PC12W cells expressing AT2R, but not AT1R, our group demonstrated that Ang II induces PPARγ expression and ligand-mediated PPARγ activity via AT2R activation, which appears to be a crucial process in the AT2R-mediated neurite outgrowth contributing to neuroprotective processes (Zhao et al. 2005). Although the AT2R is reported to induce apoptosis in unchallenged cells of neuronal origin, it seems to act in an antiapoptotic manner in neuronal cells exposed to external stressors, such as lipopolysaccharide (unpublished observation) or sodium azide, which is an inducer of hypoxia (Grammatopoulos et al. 2002). In these models, AT2R stimulation inhibits the caspase-3 apoptotic pathway.

Neuroprotection by the AT2R through antioxidative mechanisms has further been shown by our group in a stroke model in rats (McCarthy et al. 2009). In line with our data, Iwai et al. (2004) showed that stroke-induced oxidative stress is increased in AT2R-deficient mice. In addition, the involvement of delayed rectifier K^+ channel as well as of the Na^+/Ca^{2+} exchanger and Na^+/K^+ ATPase are mechanisms reported to mediate AT2R-coupled neuroprotection (Grammatopoulos et al. 2004).

Methyl methanesulfonate sensitive 2 (MMS2) belongs to a family of ubiquitin-conjugating enzyme variants and forms a complex with Ubc-13. The MMS2/Ubc-13 complex has been reported to play an important role in DNA repair through an ubiquitin-proteasome system. The group by Horiuchi described the involvement of MMS2 in the neuroprotective action of the AT2R in stroke and suggested that AT2R/MMS2-mediated neuroprotection is based on improved DNA repair (Mogi et al. 2006). As defects in DNA repair are associated with neurodegenerative diseases, such as Parkinson's disease, Alzheimer's disease, Prion disease, or amyotrophic lateral sclerosis, AT2R-coupled upregulation of MMS2 may also be beneficial in these disease entities.

10.7 INSIGHT FROM CLINICAL STUDIES

The Heart Outcome Prevention Evaluation (HOPE) study was the first large-scale study that found that ACE inhibitors significantly reduced the occurrence of first or recurrent stroke in high-risk populations, when compared to placebo (Bosch et al. 2002; Malcolm et al. 2003; Chapman et al. 2004). A similar result was obtained in the Perindopril Protection against Recurrent Stroke Study (PROGRESS) with regard to the prevention of recurrent stroke, but in this study, the reduction in strokes in patients treated with an ACE inhibitor only reached significance when the ACE inhibitor was combined with a diuretic. Moreover, in PROGRESS, BP lowering was quite robust in the ACE inhibitor/diuretic group, meaning that the hemodynamic treatment effects may have primarily accounted for the better outcome.

Regarding ARBs, the Losartan Intervention For Endpoint (LIFE) reduction in hypertension study found that in patients with isolated systolic hypertension without previous stroke, AT1R antagonist, losartan, provided a significantly greater reduction in stroke occurrence, when compared to the b-blocker, atenolol (Dahlof et al. 2002). With respect to the prevention of recurrent stroke, the Morbidity and Mortality after Stroke (MOSES; Schrader, Luders, and Kulshewski 2005) study reported superior protection in patients receiving a sartan compared to patients receiving a calcium antagonist. Interestingly, in HOPE, LIFE, and MOSES, the ACE inhibitor and AT1R antagonists provided greater protection against stroke, despite the fact that there was a similar effect on BP in treatment and control/competitor groups. Thus, clinical trials indicate that drugs that modulate the RAS offer benefits that extend beyond the control of BP.

In conjunction with the evidence that chronic inhibition of the angiotensin system can reduce the risk of stroke, the evaluation of Acute Candesartan Cilexetil thErapy in Stroke Survivors (ACCESS) study demonstrated that AT1R antagonists may also protect against stroke in an acute setting (Schrader et al. 2003). The ACCESS study found that when patients received candesartan immediately after a stroke and for the following 12 months, there was a significant reduction in cumulative 12-month mortality compared to patients who received placebo in the first 7 days poststroke. However, it must be noted that 7 days poststroke, patients in the placebo group received candesartan to treat hypertension, thus the treatment in the two study groups only differed for the first 7 days of the trial. Despite this small difference, the advantage in the candesartan group was obvious from 3 months poststroke and persisted over the subsequent 12 months follow-up period (Schrader et al. 2003; Thone-Reineke et al. 2006).

Although the evidence to support targeting the RAS in the treatment of stroke is unanimous, there is some conjecture as to whether an ACE inhibitor or an AT1R antagonist, if either, is superior. Although both classes of drugs inhibit the RAS and decrease activation of the predominant AT1R, ACE inhibitors and AT1R antagonists have opposing actions on the AT2R. The AT1R negatively regulates renin release; therefore, during AT1R blockade, circulating levels of Ang II will increase (Tea et al. 2000). An increase in the circulating levels of Ang II, in combination with the blunting of AT1R-mediated effects, results in increased AT2R stimulation during treatment with AT1R antagonists. ACE inhibitors on the other hand, reduce the conversion of angiotensin I to Ang II and, therefore, will cause a reduction in circulating levels of Ang II and subsequent AT2R stimulation.

There is also evidence from clinical studies supporting this assumption. A meta-analysis of 26 randomized clinical trials directly comparing Ang II-increasing drugs (diuretics, dihydropyridines, and ARBs) with Ang II-decreasing drugs (ACE inhibitiors and betablockers) with regard to stroke prevention revealed that Ang II-decreasing drugs are less stroke protective than Ang II-increasing drugs (Boutitie et al. 2007). Furthermore, this difference in protection provided by the two drug groups cannot entirely be explained by differing effects on BP (Boutitie et al. 2007). Thus, it is still unclear from clinical evidence as to whether blockade of the RAS using an AT1R antagonist or an ACE inhibitor provides the best protection against stroke.

Telmisartan, ramipril, or both, in Patients at High Risk for Vascular Events (ONTARGET) was the first large-scale clinical trial to directly compare an AT1R antagonist (telmisartan) with an ACE inhibitor (ramipril) in the same study (ONTARGET Investigators 2008). ONTARGET was the main of three related studies, the others being the Telmisartan Randomized Assessment Study in ACE-intolerant Subjects with Cardiovascular Disease (TRANSCEND) study, which tested telmisartan against placebo in ACE-intolerant patients excluded from ONTARGET (TRANSCEND Investigators et al. 2008), and the Prevention Regimen for Effectively Avoiding Second Strokes (PRoFESS) trial, which tested the potential benefits of telmisartan in reducing the risk of stroke in patients with a history of stroke against placebo, and two antiplatelets in a 2×2 factorial design (Diener et al. 2008).

For ONTARGET, the authors reported that telmisartan was noninferior to ACE inhibitor ramipril at reducing fatal or nonfatal cardiovascular events, with both treatment groups not achieving a statistically significant reduction in occurrence of the primary endpoint. However, there was a trend toward reducing recurrent stroke (but no other endpoints) by telmisartan versus ramipril (Diener 2009). Prevention of stroke was also present but nonsignificant in TRANSCEND and PRoFESS. However, in a combined analysis of TRANSCEND and PRoFESS, the incidence of the composite endpoint of stroke, myocardial infarction, or vascular death was significantly reduced by AT1R blockade (TRANSCEND Investigators et al. 2008). A significant effect of telmisartan treatment in reducing the number of strokes was also obtained in a *post hoc* analysis of PRoFESS, which did not include the first 6 months of treatment. When judging these data, it has to be considered that, thanks to improved background treatment, especially with statins and β-blockers, the event rate in the placebo group of TRANSCEND (which had a design similar to HOPE) was at the same level as in the treatment group in HOPE 10 years earlier. As current trials such as ONTARGET, TRANSCEND, and PRoFESS require the drug of interest to be tested against standard background treatment, the benefit of a single drug added to the background therapy of well-managed patients can generally be expected to be only moderate at best.

10.8 CONCLUSIONS

It is clear from the preceding discussions that drugs inhibiting the RAS, such as ACE inhibitors and ARBs, are neuroprotective in various animal models and provide benefit in human stroke. These effects are largely because of blockade of the central AT1R. However, much evidence was also presented that identified AT2R as a novel target in neuroprotection and neuroregeneration, particularly in animal models of stroke. Although AT2R activation may contribute to the effects seen with ARBs, it is also likely that direct central AT2R stimulation is neuroprotective. In addition, there are other ATR subtypes (Mas, AT4/IRAP) that are largely unexplored in neuroprotection, although recent data implicate both AT2R and AT4R in the neuro-protective effect of candesartan in experimental stroke (Faure et al. 2008). Thus, the involvement of various Ang II receptors in neuroprotection is still an evolving story.

ACKNOWLEDGMENT

This work (REW) was supported in part by the CASS Foundation.

REFERENCES

Albiston AL, Peck GR, Yeatman HR, Fernando R, Ye S, Chai SY. 2007. Therapeutic targeting of insulin-regulated aminopeptidase: heads and tails? *Pharmacol Ther* 116: 417–427.

Allen AM, Moeller I, Jenkins TA, Zhou J, Aldred GP, Chai SY, Mendelsohn FAO. 1998. Angiotensin receptors in the nervous system. *Brain Res Bull* 47: 17–28.

Bonnet F, Cao Z, Cooper ME. 2001. Apoptosis and angiotensin II: yet another renal regulatory system? *Exp Nephrol* 9: 295–300.

Bosch J, Yusuf S, Pogue J, Sleight P, Lonn E, Rangoonwala B, Davies R, Ostergren J, Probstfield J, on behalf of the HOPE Investigators. 2002. Use of ramipril in preventing stroke: double blind randomised trial. *BMJ* 324: 1–5.

Boutitie F, Oprisiu R, Achard JM, Mazouz H, Wang J, Messerli FH, Gueyffier F, Fournier A. 2007. Does a change in angiotensin II formation caused by antihypertensive drugs affect the risk of stroke? A meta-analysis of trials according to treatment with potentially different effects on angiotensin II. *J Hypertens* 25: 1543–1553.

Brdon J, Kaiser S, Hagemann F, Zhao Y, Culman J, Gohlke P. 2007. Comparison between early and delayed systemic treatment with candesartan of rats after ischaemic stroke. *J Hypertens* 25: 187–196.

Chabrashvili T, Kitiyakara C, Blau J, Karber A, Aslam S, Welch WJ, Wilcox CS. 2003. Effects of ANG II type 1 and 2 receptors on oxidative stress, renal NADPH oxidase, and SOD expression. *Am J Physiol Regul Integr Comp Physiol* 285: R117–R124.

Chapman N, Huxley R, Anderson C, Bousser G, Chalmers J, Colman S, Davis S, Donnan G, MacMahon S, Neal B, Warlow C, Woodward M. 2004. Effects of a perindopril-based blood pressure-lowering regimen on the risk of recurrent stroke according to stroke sub-type and medical history. *Stroke* 35: 116–121.

Chen S, Ge Y, Si J, Rifai A, Dworkin LD, Gong R. 2008. Candesartan suppresses chronic renal inflammation by a novel antioxidant action independent of AT1R blockade. *Kidney Int* 74: 1128–1138.

Chiu AT, Herblin WF, McCall DE, Ardecky RJ, Carini DJ. 1989. Identification of angiotensin II receptor subtypes. *Biochem Biophys Res Commun* 165: 196–203.

Dahlof B, Devereux RB, Kjeldsen SE, Julius S, Beevers G, de Faire U, Fyhrquist F, Ibsen H, Kristiansson K, Lederballe-Pedersen O, Lindholm LH, Nieminen MS, Omvik P, Oparil S, Wedel H. 2002. Cardiovascular morbidity and mortality in the Losartan Intervention For Endpoint reduction in hypertension study (LIFE): a randomised trial against atenolol. *Lancet* 359: 995–1003.

Dai WJ, Funk A, Herdegen T, Unger T, Culman J. 1999. Blockade of central angiotensin AT(1) receptors improves neurological outcome and reduces expression of AP-1 transcription factors after focal brain ischemia in rats. *Stroke* 30: 2391–2398.

Das UN. 2005. Is angiotensin II an endogenous pro-inflammatory molecule? *Med Sci Monit* 11:RA155–162.

De Gasparo M, Catt KJ, Inagami T, Wright JW, Unger T. 2000. International Union of Pharmacology. XXIII. The angiotensin II receptors. *Pharmacol Rev* 52: 415–472.

Diener HC. 2009. Preventing stroke: the PRoFESS, ONTARGET, and TRANSCEND trial programs. *J Hypertens Suppl* 27: S31–36.

Diener HC, Sacco RL, Yusuf S, Cotton D, Ôunpuu S, Lawton WA, Palesch Y, Martin RH, Albers GW, Bath P, Bornstein N, Chan BPL, Chen S-T, Cunha L, Dahlöf B, De Keyser J, Donnan GA, Estol C, Gorelick P, Gu V, Hermansson K, Hilbrich L, Kaste M, Lu C, Machnig T, Pais P, Roberts R, Skvortsova V, Teal P, Toni D, VanderMaelen C, Voigt T, Weber M, Yoon B-W, for the Prevention Regimen for Effectively Avoiding Second Strokes (PRoFESS) study group. 2008. Effects of aspirin plus extended-release dipyridamole versus clopidogrel and telmisartan on disability and cognitive function after recurrent stroke in patients with ischaemic stroke in the Prevention Regimen for Effectively Avoiding Second Strokes (PRoFESS) trial: a double-blind, active and placebo-controlled study. *Lancet Neurol* 7: 875–884.

Dirnagl U, Iadecola C, Moskowitz MA. 1999. Pathobiology of ischaemic stroke: an integrated view. *Trends Neurosci* 22: 391–397.

Faure S, Bureau A, Oudart N, Javellaud J, Fournier A, Achard JM. 2008. Protective effect of candesartan in experimental ischemic stroke in the rat mediated by AT(2) and AT(4) receptors. *J Hypertens* 26: 2008–2015.

Fournier A, Messerli FH, Achard JM, Fernandez L. 2004. Cerebroprotection mediated by angiotensin II: a hypothesis supported by recent randomized clinical trials. *J Am Coll Cardiol* 43: 1343–1347.

Gallinat S, Yu M, Dorst A, Unger T, Herdegen T. 1998. Sciatic nerve transection evokes lasting up-regulation of angiotensin AT2 and AT1 receptor mRNA in adult rat dorsal root ganglia and sciatic nerves. *Mol Brain Res* 57: 111–122.

Ge J, Barnes NM. 1996. Alterations in angiotensin AT1 and AT2 receptor subtype levels in brain regions from patients with neurodegenerative disorders. *Eur J Pharmacol* 297: 299–306.

Grammatopoulos T, Morris K, Ferguson P, Weyhenmeyer J. 2002. Angiotensin protects cortical neurons from hypoxic-induced apoptosis via the angiotensin type 2 receptor. *Brain Res Mol Brain Res* 99: 114–124.

Grammatopoulos TN, Johnson V, Moore SA, Andres R, Weyhenmeyer JA. 2004. Angiotensin type 2 receptor neuroprotection against chemical hypoxia is dependent on the delayed rectifier K^+ channel, Na^+/Ca^{2+} exchanger and Na^+/K^+ ATPase in primary cortical cultures. *Neurosci Res* 50: 299–306.

Greene LA, Tischler AS. 1976. Establishment of a noradrenergic clonal line of rat adrenal pheochromocytoma cells which respond to nerve growth factor. *Proc Natl Acad Sci U S A* 73: 2424–2428.

Griendling KK, Minieri CA, Ollerenshaw JD, Alexander RW . 1994. Angiotensin II stimulates NADH and NADPH oxidase activity in cultured vascular smooth muscle cells. *Circ Res* 74: 1141–1148.

Groth W, Blume A, Gohlke P, Unger, T, Culman J. 2003. Chronic pretreatment with candesartan improves recovery from focal cerebral ischemia in rats. *J Hypertens* 21: 2175–2182.

Güler A, Sahin MA, Ucak A, Onan B, Inan K, Oztaş E, Arslan S, Uysal B, Demirkiliç U, Tatar H. 2010. Protective effects of angiotensin II type-1 receptor blockade with olmesartan on spinal cord ischemia-reperfusion injury: an experimental study on rats. *Ann Vasc Surg* 24: 801–808.

HOPE. 2000. Effects of an angiotensin-converting-enzyme inhibitor, ramipril, on cardiovascular events in high-risk patients. *N Engl J Med* 342: 145–153.

Horiuchi M, Hayashida W, Akishita M, Tamura K, Daviet L, Lehtonen JY Dzau, VJ. 1999. Stimulation of different subtypes of angiotensin II receptors, AT1 and AT2 receptors, regulates STAT activation by negative crosstalk. *Circ Res* 84: 876–882.

Ito T, Yamakawa H, Bregonzio C, Terron JA, Neri AF, Saavedra JM. 2002. Protection against ischemia and improvement of cerebral blood flow in genetically hypertensive rats by chronic pretreatment with an angiotensin II AT(1) antagonist. *Stroke* 33: 2297–2303.

Iwai M, Liu HW, Chen R, Ide A, Okamoto S, Hata R, Sakanaka M, Shiuchi T, Horiuchi M. 2002. Possible inhibition of focal cerebral ischemia by angiotensin II type 2 receptor stimulation. *Circulation* 110: 843–848.

Iwasaki Y, Ichikawa Y, Igarashi O, Kinoshita M, Ikeda K. 2002. Trophic effect of olmesartan, a novel AT1R antagonist, on spinal motor neurons in vitro and in vivo. *Neurol Res* 24: 468–472.

Jones ES, Vinh A, McCarthy CA, Gaspari TA, Widdop RE. 2008. AT2 receptors: functional relevance in cardiovascular disease. *Pharmacol Ther* 120: 292–316.

Kidwell CS, Warach S. 2003. Acute ischemic cerebrovascular syndrome: diagnostic criteria. *Stroke* 34: 2995–2998.

Klee WA, Nirenberg M. 1974. A neuroblastoma times glioma hybrid cell line with morphine receptors. *Proc Natl Acad Sci U S A* 71: 3474–3477.

Kurita S, Takamura T, Ota T, Matsuzawa-Nagata N, Kita Y, Uno M, Nabemoto S, Ishikura K, Misu H, Ando H, Zen Y, Nakanuma Y, Kaneko S. 2008. Olmesartan ameliorates a dietary rat model of non-alcoholic steatohepatitis through its pleiotropic effects. *Eur J Pharmacol* 588: 316–324.

Laflamme L, Gasparo M, Gallo JM, Payet MD, Gallo-Payet N. 1996. Angiotensin II induction of neurite outgrwoth by AT2 receptors in NG108-15 cells. Effect counteracted by the AT1 receptors. *J Biol Chem* 271: 22729–22735.

Li J, Culman J, Hortnagl H, Zhao Y, Gerova N, Timm M, Blume A, Zimmermann M, Seidel K, Dirnagl U, Unger T. 2005. Angiotensin AT2 receptor protects against cerebral ischemia-induced neuronal injury. *FASEB J* 19: 617–619.

Lou M, Blume A, Zhao Y, Golhlke P, Deuschl G, Herdegen T, Culman J. 2004. Sustained blockade of brain AT1 receptors before and after focal cerebral ischemia alleviates neurological deficits and reduces neuronal injury, apoptosis and inflammatory responses in the rat. *J cereb Blood Flow Metab* 24: 536–547.

Lu Q, Zhu Y-Z, Wong PT-H. 2005. Neuroprotective effects of candesartan against cerebral ischemia in spontaneously hypertensive rats. *Neuroreport* 16: 1963–1967.

Lucius R, Gallinat S, Rosenstiel P, Herdegen T, Sievers J, Unger T. 1998. The angiotensin II 2 (AT2) receptor promotes axonal regeneration in the optic nerve of adult rats. *J Exp Med* 188: 661–670.

MacMahon S, Peto R, Collins R, Godwin J, MacMahon S, Cutler J, Sorlie P, Abbott R, Collins R, Neaton J, Abbott R, Dyer A, Stamler J. 1990. Blood pressure, stroke, and coronary heart disease : part 1, prolonged differences in blood pressure: prospective observational studies corrected for the regression dilution bias. *The Lancet* 335: 765–774.

Malcolm J, Arnold O, Yusuf S, Young J, Mathew J, Johnston D, Avezum A, Lonn E, Pogue J, Bosch J. 2003. Prevention of heart failure in patients in the heart outcomes prevention evaluation (HOPE) study. *Circulation* 107: 1284–1290.

Matsubara, H. 1998. Pathophysiological role of angiotensin II type 2 receptor in cardiovascular and renal diseases. *Circ Res* 83: 1182–1191.

McCarthy CA, Vinh A, Callaway JK, Widdop RE. 2009. Angiotensin AT2 receptor stimulation causes neuroprotection in a conscious rat model of stroke. *Stroke* 40: 1482–1489.

McKinley MJ, McAllen RM, Mendelsohn FAO, Allen AM, Chai SY, Oldfield BJ. 1990. Circumventricular organs: neuroendocrine interfaces between the brain and the hemal milieu. *Front Neuroendocrinol* 11: 91–127.

Meffert S, Stoll M, Steckelings UM, Bottari SP, Unger T. 1996. The angiotensin II AT2 receptor inhibits proliferation and promotes differentiation in PC12W cells. *Mol Cell Endocrinol* 122: 59–67.

Mogi M, Li J-M, Iwanami J, Min L-J, Tsukuda K, Iwai M, Horiuchi M. 2006. Angiotensin II type-2 receptor stimulation prevents neural damage by transcriptional activation of methyl methanesulfonate sensitive 2. *Hypertension* 48: 141–148.

Nakajima M, Hutchinson HG, Fujinaga M, Hayashida W, Morishita R, Zhang L. 1995. The angiotensin II type 2 (AT2) receptor antagonizes the growth effects of the AT1 receptor: gain-of-function study using gene transfer. *Proc Natl Acad Sci U S A* 92: 10663–10661.

Nio Y, Matsubara H, Murasawa S, Kanasaki M, Inada M. 1995. Regulation of gene transcription of angiotensin II receptor subtypes in myocardial infarction. *J Clin Invest* 95: 46–54.

Nishimura Y, Ito T, Hoe KL, Saavedra JM: 2000. Chronic peripheral administration of the angiotensin II AT(1) receptor antagonist Candesartan blocks brain AT(1) receptors. *Brain Res* 871: 29–38.

ONTARGET Investigators. 2008. Telmisartan, ramipril, or both in patients at high risk for vascular events. *N Engl J Med* 358: 1547–1559.

Paciaroni M, Caso V, Agnelli G. 2009. The concept of ischemic penumbra in acute stroke and therapeutic opportunities. *Eur Neurol* 61: 321–330.

Platten M, Youssef S, Hur EM, Ho PP, Han MH, Lanz TV, Phillips LK, Goldstein MJ, Bhat R, Raine CS, Sobel RA, Steinman L. 2009. Blocking angiotensin-converting enzyme induces potent regulatory T cells and modulates TH1- and TH17-mediated autoimmunity. *Proc Natl Acad Sci U S A* 106: 14948–14953.

Porrello ER, Delbridge LM, Thomas WG. 2009. The angiotensin II type 2 (AT2) receptor: an enigmatic seven transmembrane receptor. *Front Biosci* 1(14): 958–972.

PROGRESS 2001. Randomised trial of a perindopril-based blood-pressure-lowering regimen among 6105 individuals with previous stroke or transient ischaemic attack. *The Lancet* 358: 1033–1041.

Reinecke K, Lucius R, Reinecke A, Rickert U, Herdegen T, Unger T. 2003. Angiotensin II accelerates functional recovery in the rat sciatic nerve in vivo: role of the AT2 receptor and the transcription factor NF-kappaB. *FASEB J* 17: 2094–2096.

Rompe F, Artuc M, Hallberg A, Alterman M, Ströder K, Thöne-Reineke C, Reichenbach A, Schacherl J, Dahlöf B, Bader M, Alenina N, Schwaninger M, Zuberbier T, Funke-Kaiser H, Schmidt C, Schunck WH, Unger T, Steckelings UM. 2010. Direct angiotensin II type 2 receptor stimulation acts anti-inflammatory through epoxyeicosatrienoic acid and inhibition of nuclear factor {kappa}B. *Hypertension* 55: 924–931.

Saavedra JM, Benicky J, Zhou J. 2006. Mechanisms of the anti-ischemic effect of angiotensin II AT(1) receptor antagonists in the brain. *Cell Mol Neurobiol* 26: 1099–1111.

Santos RA, Ferreira AJ, Simões E, Silva AC. 2008. Recent advances in the angiotensin converting enzyme 2-angiotensin(1-7)-Mas axis. *Exp Physiol* 93: 519–527.

Schrader J, Luders S, Kulschewski A, Berger J, Zidek W, Treib J, Einhaupl K, Diener H, Dominiak P. 2003. The ACCESS study. Evaluation of acute candesartan cilexetil therapy in stroke survivors. *Stroke* 34: 1699–1703.

Schrader J, Luders S, Kulshewski A. 2005. Morbidity and mortality after stroke, eprosartan compared with nitrendipine for secondary prevention: principal results of a prospective randomized controlled study (MOSES). *Stroke* 36: 1218–1226.

Schupp M, Janke J, Clasen R, Unger T, Kintscher U. 2004. Angiotensin type 1 receptor blockers induce peroxisome proliferator-activated receptor-gamma activity. *Circulation* 4(109): 2054–2057.

Steckelings UM, Bottari SP, Unger T. 1992. Angiotensin receptor subtypes in the brain. *Trends Pharmacol Sci* 13: 365–368.

Steckelings UM, Kaschina E, Unger T. 2005. The AT2 receptor—a matter of love and hate. *Peptides* 26: 1401–1409.

Steckelings UM, Widdop RE, Paulis L, Unger T. 2010. The angiotensin AT2 receptor in left ventricular hypertrophy. *J Hypertens* 28 (suppl 1): S50–S55.

Stegbauer J, Lee DH, Seubert S, Ellrichmann G, Manzel A, Kvakan H, Muller DN, Gaupp S, Rump LC, Gold R, Linker RA. 2009. Role of the renin-angiotensin system in autoimmune inflammation of the central nervous system. *Proc Natl Acad Sci U S A* 106: 14942–14947.

Sugawara T, Kinouchi H, Oda M, Shoji H, Omae T, Mizoi K. 2005. Candesartan reduces superoxide production after global cerebral ischemia. *Neuroreport* 16: 325–328.

Tea B, Sarkissian D, Thian T, Hamet P, deBlois D. 2000. Proapoptotic and growth-inhibitory role of angiotensin ii type 2 receptor in vascular smooth muscle cells of spontaneously hypertensive rats in vivo. *Hypertension* 35: 1069–1073.

Telmisartan Randomised Assessment Study in ACE iNtolerant subjects with cardiovascular Disease (TRANSCEND) Investigators, Yusuf S, Teo K, Anderson C, Pogue J, Dyal L, Copland I, Schumacher H, Dagenais G, Sleight P. 2008. Effects of the angiotensin-receptor blocker telmisartan on cardiovascular events in high-risk patients intolerant to angiotensin converting enzyme inhibitors: a randomised controlled trial. *Lancet* 27(372): 1174–1183.

Tian Q, Miyazaki R, Ichiki T, Imayama I, Inanaga K, Ohtsubo H, Yano K, Takeda K, Sunagawa K. 2009. Inhibition of tumor necrosis factor-alpha-induced interleukin-6 expression by telmisartan through cross-talk of peroxisome proliferator-activated receptor-gamma with nuclear factor kappaB and CCAAT/enhancer-binding protein-beta. *Hypertension* 53: 798–804.

Thöne-Reineke C, Steckelings UM, Unger T. 2006. Angiotensin receptor blockers and cerebral protectionin stroke. *J Hypertens* 24 (suppl 1): S115–S121.

Unger T, Badoer E, Ganten D, Lang RE, Rettig R. 1988. Brain angiotensin: pathways and pharmacology. *Circulation* 77 (suppl 1): I40–I54.

Walther T, Olah L, Harms C, Maul B, Bader M, Hortnagl H, Schultheiss HP, Mies G. 2002. Ischemic injury in experimental stroke depends on angiotensin II. *FASEB J* 16: 169–176.

Whitebread S, Mele M, Kamber B, de Gasparo M. 1989. Preliminary biochemical characterization of two angiotensin II receptor subtypes. *Biochem Biophys Res Commun* 163: 284–291.

Widdop R, Jones E, Hannan R, Gaspari T. 2003. Angiotensin AT2 receptors: cardiovascular hope or hype? *Br J Pharmacol* 140: 809–824.

Wu L., Iwai M., Li Z, Shiuchi T, Min LJ, Cui TX, Li JM, Okumura M, Nahmias C, Horiuchi M. 2004. Regulation of inhibitory protein-kappaB and monocyte chemoattractant protein-1 by angiotensin II type 2 receptor activated Src homology protein tyrosine phosphatase-1 in fetal vascular smooth muscle cells. *Mol Endocrinol* 18: 666–678.

Wright JW, Harding JW. 2010. The brain RAS and Alzheimer's disease. *Exp Neurol* 223: 326–333.

Zhao Y, Foryst-ludwig A, Bruemmer D, Culman J, Bader M, Unger T, Kintscher U. 2005. Angiotensin II induces peroxisome proliferator-activated receptor gamma in PC12W cells via angiotensin type 2 receptor activation. *J Neurochem* 94: 1395–1401.

Yin J-X, Yang R-F, Li WS, Renshaw AO, Li Y-L, Schultz HD, Zimmerman MC. 2010. Mitochondria-produced superoxide mediates angiotensin II-induced inhibition of neuronal potassium current. *Am J Physiol Cell Physiol* 298: C857–C865.

Zhu YZ, Chimon GN, Zhu YC, Lu Q, Li B, Hu HZ, Yap EH, Lee HS, Wong PT. 2000. Expression of angiotensin II AT2 receptor in the acute phase of stroke in rats. *Neuroreport* 11: 1191–1194.

11 PACAP and Cellular Protection

Cellular Protection by Members of the VIP/Secretin Family with an Emphasis on Pituitary Adenylate Cyclase–Activating Polypeptide (PACAP)

Lee E. Eiden

CONTENTS

11.1 INTRODUCTION

Neuropeptides have been implicated in both the mitigation and exacerbation of cellular injury from ischemia, physical trauma, inflammation, and excitotoxicity and other physiological insults found in a wide variety of neurodegenerative diseases. One such dual-purpose neuropeptide is substance P, shown to exacerbate neuronal cell death in hippocampus after seizure [1] and yet to be neuroprotective in striatum *in vivo* [2, 2a] and in cultured cerebellar [3] and striatal [4] neurons. Of the opioid peptides, *dynorphin* is neurodamaging *in vivo* (Hauser this volume), whereas others can, with pharmacological preconditioning, mediate neuroprotection. The expression of two other neuropeptides, neuropeptide Y (NPY) and adrenomedullin (AM), are greatly increased after neurotrauma. NPY, which appears to help regulate the migration of neuroprogenitor cells into the hippocampus, is highly upregulated following hippocampus damage in rodent models of epilepsy [5, 6]. AM, which is dramatically upregulated after striatal ischemia and reperfusion, appears to be neuroprotective after central nervous system (CNS) injury [7].

Martin and coworkers have pointed out that, in neurodegenerative disease, most G-protein-coupled receptors that mediate neuroprotection are in the class II/class B receptor family and are liganded primarily by neuropeptides in the secretin subfamily [8]. Accordingly, this chapter focuses on pituitary adenylyl cyclase-activating peptide (PACAP), a prominent member of the secretin subfamily, whose role in neuroprotection was last systematically reviewed 5 years ago [9]. A summary of current progress with PACAP and related peptides reveals what we know, and what we still need to know, to develop usable neuropeptide-based drugs with neuroprotective effects.

Insults such as excitotoxicity, inflammation, trauma, and ischemia can cause cellular damage and precipitate secondary cascades of damage. Here, we consider how—following chemical, ischemic, and inflammatory insult—pharmacological or endogenous PACAP acts to mitigate cellular damage in the heart, kidney, and brain.

Progressive and acute-onset insult offer many opportunities for intervention, and it is clear that having the ability to mitigate cell damage with such interventions would lead to far better outcomes. Although we are yet to define common signal transduction pathways for such cytoprotection, an investigation of PACAP would reveal much about the cellular mechanisms by which neuropeptides and other endogenous molecules protect neurons from damage.

11.2   MODELING INJURIES REQUIRING CYTOPROTECTION

In devising therapeutic models of neuropeptide cytoprotection, it is first necessary to examine if and how such protection occurs *in vivo,* whether it is ever prophylactic (preventative) or occurs only after the fact (conditioning), and whether endogenous protective effects can be mimicked pharmacologically. It is equally necessary to determine the kinds of injuries that might benefit from pharmacological protection, whether "protection" ever interferes with endogenous healing, and which cell types can and should be protected from loss after injury to minimize degeneration without impeding regeneration. All of these considerations, therefore, need to be kept in mind when devising models to assess endogenous mechanisms of protection, testing the validity of these models, and translating findings to determine the best use of neuropeptide-based interventions in cases of human injury and disease.

11.2.1   CELL TYPE

A first consideration in assessing neuropeptides' neuroprotective effects is the type of cells damaged or killed in injury or disease. Neurons are generally considered to be the least resilient cell type and therefore the most vulnerable to CNS injury. Yet, while few drug development efforts target astrocyte, microglial, or endothelial cells following CNS insult, these cells are known to be important mediators of neuronal damage in early stages of neurotrauma and produce potentially beneficial products during healing. Their preservation following ischemic and traumatic events or neurodegenerative disease can therefore be expected to affect neuronal survival and regeneration, with significant implications for clinical outcome and quality of life.

Insofar as different cell types vary in their requirements and pathways for energy use, they are also differentially vulnerable to different insults. Some protective mechanisms are therefore also likely to be cell-type specific.

During periods of ischemic or hypoglycemic insult, for example, astrocytes are thought to conserve adenosine triphosphate (ATP) by taking advantage of enhanced uptake of glutamate released from neurons to generate lactate from glucose and divert it to neurons (through transport to the extracellular space and uptake via the MCT2 transporter) as a replacement for glucose in energy generation [10]. Protection of this so-called "lactate shuttle" may represent a point of intervention in which endogenous protective mechanisms can be modulated or enhanced to therapeutic effect.

Excitatory and oxidative stresses clearly affect types of neurons differentially. Exposure to kainic acid leads to selective loss of hippocampal neurons [11]; amyotrophic lateral sclerosis (ALS) affects predominantly motor neurons; the neurotoxin MPTP kills mainly nigrostriatal dopaminergic neurons; Alzheimer's disease (AD) results in selective loss of cholinergic cortical projections. In cases of disease, cytoprotection may therefore involve protection from cytokines, ischemia, excitotoxins (quinolinic acid; glutamate), reperfusion [reactive oxygen species (ROS) and NO synthase (NOS)], hypoglycemia (diabetic neuropathy). In addition, toxicants may differ in tempo, concentration, and combinations at different times after onset and depending on whether the insult is self-limiting, continuous, or chronic.

Can reductionist models fairly be expected to capture such complex processes? Models used in cell culture and *in vivo* to date include: oxygen–glucose deprivation, hydrogen peroxide treatment, staurosporine-induced apoptosis, glutamate and kainate excitotoxicity, serum withdrawal, ultraviolet (UV) treatment, and Aβ treatment. When assessing data gathered using cell culture models, *in vitro* experiments, and *in vivo* animal models, it is important to bear in mind the extent to which findings based on each model can be applied to human disease, and the extent to which any single treatment can be expected to result in significant improvement or salvage.

11.2.2 ACUTE VERSUS CHRONIC INJURY

In cytoprotection, it is also necessary to determine whether the insult is progressive or acute. Chemotherapeutic toxicity, for instance, is usually a discrete acutely injurious event, whereas at the other end of the injury spectrum, neurodegenerative diseases (such as Alzheimer's, Parkinson's, and Huntington's) result in neurological damage that is chronic. An interesting intermediate case is stroke, which involves both an acute-onset event and unique molecular responses, tempered by preexisting vulnerabilities in metabolic reserve and vascular resilience, in the many distinct phases of injury and recovery that occur within minutes, hours, days, weeks, and months after injury [12]. In progressive disease, where neurodegeneration increases over time, the effects become less and less reversible. Even primary traumatic brain injury in otherwise healthy young individuals will be followed by phases of secondary damage and regeneration, each uniquely susceptible to modulation by pharmacological agents. For these reasons, outcomes may benefit from the administration of therapeutic interventions in a stage-dependent way [13].

11.2.3 SINGLE AND COMBINATORIAL COMPONENTS OF BRAIN INSULT AFTER INJURY

Over the past few decades, research on neurodegeneration, ischemic-hypoxic injury, and brain trauma has identified a host of cellular stressors that contribute to the cell death thought to underlie neuropathological deficits in motor skills and cognition. These include ROS and NOS, a range of different cytokines (TNFα and IL-1β), glutamate, quinolinic acid, prostaglandins, matrix metalloproteinases, excess calcium, and amyloid peptides, such as Aβ41/43. However, to mitigate the results of adverse cellular events, these actors may need to be blocked either sequentially or in parallel. It should be noted that no single drug is likely to be a "magic bullet" for neuroprotection [14] and that determining a therapeutic window of opportunity will require identification of mediation factors, cell types affected, and stage of disease.

11.2.4 CELL PROTECTOR OR ATTACKER? KNOCKOUTS HIGHLIGHT NEUROPEPTIDES' DUAL ROLE

In cases where cells are destined to be replaced, neuropeptides that promote apoptosis, clearing unwanted cells out, may speed regeneration. By contrast, antiapoptotic neuropeptides, which are beneficial in "salvage" phases of neurodegeneration following traumatic brain injury or stroke, may be unproductive in the regenerative and healing postinjury phases.

Mice lacking expression of certain neuropeptides or their cognate receptors are valuable models for the devising and testing of potential therapies. Knockout (KO) mouse models have therefore been created that are deficient in neurokinin 1, AM, calcitonin gene-related peptide (CGRP), or NPY. In addition, KO mouse models have been made lacking vasoactive intestinal polypeptide (VIP) and PACAP, their cognate receptors, or the receptor-associated modulatory proteins (RAMPs) that support specificity of receptor function for AM and CGRP.

In many cases, particularly where a single peptide uses multiple receptors, phenocopying of receptor and peptide KOs has provided information needed for the development of drug-like molecules for specific applications and without specific side-effects. However, given that peptides are generally widely distributed and play many different roles, KO models must also be temporally and regionally specific if their beneficial effects are not to be masked or otherwise compensated for. Furthermore, the pharmacological effects of neuropeptides are unlikely to be fully predictable based on negative effects of peptide or receptor deficiency. The chief use of neuropeptide-deficient animal models may therefore be, as with PACAP, to show the total *in vivo* biology of peptide activity and metabolism as a guide to the level of specificity required in a peptidomimetic drug. In addition, KO animal models are essential for uncovering off-target effects and designing proof-of-concept tests in drug development.

11.2.5 CELL DEATH BY NECROSIS AND APOPTOSIS

The journey from injury to cell death can follow many different pathways. Cell death from necrosis, for example, can involve calcium overload and subsequent ion transporter overactivity, ATP depletion, ROS, and cellular destruction [12]. However, at several points in this pathway, necrosis can cross over to apoptotic cell death— usually when mitochondrial stress leads to caspase activation and "programmed" DNA scission results in complete loss of cellular function. The question then arises as to whether cells protected from apoptosis will simply die more slowly by other means. Leker and Shohami [15] have suggested that specific blockade targeted to penultimate steps in cell death may simply redirect the process to the other (necrotic or apoptotic) pathway.

11.3 NEUROPROTECTANT VIP

The first member of the secretin superfamily to be implicated in neuroprotection was VIP. VIP was initially found to promote the survival of cultured neurons by elevating extracellular potassium, which mimics the stimulation of electrical activity [16]. Brenneman and coworkers established that VIP produces these neurotrophic effects indirectly [17] by causing astrocytes to release a factor later called activity-dependent neurotrophic factor (ADNF) [18]. In this scenario, endogenous VIP released from neurons during enhanced electrical activity interacts with astrocyte receptors to allow increased expression and release of ADNF. NAP, an active peptide fragment of ADNF, has since been shown to have a variety of neuroprotective effects, including blockade of glutamate excitotoxicity, Aβ toxicity, and HIV gp120 cytotoxicity, in neuronal cultures and blockade of cell loss *in vivo* after various neurological insults (including gp120 administration) [19].

Peptide histidine isoleucine (PHI), a second member of the secretin super-family released on processing of the VIP/PHI prohormone, also appears to be neuroprotective. Evidence for this assumption comes mainly from the activation of a class of VIP-binding, guanosine triphosphate-independent and non-cAMP-generating receptors expressed in early embryonic development. However, the existence and neuroprotective role of these VIP/PHI receptors is still controversial (see review by Dejda and others [9]) and awaits experimental resolution.

PACAP—discovered by Miyata and coworkers in 1989 [20], almost 20 years after VIP [21]—rapidly surpassed both VIP and PHI in scope of physiological and pharmacological neuroprotective actions [22]. (PACAP-38 is the predominant form of the peptide, although PACAP-27 and -38 are bioequivalent in most assays.)

Both VIP and PACAP, which have similar sequences (VIP-28 and PACAP-27 have 18 amino acid residues in common, see Table 11.1) operate with predictably similar bioactivity. However, the two peptides are not identical physiologically. VIP interacts with the receptors VPAC1 and VPAC2, whereas PACAP interacts with PAC1, VPAC1, and VPAC2. They also differ in sites of production and release.

TABLE 11.1
Neuroprotective Neuropeptides and Their Cognate G-Protein-Coupled Receptors

Peptide	Sequence*	Receptor
PACAP-38	HSDGIFTDSYS**R**Y**RK**QMAV**KK**YLAAVLG**KRY** **KQRVKNK**-amide	PAC1, VPAC1, VPAC2
VIP	HSDAVFTDNYT**R**L**RK**QMAV**KK**YLNSILN- amide	VPAC1, VPAC2
PHM/PHI	HADGVFTSD<u>F</u>S<u>K</u>LLGQ<u>L</u>SA**KK**YLESL<u>M</u>-amide	?
Substance P	**R**P**K**PQQFFGLM-amide	NK-1
Adrenomedullin	GC**R**FGTCTV<u>V</u>Q**K**LAHQIYQFTD**K**D**K**D<u>N</u>VA P**R**N**K**ISPQGY-amide	CLRL+RAMP2/3
α-CGRP	A<u>C</u>DTATCVTH**R**LAGLLS**R**SGGVV**K**<u>N</u>NFV PTNVGS<u>K</u>AF-amide	CLRL+RAMP1
Opioid peptides	YGGFM, YGGFL, YGGFM**R**F-amide, YGGFL**RR**I**R**P**K**L**K**	μ, κ, δ

*Sequences of human with rodent nonconcordance underlined, and basic residues in bold.

I, ischemia; I/R, ischemia-reperfusion; TBI, traumatic brain injury; O, others, including gp120 toxic-ity, excitotoxicity, survival in development, and pharmacological preconditioning. Does not include cytoprotection of nonneuronal tissue, although some have that activity.

Where activity is documented by only a few examples in the primary literature, reference is supplied.

Though this discussion focuses on PACAP, it is important to keep in mind that its activity largely overlaps with that of VIP as a pharmacological agent. For this reason, therapeutic development will be guided by phenocopying of PACAP and PAC1 KO mice and the ability to generate compounds truly selective for the three types of VIP/PACAP G-protein-coupled receptors.

11.4 PACAP PROTECTION IN NEURONAL SYSTEMS

Systematic consideration of PACAP's cytoprotective actions on a tissue-by-tissue basis allows us to build a picture of how PACAP, as a potential drug, will affect the whole organism. *In vivo* studies give us a further appreciation of the range of cellular actions involved in producing PACAP's cytoprotective effects. It is important to note that not all effects registered in cell culture systems as "neuroprotective" will be so in any given *in vivo* therapeutic situation. For example, effects on neurite extension and cell survival may be limited to certain developmental windows and therefore not applicable to adult tissue. Yet, even effects confined to specific stages of development in model systems may shed useful light on general cellular properties found in adult animals and humans. It is also possible that the regenerative phase of injury episodes (such as traumatic brain injury, stroke, and Alzheimer's disease) may in part recapitulate development. Careful attention must therefore be paid to each feature of PACAP action, whether measured in a particular cell culture model or by *in vivo* rodent neuroprotection assay, so that results can later be applied in different therapeutic contexts.

11.4.1 RETINA

The retina is a collection of highly accessible CNS neurons. As a number of diseases involve the progressive degeneration of retinal neurons and subsequent loss of vision, the retina is a logical starting point for the consideration of PACAP neuroprotection as a potential therapy.

As a result of hypoperfusion of the retina in type II diabetes, retinal degeneration can occur in combination with dysangiogenesis. Loss of vision presumably occurs as a result of obstructive vascular proliferation and ischemic death of ganglion cells and other retinal neurons that organize and convey visual information. Glutamatergic excitotoxicity leading to ganglion cell death is also postulated as a mechanism of retinopathy in glaucoma, although this view has recently been challenged [23].

PACAP is present in a subpopulation of retinal ganglion cells that projects to the suprachiasmatic nucleus (SCN) of the hypothalamus. There it serves the highly specific purpose of enabling circadian phase advance in response to light stimulation during the late night hours [24, 25]. Although the physiological function of PACAP in the retina thus appears to be primarily retinofugal, PAC1 receptors are nevertheless present in the retina. PACAP is also present in retinal amacrine cells.

As evidenced by exacerbation of retinal damage after application of the PAC1 antagonist PACAP(6-38), both endogenous and pharmacologically applied (exogenous) PACAP exert neuroprotective effects in retina [26]. PACAP-27 and -38 were shown to decrease cell death induced by 1 mM glutamate in cultured retinal neurons at doses from 10–100 nM [27]. Subsequently, based on the expectation that

PACAP might be neuroprotective in anoxia given its exceptionally high concentrations (compared with mammals) in the brains and retinas of diving turtles resistant to long periods of anoxia [28–32], Reglodi and coworkers identified PACAP-38's ability to spare degraded horizontal cell function in anoxic turtle retinal slices [28, 32]. Direct injection of 100 pmol of PACAP-38 into the vitreous humor was further shown to be protective against subcutaneous monosodium glutamate treatment, which causes extensive retinal damage over several days [33]. Further experiments have shown PACAP treatment to protect the retina in rodents against a range of insults (diabetic, excitotoxic, UV-light, and optic nerve transection and ischemia following carotid artery blockade) and cultured retinal neurons (reviewed in [26]) against apoptotic cell death induced by hypoxia, glutamate, N-methyl-D-aspartate (NMDA), endoplasmic reticulum (ER) stress, and chemicals. Moreover, PACAP neuroprotection appears to extend to all types of retinal neuronal cells [34] and to reverse functional, as well as cytological, excitotoxic damage to the retina [35].

11.4.2 CEREBELLUM

PACAP's role in neuronal development and survival and its implications for neuroprotection after toxic insult are clearly apparent in the cerebellum, particularly in the neurons of its inner granule layer (IGL). IGL granule cells account for the majority of neurons in the brain. They function by receiving inputs (signals) from other brain regions through glutamatergic (excitatory) mossy fiber projections, which are processed by excitatory glutamatergic synaptic convergence on Purkinje cells.

Purkinje cells control neurons deep in the cerebellar nuclei that represent processed output from cerebellum to control muscle movement, and learning associated with movement coordination and conditioning. When cultured from 8-day-old rats, the survival of granule cells—in keeping with their great number and principal excitatory input from precerebellar nuclei—is almost completely dependent on activity. The development of activity-dependent survival over time corresponds with the *in vivo* development of synaptic connections and requires calcium influx. In this way, Purkinje cells mimic the postsynaptic effects of glutamate, the natural *in vivo* excitatory transmitter of mossy fiber inputs to granule cells [36]. For these reasons, culturing activity-dependent cerebellar granule cells is now a standard technique for assessing the effects of putative neurotrophins on neuronal development; of putative developmental neurotoxins on neuronal survival; and of putative neuroprotectants on neurogenesis, neuronal differentiation, and neuronal survival.

Research interest in PACAP's neurotrophic and neuroprotective effects on cerebellar granule cells derives primarily from the observation that in rat and human cerebellar neurons, PACAP-38 levels are much higher than those of other neuropeptides [29, 31] and that rat and human cerebellum contain a high number of PACAP receptors [37, 38]. Importantly, PACAP receptor binding is high not only in adult cerebellum but also in the germinal matrix of the cerebellum during neurogenesis [37]. This implies a developmental role for PACAP, confirmed in cell culture by the observation that in the absence of elevated potassium or after its withdrawal, PACAP enhances granule cell survival and neuritogenesis [39–42]. This was further confirmed *in vivo* by elegant neonatal rat experiments in which microinjection of PACAP into the subarachnoid space was shown to increase the number of granule cells and thickness of the IGL [43].

Cerebellar granule cells cultured in complete medium require elevated potassium for survival. In this state, the cells are considered mature but susceptible—through apoptotic mechanisms—to oxidative (hydrogen peroxide) [44], endogenous lipotoxin (ceramide) [45], and neurotoxic [46] insult. Protection against these insults has been shown to be afforded by the administration of PACAP-38 (and -27) at nanomolar concentrations.

Endogenous PACAP is also implicated in the neuroprotection of granule cells. Tabuchi and coworkers report that a novel splice variant of the PACAP gene, induced by activity (25 mM KCl), seems to be involved in activity-dependent neuroprotection. Thus, inhibition of voltage-gated calcium channel function blocks both neuronal survival and PACAP induction. In addition, the PACAP antagonist PACAP (6-39) blocks granule cell survival elicited by the application of 25 mM KCl and also blocks rescue by PACAP itself [47, 48].

Endogenous PACAP released from cerebellar neurons in culture has also been shown to protect these cells from ethanol or hydrogen peroxide. By the same token, after exposure to ethanol or hydrogen peroxide, cerebellar granule cells cultured from PACAP-deficient donor mice are more susceptible to cell death than are wild-type neurons, which contain relatively high levels of PACAP [49].

PACAP is important not only in enhancing the survival of granule cells but also in their initial generation during development. Cameron and coworkers report that migration of prospective granule cells to the IGL during development depends on PACAP [50]. In fact, by postnatal day 7, PACAP-deficient mice show an abnormal decrease in IGL thickness [51], which appears to be completely corrected in all cortical layers by postembryonic day 11 [49]. The brain's ability to correct the transient deviation from normal cerebellar development induced by the absence of PACAP suggests a complex set of nested compensatory controls to ensure that the brain arrives at the correct number and configuration of granule cells, which are the majority of neurons in the mammalian brain.

PACAP's role in compensating for abnormal cell number must also be considered in the context of the unregulated cell growth that leads to cancer. For instance, when removal of a single neurotrophin or growth factor during development from neuroblast to mature neuronal cell leads to abnormal cell growth, PACAP functions as an important protector of the cerebellum. During development, PACAP signaling (through protein kinase A, PKA) appears to hold in check sonic hedgehog signaling (through the smoothened–patched receptor-activated pathway) that controls granule cell proliferation and—in cases of compromised negative feedback of sonic hedgehog signaling—leads to medulloblastoma [52].

In mice lacking the patched receptor that acts as a brake on sonic hedgehog signaling in humans, the incidence of medulloblastoma is low. However, that incidence is greatly increased in patched receptor-deficient plus PACAP-deficient mice, which lack a second critical "emergency brake" on uncontrolled proliferation [53, 54]. Findings such as these show that, in designing PACAP-related therapeutics whose primary focus is neuroprotection, researchers must also take into account the neuropeptide's many widespread and regionally and temporally specific actions.

In addition, although rarely discussed, the question of whether so-called neuro-protectants identified using cerebellar granule cell cultures are relevant to both adult and neonatal neuroprotection *in vivo* needs to be considered when contemplating therapeutic applications for neuroprotective compounds identified using such screens [38].

11.4.3 HIPPOCAMPUS AND CORTEX

Stroke, neurodegenerative diseases (such as Alzheimer's), major traumatic brain injury, and minor traumatic brain injury (due to blast exposure or physical concussion) frequently result in neuronal cell death in diencephalon and telencephalon. In stroke ischemia, the interruption of blood flow to the brain frequently results in neuronal damage. Where transient cerebral ischemia is followed by blood reperfusion, the inflow of ROS, cytokines, calcium, and extracellular ions can cause additional secondary damage, exacerbating the initial ischemic injury [55, 56].

To simulate toxicity in assays of neuroprotective agents against this type of neuronal damage, cell culture systems are treated with glutamate or NMDA; deprived of glucose, oxygen, serum, or trophic substances; or subjected to chemical stressors, (such as chemokines, cytokines, and generators of oxidative or ER stress. In many cases, cells cultured from KO mice allow testing for the involvement of specific gene products hypothesized to be involved in cell death and survival. Assays such as these are used to test for the many different components contributing to cell death after insult and for the enzymatic activity and gene products mobilized either to effect cell death after injury or to minimize damage to bystander cells.

Using four-vessel transient occlusion in rats, Uchida and coworkers found that PACAP-38 infused into the brain prevents the death of CA1 hippocampal neurons [57]. Moreover, PACAP, whether administered before or after ischemia onset or given intracerebro-ventricularly (i.c.v.) or systemically, was beneficial. Issues of timing and method of administration are crucial, as neither pretreatment nor i.c.v. administration are viable therapeutic options for human stroke.

After immunostaining studies of PACAP receptors in astrocytes, Uchida and coworkers hypothesized that PACAP's neuroprotective actions involve astrocytes. The fact that PAC1 receptors on astrocytes are coupled to cAMP signaling [58] does, indeed, suggest a mechanism by which PACAP might indirectly protect cells in the hippocampus.

Shioda and coworkers further suggest that PACAP protection against hippocampal neuronal apoptosis following an ischemic episode requires the presence of IL-6, which may be supplied *in vivo* by PACAP-stimulated astrocytes [59]. In later experiments, Reglodi and coworkers showed that PACAP infused i.c.v. even 4 hours after onset of transient middle cerebral artery occlusion (MCAO) in rats decreased the size of cortical infarct as measured 48 hours after MCAO. These results confirm that PACAP neuroprotection of the hippocampus after an ischemic episode also extends to the cerebral cortex [60].

The first reports indicating neuroprotective effects of PACAP-38 in cultured neurons were published in 1996 by Morio and others using cortical neurons [61]. These results were later confirmed by Shintani and others [62] and Frechilla and others [63], who used slightly different concentrations of PACAP and methods of cortical neuron culture. In these cases, it was assumed that astrocyte contamination of cortical cultures was minimal and that PACAP was therefore likely to affect the neurons directly. In fact, when Frechilla and others compared PACAP neuroprotective effects in mixed neuronal-glial cultures and neuron-enriched cultures, they found no differences in PACAP's protective effects against glutamate excitotoxicity [63].

Insofar as PACAP neuroprotection in culture can be mimicked by dibutyryl cAMP, it is likely to involve activation of adenylate cyclase through Gs [61]. However, the effect of PACAP may not be direct, as PACAP upregulated brain-derived neurotrophic factor (BDNF) mRNA levels in cortical cultures treated with glutamate, and PACAP's neuroprotective effects could be blocked by co-administration of a BDNF-neutralizing antibody [63].

Several lines of evidence suggest that PACAP's neuroprotective action in cultured neurons and *in vivo* is not only pharmacological but also reflects neuroprotective action by endogenous PACAP. These results come primarily from studies in PACAP-deficient mice subjected to permanent MCAO, in which PACAP-deficient mice show greater infarct size and cognitive impairment, both of which are ameliorated when PACAP-38 is administered 1 hour after MCAO [64, 65].

In an independent study, Ohtaki and others demonstrated protective effects of PACAP and deleterious effects of PACAP deficiency in both transient and permanent MCAO [66]. They further reported that PACAP's neuroprotective effects against MCAO were abrogated in IL6-deficient mice, consistent with Shioda and coworkers' earlier suggestion that in hippocampus, IL-6 may mediate PACAP's neuroprotective action [67]. Shintani and others have also reported that, while exogenous PACAP enhances BDNF mRNA induction after glutamate administration in cultured cortical neurons, treatment with the PAC1 antagonist PACAP(6-38) alone decreases BDNF mRNA levels in glutmate-treated cultures. This suggests that cortical neurons possess some level of autoprotection against excitotoxicity, and that it is mediated by endogenous PACAP.

Although most reports of PACAP neuroprotection in telecephalon focus on ischemic brain injury, PACAP-38 has also been reported as beneficial in reversing, or even preventing, certain deleterious effects of physical trauma to the brain. PACAP-38 administration can, for instance, ameliorate axonal damage caused by impact acceleration [68]. Wu and coworkers also report decreased PACAP and BDNF expression in human AD cortical tissue and in the cortex of the familial-associated mutant mouse model for AD [69]. This observation is particularly intriguing, as PACAP-38 is protective against diffuse axonal injury (DAI) in which amyloid precursor protein transport is impeded, suggesting a link between PACAP's action in DAI and in amyloid protein dysregulation and conversion to $A\beta(1-42)$ in AD.

As described earlier, PACAP's actions in telecephalic neuroprotection appear to be a complex mix of mechanisms that affect neurons directly, as well as indirectly through astrocyte activation. It is therefore not surprising that studies report beneficial effects of PACAP at widely varying concentrations: at less than 1 nM but not above 10 nM in one study [61] and at up to 100 nM in another [63]. Deleterious as well as protective effects have also been reported when PACAP is administered after MCAO, both at high (100 nM) and low (1–5 nM) concentrations of PACAP-27 or -38 [70].

Yet a third mechanism of PACAP neuroprotective action appears to be promotion of hippocampal neurogenesis in the adult mouse [71] by enhancing the survival of hippocampal progenitor cells mobilized in response to injury. This type of neuroprotection takes place in the regenerative phase following brain injury indicating that different mechanisms of PACAP neuroprotection take place at different times. As each mechanism is mediated by different concentrations of the neuropeptide, its time course is critically important in the application of exogenous PACAP as a treatment for brain injury.

11.4.4 Dopaminergic Neurons

The substantia nigra, in which the progressive loss of dopaminergic neurons leads to Parkinson's disease, has the brain's highest concentration of PACAP-38 [29]. Reglodi and coworkers recently reported that PACAP-38 given prior to unilateral 6-hydroxydopamine (6-OHDA) treatment in the rat blocked 50–95% of loss of dopaminergic neurons and also of the hypokinesia associated with 6-OHDA-induced hypodopaminergic innervation of the striatum. PACAP is further reported to help reverse dopaminergic neuronal apoptosis after administration of the Parkinson's-inducing drug, MPTP [72].

Intriguingly, the mechanism of PACAP's action here appears to be, at least in part, reversal of MPP+-induced decreases in eIF4E and other components critical for new protein synthesis in dopaminergic neurons. This observation is consistent with PACAP's reported upregulation of eIF4E mRNA production in PC12 and primary chromaffin cells, both of which are closely related to dopaminergic cells of the substantia nigra [73].

11.4.5 Damage to Axons, Spinal Cord, Peripheral Neurons

As mentioned in section 11.4.3, PACAP treatment ameliorates resulting axonal damage in the impact acceleration model of DAI [68, 74]. Spinal cord damage is also reportedly responsive to PACAP-38 in the context of co-treatment with human mesenchymal stem cells, emphasizing PACAP's indirect actions in regeneration after injury [75].

Waschek and coworkers have examined in detail the effects of PACAP following facial nerve crush as a well-defined model for axonal regeneration in which the contributions of the area of the primary injury, and the response of the soma of the injured cell, can be individually examined across the temporal course of nerve repair and axonal regeneration. After facial nerve axotomy, both VIP and PACAP are strongly induced in the facial motor nucleus. The induction of PACAP mRNA in particular requires a lymphocytic response, which significantly does not occur in severe combined immunodeficiency disease (SCID) mice [76]. CD4-positive helper cell conversion to the Th2 phenotype appears to be most important in the response, which is required for nerve regeneration [77]. The fact that nerve regeneration is impaired in PACAP-deficient mice [78] indicates a role for endogenous PACAP in nerve regeneration. These results further confirm that pharmacological PACAP's effects in axonal regeneration after injury are worthy of further investigation.

11.5   PACAP PROTECTION IN NONNEURONAL SYSTEMS

PACAP cytoprotection of nonneuronal cells is an area of investigation that has mushroomed in recent years. This is in part due to the realization that PACAP is distributed ubiquitously in the autonomic nerves of virtually all target organs. In addition, we now know that under certain conditions, such as ischemia, in which "vegetative" signaling is favored, it is likely that PACAP is released from parasympathetic nerves. Moreover, it now appears that several of the pleiotropic cellular actions elucidated for PACAP-induced neuroprotection (see section 11.6) likely apply to the protection of all cells under stress from toxins or lack of oxygen, regardless of cell type.

11.5.1   HEART

In cases of ischemia followed by reperfusion (I/R injury), temporary cessation of coronary blood flow to the heart results in myocardial infarction (heart attack). Lack of myocardial oxygenation causes a state of hibernation in which work performed by the affected myocardial cells is decreased; this also decreases their vulnerability to further anoxic damage and death. However, the performance of the heart muscle is also decreased, enhancing vulnerability to further myocardial damage [79]. Paradoxically, upon reperfusion following resolution of embolic blockade of blood flow, the reintroduction of oxygen and calcium ions into affected tissue can result in still more injury, as hypoperfused tissue is more vulnerable to the formation of ROS and calcium overload [55, 56].

Gasz and coworkers have reported that following exposure to 1 mM hydrogen peroxide—an oxidative stress thought to mimic the generation of ROS during I/R—even nanomolar concentrations of PACAP-38 protect cardiomyocytes against apoptosis. This protection was accompanied by decreases in cleavage of caspase-3 and decreased Bad phosphorylation, indicating increased activity of the antiapoptotic proteins Bcl-2 and Bcl-xL [80]. Simulating I/R with an ischemic buffer (an acidic and lactate-rich medium lacking glucose and oxygen) followed by a neutral medium containing glucose and oxygen, this group further demonstrated that PACAP-38 can induce increased activation in the PKA and Akt signaling pathways and protect cardiomyocytes from apoptosis during perfusion [81].

To date, PACAP has not been shown to ameliorate I/R injury *in vivo*, and PACAP(6-38) has not been shown to exacerbate it. Rather, in protecting cultured myocytes from ischemic cell death, PACAP mimicks ischemic preconditioning [82], and in myocardial tissue of mice subjected to I/R, PACAP-38 levels rise [83]. These observations highlight PACAP's high potential for therapeutic applications and underscore the need for further investigations in the near future.

Chemotherapeutic agents that trigger apoptotic cell death are another major type of myocardial insult in which PACAP has been found to have cardioprotective effects. Mori and coworkers found, for instance, that in doxorubicin-induced cardiomyopathy, PACAP-deficient mice showed more doxorubicin cardiotoxicity than did wild-type mice [84]. In cardiomyocyte cultures treated with 20 nM PACAP-28 and then challenged with doxorubicin, there is a reduction in doxorubicin-induced caspase-3 activation and Bad phosphorylation—both clear cytoprotective effects [85]. Insofar as doxorubicin's tendency to induce cardiomyopathy limits its usefulness as an anticancer agent, it is important to find out whether and how PACAP might be used clinically to counteract its toxic effects.

11.5.2 KIDNEY

PACAP's cytoprotective effect in kidney disease was first reported by Li and coworkers, who associated it with the secondary activation of NF-kB in the renal tubules after light-chain uptake and protein overload [86]. Reports of PACAP's cytoprotective effects against I/R injury in brain and heart spurred new interest in its potential for protecting the kidney.

Accordingly, investigations by the Reglodi laboratory, using a renal artery clamp and reperfusion in the rat, showed that PACAP administered intravenously prior to onset of ischemia increased survival, even after 60 minutes of ischemia [87]. Although kidney cells cultured from PACAP-deficient mice were only slightly more susceptible than wild-type cells to death from cobalt chloride-induced ischemia [88], doses of 0.1–10 nM PACAP-38 made cultured kidney cells less vulnerable to oxidative stress (1–6 mM hydrogen peroxide) [89]. Further studies showed that, following a 60-minute ischemia-reperfusion, endogenous PACAP-38 expression was increased in kidney cortex for several hours [90]. In addition, this same group reported that following I/R, PACAP-deficient mice exhibited increased kidney damage [91].

Thus, the kidney, like the brain and the heart, is endowed with a source of endogenous PACAP that provides some measure of protection against ischemia. The cellular compartment that releases endogenous PACAP for cytoprotection of the kidney has not been identified. A suggested mechanism for PACAP action is a decrease in cytokine-dependent apoptosis, but compared with PACAP's demonstrated cytoprotective effects in this tissue, PACAP-38's ability to blunt cytokine production was unimpressive [92].

11.5.3 IMMUNE SYSTEM IN SEPSIS AND AUTOIMMUNITY

In cases of ischemic or oxidative challenge, excessive calcium influx, or metabolic stress, PACAP cytoprotection of nervous system, heart, and kidney tissues appears to be largely direct—that is, mediated through PACAP receptors present on the cells that exhibit PACAP-dependent protection. In cases of multiorgan system failure secondary to septic shock or rheumatoid arthritic joint inflammation, PACAP is thought to protect by deactivating endogenous toxin-producing cells, rather than by activating antiapoptotic or antioxidative signaling mechanisms in target cells (see section 11.6 for a review of the molecular mechanisms and subcellular loci of PACAP cytoprotection.)

In animal models of septic shock, multi-organ system failure secondary to prolonged inflammation is usually mimicked by perforation of the intestinal wall or by the administration of lipopolysaccharide (LPS), which profoundly elevates levels of the inflammatory cytokines TNF-α and IL-1β [93].

In vitro experiments by Delgado and coworkers were the first to implicate VIP and PACAP in the regulation of inflammation secondary to endotoxemia. This neuroprotective effect was achieved by inhibiting the production and secretion of TNF-α and IL-1β in LPS-activated peritoneal macrophages [94]. These experiments showed that in LPS-treated mice, i.p. administration of 5 nmol of either PACAP or VIP attenuated TNF-α and IL-6 production in serum and peritoneal fluid, thereby protecting mice from lethal endotoxemia [95]. Martinez and others further reported that bolus i.p. injection of VIP and PACAP at the time of LPS administration could decrease lethality of septic shock in the LPS model. The fact that the protective effects of both VIP and PACAP were reduced (but not abolished) in PAC1-deficient mice suggests that these peptides activate other receptors in addition to PAC1 receptors (see section 11.7) [96].

Rheumatic arthritis is caused by the destruction of joint tissue through chronic autoimmune inflammation, in which TNF-α is locally overproduced. In endotoxemia, this occurs systemically, along with chronic overproduction of tissue-destroying tissue metalloproteinases [97]. As VIP and PACAP suppress TNF-α production from activated macrophages in endotoxemia, it is perhaps not surprising that both peptides are reported to decrease the severity of tissue destruction from rheumatoid arthritis (RA), and, in rodents, collagen-induced arthritis [98, 99].

In endogenous PACAP protection of the kidney, in endotoxemia and in RA, the cellular and anatomical sources of PACAP release for vascular and synovial cytoprotection are not yet clear. PACAP is reported to be present in lymphocytes in thymus, bone marrow, spleen, lymph nodes, intestinal mucosa in the rat [100], and in both T and B lymphocytes isolated from lymphoid tissues (reviewed by Delgado et al. [101]). Given PACAP's presence in the autonomic nervous system and the emerging connection between immune/inflammatory suppression and cholinergic autonomic innervation, the possibility of a neural-immune connection in which PACAP is supplied by parasympathetic innervation of inflamed tissues at various sites should also be considered [102].

11.6 SIGNALING MECHANISMS AND SUBCELLULAR LOCI IN PACAP CYTOPROTECTION

Mechanisms of cell injury ultimately involve the activation of either necrotic or apoptotic cell death cascades [103]. Various types of neuronal injury trigger distinct signaling pathways and can be attenuated or blocked only by highly specific modes of cytoprotection.

In stroke, for example, though oxidative cellular stress and excitotoxicity (glutamate) lead to activation of cyclic AMP response element binding protein (CREB) [104], only oxidative stress induces the CREB-dependent protein PGC-1α in neurons [102, 103]. Signaling pathways and combinatorial activation of such pathways thus clearly play an important role in tailoring the cytoprotective response to the specific stressor and type of cell stressed. Therefore, not surprisingly, comparisons of the molecular and cellular mechanisms involved in PACAP signaling in cerebellar and hippocampal granule cells, cortical neurons, and PC12 cells reveal a diversity of pathways that are potentially activated when cell-death cascades are initiated to protect neurons.

In the absence of neuronal activity, PACAP-dependent survival of cerebellar granule cells requires the induction of fos, independent of PKC, through the cAMP/PKA pathway [105]. This pathway is distinct from that in which PACAP causes neuritogenesis in PC12 cells, in which cAMP elevation leads to extracellular regulated kinase (ERK)-dependent activation of egr-1 through a pathway independent of both PKC and PKA [106–108].

Mechanisms and pathways by which PACAP acts to block cell death also differ. Cerebellar granule cell death, induced by oxidative stress (hydrogen peroxide) occurring through a caspase-3-dependent mechanism, is blocked by inhibition of ERK activation rather than inhibition of PKA [44], as is PACAP-dependent protection against ceramide [45]. Ethanol-induced granule cell neurotoxicity is accompanied by hyperpolarization and enhanced delayed rectifier potassium channel (I_K), both of which are reversed by PACAP application, which is neuroprotective for ethanol cytotoxicity [109].

It must be noted that these mechanisms are not necessarily exclusive, as it is not yet established whether PACAP protection against ceramide and H_2O_2 is also PKA-dependent, whether PACAP-dependent enhanced survival in low-potassium medium is independent of ERK, or whether modulation of I_K is involved in cytoprotective effects of PACAP in addition to ethanol cytotoxicity.

In most reported examples, PACAP cytoprotection occurs through PAC1 activation on neuronal cells undergoing calcium-, ER-, or metabolic (ischemic) stress. But neuroprotection, especially in the brain, may also be mediated through PACAP-dependent secretion of neuroprotective factors from astrocytes or the inhibition of cytotoxic (especially cytokine) factors from inflammatory cells (such as macrophages, microglial cells, and perhaps even endothelial cells).

PACAP actions on mouse peritoneal macrophages, for example, involve cAMP/PKA-dependent inhibition of TNF-α induction via NF-kB signaling [94]. Human THP-1 cells were used to demonstrate that both VIP and PACAP inhibit NF-kB, in part through CREB-binding protein (CBP) recruitment to CREB at the expense of p65. This attenuates p65-dependent transcription of TNF-α, which requires co-activation by CBP [110]. Based on a large body of work from their own laboratories, Leceta and others suggest, though this is controversial, that both VIP and PACAP act through the PAC1 receptor, expressed on macrophages, to reduce NF-kB binding to promoters for various cytokines (including TNF-α, IL-6, IL-12) and also NOS through a PKA-dependent mechanism [111].

The following is a summary of current knowledge about the mechanism of PACAP cytoprotection:

- The mechanisms of PACAP actions on cerebellar granule cells are generally independent of p38 and c-Jun kinase (JNK) activation and include PKA-dependent survival (trophic effects) and PKA-independent caspase inhibition (neuroprotective effects).
- In hippocampus, PACAP neuroprotection against ischemia likely depends on the synergistic neuroprotective effects of its direct neuronal actions and those of IL-6 secreted from astrocytes in response to PACAP [59, 67, 112].

- PACAP neuroprotection of hippocampus against ischemia seems to involve downregulation of activated neuronal JNK and p38 and concomitant activation of neuronal ERK [59, 67, 112].
- Cortical neuroprotection against excitotoxicity also appears to be mediated in part through PACAP-stimulated BDNF production and signaling, through the tyrosine kinase receptor (TrkB), in (not yet fully characterized) pathways leading to PACAP induction of BDNF [63].

Thus, at least four modes of signaling may therefore be involved in PACAP's cytoprotective effects: generation of cAMP/PKA (to inhibit oxidative cytotoxicity in cerebellar granule cells), activation of ERK (to promote survival in low-potassium medium), reversal of I_K activation (to reverse the effects of ethanol neurotoxicity), and inhibition of p38 and JNK (to reverse hypoxia-induced cell death in hippocampal neurons). It is not yet clear whether these are all distinct pathways. In hippocampus injured by ischemia, for example, the activation of PAC1 receptors on astrocytes to generate IL-6 introduces a second level of indirect neuroprotection by PACAP [67]. Molecular elucidation of these pathways is vital if we are ever to generate pharmacological neuroprotectants with highly specific therapeutic targets (see section 11.10).

Another emerging hypothesis is that PACAP neuroprotection occurs through induction of proteins to limit damage caused by the excessive generation of intracellular calcium. These potentially neuroprotective, PACAP-induced proteins include selenoprotein T and stanniocalcin 1 (Stc1), both encoded by PACAP-responsive genes in PC12 cells [113–115] and later verified *in vivo* as PACAP targets [116, 117]. Interestingly, induction of Stc1 itself appears to be under the regulation of calcium influx, which is stimulated by PACAP signaling through the PAC1 "hop" receptor variant. This variant contains a sequence in the third intracellular loop that couples calcium influx to activation of the PAC1 receptor [118].

Enhanced transcription of the Stc1 gene may be related to PACAP's ability to protect cells against calcium overload in several different scenarios. In addition to protecting against glutamate-induced excitotoxicity, it may also protect neurons and neuroendocrine cells against potential calcium overload during prolonged periods of stress, which induces PACAP transmission through PAC1hop [117]. The possibility that PACAP can protect cells against the calcium overload caused by its own signaling is intriguing, because it suggests that pharmacological doses of PACAP may have similar cytoprotective effects during prolonged stress, which causes hippocampal and hypothalamic neuronal loss.

11.7 PACAP AND PACAP-RECEPTOR DISTRIBUTION ACROSS ORGANS AND SPECIES: IMPLICATIONS FOR CYTOPROTECTIVE PHARMACOLOGY

The distribution of PACAP and its receptors in the context of its physiological functions has been recently reviewed [119, 120]. In this section, we will consider its distribution in the context of translational opportunities for PACAP-directed drug development for cytoprotection.

In vitro studies have shown that PACAP binds to the PAC1, VPAC1, and VPAC2 receptors, whereas VIP binds to VPAC1 and VPAC2 [121]. *In vivo* studies, using KOs of both peptides and of the PAC1 and VPAC2 receptors, have significantly clarified these receptor interactions and overlaps, for example, in elucidating the respective roles of VIP, PACAP, and their receptors in the regulation of circadian function mediated by the suprachiasmatic nucleus (SCN). Results from these studies suggest a level of specificity beyond ligand–receptor interaction, with VPAC2 activation seemingly private to VIP even though pharmacologically PACAP also recognizes this receptor. Thus, the role of VIP as an internal organizer for the endogenous circadian pacemaker is phenocopied in VIP and VPAC2 KO mice [122–125]. By contrast, PACAP's clock resetting function is impaired in parallel in PACAP and PAC1 KO mice [24, 25, 126, 127]. This is true even though both peptides are released within the SCN and both receptors are present on cells within it. PACAP and PAC1 KOs also show similar cognitive impairments that have not been detected in VIP and VPAC2 KOs [128, 129].

On the other hand, both PACAP and VIP appear to affect metabolic rate and insulin levels, and both VPAC2 and PAC1 receptors appear to be important in the regulation of insulin secretion and growth rate [130]. PACAP and VIP KO mice are both smaller and lighter than wild-type mice [131]. (The KO phenotype for VPAC1, found predominantly within the vasculature [132], has not yet been fully described.)

As mentioned earlier, the neuroprotective effects of PACAP in ischemia, which are mediated through both astrocytes and neurons in hippocampus, are thought to be PAC1-dependent. If so, a PAC1-specific agonist that would activate both cell types should provide full PACAP protection in ischemia [67]. In fact, pharmacological protection by PACAP in stroke has been shown to be greater than that afforded by endogenous PACAP. Thus, treatment with exogenous PACAP-38 not only reverses the increased stroke damage seen in PACAP-deficient mice compared to wild-type mice, but it also ameliorates stroke damage in wild-type and PACAP-deficient animals [64, 65]. This result implies that PACAP's cytoprotective actions may extend beyond cells innervated by PACAPergic terminals in the brain. Moreover, it is consistent with the findings of Shioda and others that PACAP enhances release of the antiinflammatory cytokine IL-6 from brain astrocytes [59], and with Ohtaki and coworkers' finding that PACAP neuroprotection in stroke is blunted in IL6-deficient mice [66].

PACAP cytoprotection appears to hold particular promise for the treatment of RA and sepsis. Both are common and devastating immune complications in humans for which there are few effective treatments. Moreover, both RA and sepsis involve a complex interplay between PACAP and VIP, and PAC1 and VPAC2 receptors.

Delgado and coworkers, for example, describe the ameliorative actions of both VIP and PACAP in lethal endotoxic shock as involving nonadditive suppression of TNF-α and IL-6 production. This implies that VPAC1/VPAC2—but not PAC1 receptors—mediate this effect [95]. Further work in rodent models using specific VPAC1 [(K15, R16, L27) VIP (1–7)-GRF (8–27)] and VPAC2 (Ro25–1553) agonists led to the conclusion that VPAC1 receptors mediate the antiinflammatory effects of VIP and PACAP in the periphery [133].

In contrast to most other findings concerning the specificity of VIP for the VPAC1 and VPAC2 receptors and PACAP's ability to activate PAC1, VPAC1, and VPAC2, Leceta and coworkers found that, at least in human macrophage cell lines, both VIP and PACAP apparently act through the PAC1 receptor [111].

This finding suggests that the specificity of this PAC1 receptor differs from those found in virtually all other tissues studied so far. It also underscores that—in the development of receptor-specific agonists of the PAC1 receptor, whose expression entails extensive alternative splicing—it will be necessary to take many different receptor signaling properties in different cell types into account [119, 134–136]. The current lack of a VPAC1 receptor KO mouse and (with some exceptions) definitive anatomical descriptions of the distribution of PAC1, VPAC1, and VPAC2 receptors in human organs and tissues represents another serious impediment to progress in the development of PACAP-based drug therapies.

11.8 THERAPEUTIC APPLICATIONS AND PROTOTYPE COMPOUNDS

Although it is a neuropeptide with an isoelectric point of 11 and a net charge of 10 at pH 7, PACAP's specific transport system across the blood–brain barrier gives it access to the brain from peripheral sites of administration [137]. This property makes proof-of-concept experiments concerning PACAP's neuroprotective properties especially convenient to pursue in animal models of Alzheimer's disease, retinal degeneration, cytoprotection against off-target effects of cancer chemotherapeutic agents, and other types of neural injury or degeneration. At the same time, consideration of PACAP as a drug target in these diseases and conditions would be accelerated by the development of a PAC1 receptor agonist with drug-like characteristics (such as oral absorption, moderately long half-life, predictable bioavailability, some selectivity among PACAP receptor subtypes, and even selectivity for specific PAC1 variants).

The sand fly peptide maxadilan represents a "'bioequivalent" of such a drug. It is a 61 amino acid C-terminally amidated peptide that is a potent vasodilator and immunosuppressant in mammals. These properties presumably facilitate host infection by Leishmania Major, a protozoan parasite for which the sand fly is a vector [138]. Deletion of residues 24-42 generates an antagonist that is also specific for the PAC1 receptor. Consistent with its immunosuppressive effects, maxadilan treatment of mouse macrophages suppresses TNF-α and promotes IL-6 production [139].

Vaudry and coworkers have recently reported a modified peptide analog of PACAP that potently activates the PAC1 and VPAC1 receptors and has neuroprotective activity against MPP+ in cultured neuroblastoma cells [140]. This compound complements the VPAC1 receptor-specific agonist Ro 25-1392. However, a potent "drug-like" agonist for the PAC1 receptor does not yet exist, nor do specific antagonists for either PAC1 or VPAC2.

In vivo, PACAP generally acts to antagonize rather than facilitate neurode-generation, inflammation, and cell death. For this reason, PACAP agonists rather than antagonists will be the targets for PACAP-related cytoprotective drugs. Hurdles, in addition to the absence of a drug-like, receptor-specific PACAP agonist, must be overcome. Some of the "paradoxical" effects of PACAP, VIP, and maxadilan, for instance, may be explained by as yet undiscovered RAMPs and splice variants of PAC1, which are employed or even specifically induced in certain pathophysiological conditions. Even if an effective drug-like molecule or stable and bioavailable peptoid drug is developed as a cytoprotectant, emerging evidence of PACAP's role as an endogenous anxiogenic transmitter in the amygdala [141] suggests that anxiety may be an unwanted off-target effect. (For potential therapeutic applications for PACAP-related cytoprotective drugs, see Table 11.2.)

11.9 OTHER NEUROPEPTIDES AND PACAP'S ROLE IN COMBINATORIAL TREATMENT FOR CYTOPROTECTION IN PROGRESSIVE DISEASE

Injuries leading to neuronal damage in brain are best characterized as physiological and pathophysiological *processes*, rather than *events*. Stroke, for example, proceeds through an initial stage of ischemia and primary neuronal cell death (infarction), secondary neuronal death involving first excitotoxicity (caused by oxygen free radicals and glutamate), and later, within the penumbra surrounding the infarct, inflammation and cytokine generation followed by a phase of healing (with glial scarring, axonal regeneration, and neurogenesis) during recovery [12].

The efficacy of PACAP neuroprotection has mainly been examined for the period 24–48 hours after MCAO or global ischemia—the phase in which cytokines, ROS, and glutamate are likely to be the major mediators of cell damage.

Rat and mouse experiments confirm that, when administered within several hours of MCAO, PACAP effectively decreases the size of the infarct from initial ischemia and secondary neuronal death and markedly improves neurological status (primarily measured as motor performance) [65, 142]. Further study is needed to determine whether PACAP's pronounced lymphocyte-mediated effects on axonal repair (observed to date mainly in brain stem and spinal cord but likely to occur also in cortex and hippocampus) continue in the weeks following stroke [75, 77, 78].

PACAP has also been tested in combination with other treatments. In one experiment, stem cell therapy and PACAP were used together to treat spinal cord injury. In another experiment, VIP and PACAP were applied in cases of experimental allergic encephalitis and facial motor nerve damage and regeneration [78, 143]. Based on these experiments, PAC1 and VPAC2 agonist drugs offer many opportunities for use in therapeutic interventions at different times following traumatic brain injury, spinal cord injury, or stroke—first as cytoprotective agents and later, during recovery, as catalysts of neurogenesis and axonal repair.

In animal and cell culture models, other class II neuropeptides have also been shown to have neuroprotective properties. Substance P, for instance, may protect cerebellar granule cells from apoptosis by inhibiting I_k currents—a mechanism it shares with PACAP. However, as part of its mechanism for promoting neuronal survival during serum and potassium deprivation-induced cell death, substance P inhibits calpain activation rather than (as for PACAP) caspase-3 activity [3].

The fact that substance P also promotes the survival of dopaminergic neurons in culture suggests that it protects neurons that die when deprived of excitatory

TABLE 11.2
PACAP Cytoprotective Therapeutic Applications

Syndrome/ Disease	PACAP/VIP Agonist/ Antagonist	Anticipated Therapeutic Effect	Cell Target	Representative Report/Review
Septic shock	PAC1/ VPAC1 agonist	Decrease TNF, IL-6 production; survival	Macrophages/ vascular endothelium	Delgado et al., 2003
Rheumatoid arthritis	PAC1/ VPAC1 agonist	Decrease TNF, IL-6 production; improved joint function	Macrophages/ synovial fibroblasts	Delgado et al., 2003
Stroke	PAC1 agonist	Enhanced IL-6; direct neuroprotection; enhanced cognition and motor function	Cortical and hippocampal neurons	Stumm et al., 2007
Cardiac I/R injury	PAC1/ VPAC1 agonist?	Direct myocardiocyte protection (antiapoptotic); improved cardiac function	Cardiac myocytes	Roth et al., 2009
Kidney I/R injury	PAC1/ VPAC1 agonist?	Direct renal cell protection (antiapoptotic?); improved kidney function	Cells of renal tubules	Szakaly et al., 2011
Chronic stress	PAC1 agonist?	Neuroprotective gene induction	Brain	Stroth et al., 2011
Neonatal ethanol toxicity	PAC1 agonist	Granule cell protection (antiapoptotic); improved motor function	Cerebellum	Vaudry et al., 2002
Crohn's disease	VPAC1 agonist?	Antiinflammatory; improved GI function	Gut macrophages? Th2 lymphocytes?	Abad et al., 2006
Multiple sclerosis	PAC1/ VPAC1 agonist	Decrease TNF, IL-6 production; improved motor (and cognitive) function	Th2 lymphocytes?	Tan et al., 2009
Chemotherapy adjunct	PAC1 agonist	Inhibited apoptosis; improved cardiac function	Cardiac myocytes	Mori et al., 2010
Retinal degeneration	PAC1 agonist	Mitigate loss of vision	Ganglion cells, horizontal cells	Atlasz et al., 2010

Receptor targets inferred from effects of VIP versus PACAP; PAC1 versus VPAC1 KO; VIP versus PACAP antagonist or agonist pharmacological effects.
I/R injury, ischemia/reperfusion injury.

synaptic input [4]. Although substance P protects striatal neurons from quinolinic acid excitotoxicity [2], resistance of substance-P-deficient mice to kainate-induced hippocampal cell death suggests that in the hippocampus, substance P may also contribute to seizure-induced excitotoxicity [1].

During global cerebral ischemia, expression of AM, another class II GPCR neuropeptide ligand, is upregulated in the brain [7]. The activity of exogenous AM *in vivo*, whether to exacerbate or ameliorate ischemic neuronal cell death, may depend on the extent of astroglial activation and the brain region examined [144, 145]. During reperfusion following ischemia, the neuropeptide *urocortin* acting at the CRF-R2 receptor is proposed as a potent cardioprotective factor [146]. CGRP ameliorates I/R injury in liver [147].

Acting through Y1 receptors, NPY plays an important role in neurogenesis in the dentate gyrus of the hippocampus. Acting through Y2 receptors on synaptic transmission in CA3, NPY acts as an antiepileptic as well. Moreover, NPY expression *in vivo* is rather robustly increased by kainate treatment, suggesting an important role in homeostatic responses to excitotoxic stimuli [5]. Despite these findings, a neuroprotective role for NPY—either *in vivo* or in cell culture—is yet to be established. One report describes decreased neuronal apoptosis following kainate treatment of mice given NPY i.c.v. within 8 hours of kainic acid administration [6].

Remarkably, PACAP strongly induces the expression of mRNA encoding the prohormones for NPY, CGRP, and substance P in neuroendocrine cells [114], which may in part account for PACAP's neuroprotective effects.

PACAP transactivation of the neuronal insulin-like growth factor (IGF) receptor is another potential mechanism by which PACAP could exert neuroprotective effects in the CNS [148]. As IGF primarily affects regeneration and repair, this mechanism may provide a route for PACAP use in long-term treatment following neurotrauma.

In summary, PACAP's varied neuroprotective effects suggest that it has great therapeutic potential, which might be increased even further in long-term and preconditioning treatment paradigms. Although PACAP's neuroprotective effects have already been demonstrated in the initial traumatic phases of I/R injury in brain, heart, and other organs, its ability to act on the expression of other neuroprotective peptides and proteins in the brain suggest that it may also be therapeutic in prolonged brain injury recovery and chronic neurodegenerative disease treatments.

11.10 TRANSLATION TO CLINICAL PRACTICE: GAPS AND OPPORTUNITIES

PACAP's future in the clinic rests primarily on the identification of human PACAP receptors operative for cytoprotection in joint synovia, kidney tubules, cardiomyocytes, and hippocampal and cortical neurons. At the most basic level, drug therapies against injury in RA, septic shock, stroke, I/R injury in heart, chemotherapeutic cardiotoxicity, traumatic brain injury, and ethanol neurotoxicity cannot be developed until uncertainty about the involvement of PAC1 versus VPAC1 and VPAC2 in cytoprotection is resolved.

Another major impediment to therapeutic PACAP is the current failure to develop a PACAP receptor agonist with drug-like characteristics (bioavailability, stability, potency, specificity). Such a compound could be a as diseases in which neurons are lost are now seen as progressing in stages that may each have a different target for intervention, the idea of a "one-hit'" drug for neuroprotection has largely been abandoned, and pleiotropic neuroprotectants are being sought.

Far from the "one-hit" model, PACAP neuroprotection employs many different signal transduction pathways. In cancer, granule cell oxidative stress and activity-dependent survival of developing neurons, for example, PACAP signaling, uses a canonical cAMP pathway involving PKA. At the same time, other PACAP-mediated actions (regulating transcription of a wide variety of other peptides, neuroprotective proteins, and transactivators of gene transcription, including Egr1, Ier3, and Nr4a) are accomplished through a parallel but quite distinct nonclassical cAMP-signaling pathway to ERK. Continued attention to the pharmacological properties of these unique downstream components of PACAP signaling may eventually yield compounds that allow us to leap over the stubborn pharmacology of the receptors themselves to harness (pharmacologically) their uniquely selective mechanisms that affect cellular decisions to die, repair, and regenerate.

At the present time, high-throughput screening is needed to identify each component of each signal transduction pathway leading from the activation of the PAC1 receptor to neuroprotection. This is the necessary first step toward the creation of drugs for neuroprotection, which have the potential to tip the balance away from formerly inevitable outcomes in cases of neurotrauma and autoimmune and neurodegenerative diseases. Table 11.2 lists several other practical therapeutic targets and pathways for the development of drug therapies, which are imperative in the next decade of translational PACAP research.

REFERENCES

1. Liu, H., et al., Resistance to excitotoxin-induced seizures and neuronal death in mice lacking the preprotachykinin A gene. *Proc Natl Acad Sci U S A*, 1999. **96**(21): 12096–101.
2. Sanberg, P.R., et al., Substance P containing polymer implants protect against striatal excitotoxicity. *Brain Res*, 1993. **628**(1–2): 327–9.
2a. Kowall, N., et al., An in vivo model for the neurodegenerative effectsof beta amyloid and protection by substance P. *Proc Natl Acad Sci U S A*, 1991. **88**: 7247–51.
3. Amadoro, G., et al., Substance P provides neuroprotection in cerebellar granule cells through Akt and MAPK/Erk activation: evidence for the involvement of the delayed rectifier potassium current. *Neuropharmacology*, 2007. **52**(6): 1366–77.
4. Salthun-Lassalle, B., et al., Substance P, neurokinins A and B, and synthetic tachykinin peptides protect mesencephalic dopaminergic neurons in culture via an activity-dependent mechanism. *Mol Pharmacol*, 2005. **68**(5): 1214–24.
5. Sperk, G., T. Hamilton, and W.F. Colmers, Neuropeptide Y in the dentate gyrus. *Prog Brain Res*, 2007. **163**: 285–97.
6. Wu, Y.F. and S.B. Li, Neuropeptide Y expression in mouse hippocampus and its role in neuronal excitotoxicity. *Acta Pharmacol Sin*, 2005. **26**(1): 63–8.
7. Encinas, J.M., et al., Adrenomedullin over-expression in the caudate-putamen of the adult rat brain after ischaemia-reperfusion injury. *Neurosci Lett*, 2002. **329**(2): 197–200.

8. Martin, B., et al., Class II G protein-coupled receptors and their ligands in neuronal function and protection. *Neuromolecular Med*, 2005. **7**(1–2): 3–36.
9. Dejda, A., P. Sokolowska, and J.Z. Nowak, Neuroprotective potential of three neuropeptides PACAP, VIP and PHI. *Pharmacol Rep*, 2005. **57**(3): 307–20.
10. Pellerin, L., et al., Activity-dependent regulation of energy metabolism by astrocytes: an update. *Glia*, 2007. **55**(12): 1251–62.
11. McLin, J.P. and O. Steward, Comparison of seizure phenotype and neurodegeneration induced by systemic kainic acid in inbred, outbred, and hybrid mouse strains. *Eur J Neurosci*, 2006. **24**(8): 2191–202.
12. Dirnagl, U., R.P. Simon, and J.M. Hallenbeck, Ischemic tolerance and endogenous neuroprotection. *Trends Neurosci*, 2003. **26**(5): 248–54.
13. Beauchamp, K., et al., Pharmacology of traumatic brain injury: where is the "golden bullet"? *Mol Med*, 2008. **14**(11–12): 731–40.
14. Witkop, B., Paul Ehrlich and his magic bullets—revisited. *Proc Am Philos Soc*, 1999. **143**(4): 540–57.
15. Leker, R.R. and E. Shohami, Cerebral ischemia and trauma-different etiologies yet similar mechanisms: neuroprotective opportunities. *Brain Res Brain Res Rev*, 2002. **39**(1): 55–73.
16. Brenneman, D.E. and L.E. Eiden, Vasoactive intestinal peptide and electrical activity influence neuronal survival. *Proc Natl Acad Sci U S A*, 1986. **83**: 1159–62.
17. Brenneman, D.E., et al., Nonneuronal cells mediate neurotrophic action of vasoactive intestinal peptide. *J Cell Biol*, 1987. **104**(6): 1603–10.
18. Brenneman, D.E. and I. Gozes, A femtomolar-acting neuroprotective peptide. *J Clin Invest*, 1996. **97**(10): 2299–307.
19. Gozes, I. and S. Furman, Clinical endocrinology and metabolism. Potential clinical applications of vasoactive intestinal peptide: a selected update. *Best Pract Res Clin Endocrinol Metab*, 2004. **18**(4): 623–40.
20. Miyata, A., et al., Isolation of a novel 38 residue-hypothalamic polypeptide which stimulates adenylate cyclase in pituitary cells. *Biochem Biophys Res Commun*, 1989. **164**: 567–74.
21. Said, S.I. and V. Mutt, Polypeptide with broad biological activity: isolation from small intestine. *Science*, 1970. **169**(951): 1217–18.
22. Brenneman, D.E., Neuroprotection: a comparative view of vasoactive intestinal peptide and pituitary adenylate cyclase-activating polypeptide. *Peptides*, 2007. **28**(9): 1720–6.
23. Lotery, A.J., Glutamate excitotoxicity in glaucoma: truth or fiction? *Eye (Lond)*, 2005. **19**(4): 369–70.
24. Beaule, C., et al., Temporally restricted role of retinal PACAP: integration of the phase-advancing light signal to the SCN. *J Biol Rhythms*, 2009. **24**(2): 126–34.
25. Colwell, C.S., et al., Selective deficits in the circadian light response in mice lacking PACAP. *Am J Physiol Regul Integr Comp Physiol*, 2004. **287**(5): R1194–R1201.
26. Atlasz, T., et al., Pituitary adenylate cyclase activating polypeptide in the retina: focus on the retinoprotective effects. *Ann N Y Acad Sci*, 2010. **1200**: 128–39.
27. Shoge, K., et al., Attenuation by PACAP of glutamate-induced neurotoxicity in cultured retinal neurons. *Brain Res*, 1999. **839**(1): 66–73.
28. Reglodi, D., et al., Pituitary adenylate cyclase activating polypeptide is highly abundant in the nervous system of anoxia-tolerant turtle, *Pseudemys scripta* elegans. *Peptides*, 2001. **22**(6): 873–8.
29. Ghatei, M.A., et al., Distribution, molecular characterization of pituitary adenylate cyclase-activating polypeptide and its precursor encoding messenger RNA in human and rat tissues. *J Endocrinol*, 1993. **136**: 159–66.
30. Teuchner, B., et al., VIP, PACAP-38, BDNF and ADNP in NMDA-induced excitotoxicity in the rat retina. *Acta Ophthalmol*, 2011. **89**(7): 670–5.

31. Arimura, A., et al., Tissue distribution of PACAP as determined by RIA: highly abundant in the rat brain and testes. *Endocrinology*, 1991. **129**(5): 2787–9.
32. Rabl, K., et al., PACAP inhibits anoxia-induced changes in physiological responses in horizontal cells in the turtle retina. *Regul Pept*, 2002. **109**(1–3): 71–4.
33. Babai, N., et al., Search for the optimal monosodium glutamate treatment schedule to study the neuroprotective effects of PACAP in the retina. *Ann N Y Acad Sci*, 2006. **1070**: 149–55.
34. Atlasz, T., et al., Evaluation of the protective effects of PACAP with cell-specific markers in ischemia-induced retinal degeneration. *Brain Res Bull*, 2010. **81**(4–5): 497–504.
35. Varga, B., et al., PACAP improves functional outcome in excitotoxic retinal lesion: an electroretinographic study. *J Mol Neurosci*, 2011. **43**(1): 44–50.
36. Gallo, V., et al., The role of depolarization in the survival and differentiation of cerebellar granule cells in culture. *J Neurosci*, 1987. **7**(7): 2203–13.
37. Basille, M., et al., Localization and characterization of PACAP receptors in the rat cerebellum during development: evidence for a stimulatory effect of PACAP on immature cerebellar granule cells. *Neuroscience*, 1993. **57**: 329–38.
38. Basille, M., et al., Ontogeny of PACAP receptors in the human cerebellum: perspectives of therapeutic applications. *Regul Pept*, 2006. **137**(1–2): 27–33.
39. Gonzalez, B.J., et al., Pituitary adenylate cyclase-activating polypeptide promotes cell survival and neurite outgrowth in rat cerebellar neuroblasts. *Neuroscience*, 1997. **78**: 419–30.
40. Kienlen Campard, P., et al., PACAP type I receptor activation promotes cerebellar neuron survival through the cAMP/PKA signaling pathway. *DNA Cell Biol*, 1997. **16**: 323–33.
41. Cavallaro, S., et al., Pituitary adenylate cyclase activating polypeptide prevents apoptosis in cultured cerebellar granule neurons. *Mol Pharmacol*, 1996. **50**(1): 60–6.
42. Chang, J.Y., V.V. Korolev, and J.Z. Wang, Cyclic AMP and pituitary adenylate cyclase-activating polypeptide (PACAP) prevent programmed cell death of cultured rat cerebellar granule cells. *Neurosci Lett*, 1996. **206**(2–3): 181–4.
43. Vaudry, D., et al., Neurotrophic activity of pituitary adenylate cyclase-activating polypeptide on rat cerebellar cortex during development. *Proc Natl Acad Sci U S A*, 1999. **96**: 9415–20.
44. Vaudry, D., et al., PACAP protects cerebellar granule neurons against oxidative stress-induced apoptosis. *Eur J Neurosci*, 2002. **15**(9): 1451–60.
45. Vaudry, D., et al., Pituitary adenylate cyclase-activating polypeptide prevents C2-ceramide-induced apoptosis of cerebellar granule cells. *J Neurosci Res*, 2003. **72**(3): 303–16.
46. Vaudry, D., et al., Pituitary adenylate cyclase-activating polypeptide protects rat cerebellar granule neurons against ethanol-induced apoptotic cell death. *Proc Natl Acad Sci U S A*, 2002. **99**(9): 6398–403.
47. Tabuchi, A., et al., Involvement of endogenous PACAP expression in the activity-dependent survival of mouse cerebellar granule cells. *Neurosci Res*, 2001. **39**: 85–93.
48. Tabuchi, A., M. Koizumi, and M. Tsuda, Novel splice variants of PACAP gene in mouse cerebellar granule cells. *Neuroreport*, 2001. **12**(6): 1181–6.
49. Vaudry, D., et al., Endogenous PACAP acts as a stress response peptide to protect cerebellar neurons from ethanol or oxidative insult. *Peptides*, 2005. **26**(12): 2518–24.
50. Cameron, D.B., et al., Cerebellar cortical-layer-specific control of neuronal migration by pituitary adenylate cyclase-activating polypeptide. *Neuroscience*, 2007. **146**(2): 697–712.
51. Allais, A., et al., Altered cerebellar development in mice lacking pituitary adenylate cyclase-activating polypeptide. *Eur J Neurosci*, 2007. **25**(9): 2604–18.
52. Nicot, A., et al., Pituitary adenylate cyclase-activating polypeptide and sonic hedgehog interact to control cerebellar granule precursor cell proliferation. *J Neurosci*, 2002. **22**(21): 9244–54.

53. Waschek, J.A., et al., Hedgehog signaling: new targets for GPCRs coupled to cAMP and protein kinase A. *Ann N Y Acad Sci*, 2006. **1070**: 120–8.
54. Lelievre, V., et al., Disruption of the PACAP gene promotes medulloblastoma in ptc1 mutant mice. *Dev Biol*, 2008. **313**(1): 359–70.
55. Dorweiler, B., et al., Ischemia-reperfusion injury. Pathophysiology and clinical implications. *Eur J Trauma Emerg Surg*, 2007. **33**: 600–12.
56. Hess, M.L. and N.H. Manson, Molecular oxygen: friend and foe. The role of the oxygen free radical system in the calcium paradox, the oxygen paradox and ischemia/reperfusion injury. *J Mol Cell Cardiol*, 1984. **16**(11): 969–85.
57. Uchida, D., et al., Prevention of ischemia-induced death of hippocampal neurons by pituitary adenylate cyclase activating polypeptide. *Brain Res*, 1996. **736**(1–2): 280–6.
58. Grimaldi, M. and S. Cavallaro, Functional and molecular diversity of PACAP/VIP receptors in cortical neurons and type I astrocytes. *Eur J Neurosci*, 1999. **11**(8): 2767–72.
59. Shioda, S., et al., PACAP protects hippocampal neurons against apoptosis: involvement of JNK/SAPK signaling pathway. *Ann N Y Acad Sci*, 1998. **865**: 111–17.
60. Reglödi, D., et al., Delayed systemic administration of PACAP38 is neuroprotective in transient middle cerebral artery occlusion in the rat. *Stroke*, 2000. **31**: 1411–17.
61. Morio, H., et al., Pituitary adenylate cyclase-activating polypeptide protects rat-cultured cortical neurons from glutamate-induced cytotoxicity. *Brain Res*, 1996. **741**(1–2): 82–8.
62. Shintani, N., et al., Neuroprotective action of endogenous PACAP in cultured rat cortical neurons. *Regul Pept*, 2005. **126**(1–2): 123–8.
63. Frechilla, D., et al., BDNF mediates the neuroprotective effect of PACAP-38 on rat cortical neurons. *Neuroreport*, 2001. **12**(5): 919–23.
64. Chen, Y., et al., Expression profiling of cerebrocortical transcripts during middle cerebral artery occlusion and treatment with pituitary adenylate cyclase-activating polypeptide (PACAP) in the mouse, in *Pharmacology of Cerebral Ischemia*, J. Krieglstein and S. Klumpp, Editors. Stuttgart, Germany: Medpharm Scientific Publishers; 2004:267–77.
65. Chen, Y., et al., Neuroprotection by endogenous and exogenous PACAP following stroke. *Regul Pept*, 2006. **137**(1–2): 4–19.
66. Ohtaki, H., et al., Pituitary adenylate cyclase-activating polypeptide (PACAP) decreases ischemic neuronal cell death in association with IL-6. *Proc Natl Acad Sci U S A*, 2006. **103**(19): 7488–93.
67. Shioda, S., et al., Pleiotropic functions of PACAP in the CNS: neuroprotection and neurodevelopment. *Ann N Y Acad Sci*, 2006. **1070**: 550–60.
68. Farkas, O., et al., Effects of pituitary adenylate cyclase activating polypeptide in a rat model of traumatic brain injury. *Regul Pept*, 2004. **123**(1–3): 69–75.
69. Wu, Z.L., et al., Comparative analysis of cortical gene expression in mouse models of Alzheimer's disease. *Neurobiol Aging*, 2006. **27**(3): 377–86.
70. Stumm, R., et al., Pituitary adenylate cyclase-activating polypeptide is up-regulated in cortical pyramidal cells after focal ischemia and protects neurons from mild hypoxic/ischemic damage. *J Neurochem*, 2007. **103**(4): 1666–81.
71. Ago, Y., et al., Role of endogenous pituitary adenylate cyclase-activating polypeptide in adult hippocampal neurogenesis. *Neuroscience*, 2011. **172**: 554–61.
72. Deguil, J., et al., Neuroprotective effect of PACAP on translational control alteration and cognitive decline in MPTP parkinsonian mice. *Neurotox Res*, 2010. **17**(2): 142–55.
73. Samal, B., et al., Meta-analysis of microarray-derived data from PACAP-deficient adrenal gland in vivo and PACAP-treated chromaffin cells identifies distinct classes of PACAP-regulated genes. *Peptides*, 2007. **28**(9): 1871–82.
74. Tamas, A., et al., Postinjury administration of pituitary adenylate cyclase activating polypeptide (PACAP) attenuates traumatically induced axonal injury in rats. *J Neurotrauma*, 2006. **23**(5): 686–95.

75. Fang, K.M., et al., Effects of combinatorial treatment with pituitary adenylate cyclase activating Peptide and human mesenchymal stem cells on spinal cord tissue repair. *PLoS One*, 2010. **5**(12): e15299.
76. Armstrong, B.D., et al., Lymphocyte regulation of neuropeptide gene expression after neuronal injury. *J Neurosci Res*, 2003. **74**(2): 240–7.
77. Armstrong, B.D., et al., Impairment of axotomy-induced pituitary adenylyl cyclase-activating peptide gene expression in T helper 2 lymphocyte-deficient mice. *Neuroreport*, 2006. **17**(3): 309–12.
78. Armstrong, B.D., et al., Impaired nerve regeneration and enhanced neuroinflammatory response in mice lacking pituitary adenylyl cyclase activating peptide. *Neuroscience*, 2008. **151**(1): 63–73.
79. Depre, C. and S.F. Vatner, Cardioprotection in stunned and hibernating myocardium. *Heart Fail Rev*, 2007. **12**(3–4): 307–17.
80. Gasz, B., et al., Pituitary adenylate cyclase activating polypeptide protects cardiomyocytes against oxidative stress-induced apoptosis. *Peptides*, 2006. **27**(1): 87–94.
81. Racz, B., et al., PKA-Bad-14-3-3 and Akt-Bad-14-3-3 signaling pathways are involved in the protective effects of PACAP against ischemia/reperfusion-induced cardiomyocyte apoptosis. *Regul Pept*, 2008. **145**(1–3): 105–15.
82. Roth, E., et al., Effects of PACAP and preconditioning against ischemia/reperfusion-induced cardiomyocyte apoptosis in vitro. *Ann N Y Acad Sci*, 2009. **1163**: 512–6.
83. Alston, E.N., et al., Cardiac ischemia-reperfusion regulates sympathetic neuropeptide expression through gp130-dependent and independent mechanisms. *Neuropeptides*, 2011. **45**(1): 33–42.
84. Mori, H., et al., Cardioprotective effect of endogenous pituitary adenylate cyclase-activating polypeptide on Doxorubicin-induced cardiomyopathy in mice. *Circ J*, 2010. **74**(6): 1183–90.
85. Racz, B., et al., Protective Effect of PACAP Against Doxorubicin-Induced Cell Death in Cardiomyocyte Culture. *J Mol Neurosci*, 2010. **42**(3): 419–27.
86. Li, M., et al., Renoprotection by pituitary adenylate cyclase-activating polypeptide in multiple myeloma and other kidney diseases. *Regul Pept*, 2008. **145**(1–3): 24–32.
87. Szakaly, P., et al., Effects of PACAP on survival and renal morphology in rats subjected to renal ischemia/reperfusion. *J Mol Neurosci*, 2008. **36**(1–3): 89–96.
88. Horvath, G., et al., Mice deficient in neuropeptide PACAP demonstrate increased sensitivity to in vitro kidney hypoxia. *Transplant Proc*, 2010. **42**(6): 2293–5.
89. Horvath, G., et al., Effects of PACAP on Oxidative Stress-Induced Cell Death in Rat Kidney and Human Hepatocyte Cells. *J Mol Neurosci*, 2011. **43**(1): 67–75.
90. Szakaly, P., et al., Changes in pituitary adenylate cyclase-activating polypeptide following renal ischemia-reperfusion in rats. *Transplant Proc*, 2010. **42**(6): 2283–6.
91. Szakaly, P., et al., Mice deficient in pituitary adenylate cyclase activating polypeptide (PACAP) show increased susceptibility to in vivo renal ischemia/reperfusion injury. *Neuropeptides*, 2011. **45**(2): 113–21.
92. Horvath, G., et al., Effects of PACAP on mitochondrial apoptotic pathways and cytokine expression in rats subjected to renal ischemia/reperfusion. *J Mol Neurosci*, 2010. **42**(3): 411–8.
93. Movat, H.Z., Tumor necrosis factor and interleukin-1: role in acute inflammation and microvascular injury. *J Lab Clin Med*, 1987. **110**: 668–81.
94. Delgado, M., et al., Vasoactive intestinal peptide and pituitary adenylate cyclase-activating polypeptide inhibit endotoxin-induced TNF-alpha production by macrophages: in vitro and in vivo studies. *J Immunol*, 1999. **162**(4): 2358–67.
95. Delgado, M., et al., Vasoactive intestinal peptide (VIP) and pituitary adenylate cyclase-activation polypeptide (PACAP) protect mice from lethal endotoxemia through the inhibition of TNF-alpha and IL-6. *J Immunol*, 1999. **162**: 1200–5.

96. Martinez, C., et al., Anti-inflammatory role in septic shock of pituitary adenylate cyclase-activating polypeptide receptor. *Proc Natl Acad Sci U S A*, 2002. **99**(2): 1053–8.
97. Firestein, G.S., Mechanisms of tissue destruction and cellular activation in rheumatoid arthritis. *Curr Opin Rheumatol*, 1992. **4**(3): 348–54.
98. Delgado, M., et al., Vasoactive intestinal peptide prevents experimental arthritis by downregulating both autoimmune and inflammatory components of the disease. *Nature Med*, 2001. **7**: 563–8.
99. Abad, C., et al., Pituitary adenylate cyclase-activating polypeptide inhibits collagen-induced arthritis: an experimental immunomodulatory therapy. *J Immunol*, 2001. **167**(6): 3182–9.
100. Gaytan, F., et al., Pituitary adenylate cyclase-activating peptide (PACAP) immunolocalization in lymphoid tissues of the rat. *Cell Tissue Res*, 1994. **276**(2): 223–7.
101. Delgado, M., et al., PACAP in immunity and inflammation. *Ann N Y.Acad Sci*, 2003. **992**: 141–57.
102. Tracey, K.J., Physiology and immunology of the cholinergic antiinflammatory pathway. *J Clin Invest*, 2007. **117**(2): 289–96.
103. Sapolsky, R.M., Neuroprotective gene therapy against acute neurological insults. *Nat Rev Neurosci*, 2003. **4**(1): 61–9.
104. Witte, O.W. and G. Stoll, Delayed and remote effects of focal cortical infarctions: secondary damage and reactive plasticity. *Adv Neurol*, 1997. **73**: 207–27.
105. Vaudry, D., et al., Pituitary adenylate cyclase-activating polypeptide stimulates both c-Fos gene expression and cell survival in rat cerebellar granule neurons through activation of the protein kinase A pathway. *Neuroscience*, 1998. **84**: 801–12.
106. Ravni, A., et al., The neurotrophic effects of PACAP in PC12 cells: control by multiple transduction pathways. *J Neurochem*, 2006. **98**(2): 321–29.
107. Ravni, A., et al., A cAMP-dependent, PKA-independent signaling pathway mediating neuritogenesis through Egr1 in PC12 cells. *Mol Pharmacol*, 2008. **73**: 1688–708.
108. Vaudry, D., et al., Signaling pathways for PC12 cell differentiation: making the right connections. *Science*, 2002. **296**(5573): 8–1649.
109. Castel, H., et al., The delayed rectifier channel current IK plays a key role in the control of programmed cell death by PACAP and ethanol in cerebellar granule neurons. *Ann N Y Acad Sci*, 2006. **1070**: 173–9.
110. Delgado, M. and D. Ganea, Vasoactive intestinal peptide and pituitary adenylate cyclase-activating polypeptide inhibit nuclear factor-kappa B-dependent gene activation at multiple levels in the human monocytic cell line THP-1. *J Biol Chem*, 2001. **276**(1): 369–80.
111. Leceta, J., et al., Receptors and transcriptional factors involved in the anti-inflammatory activity of VIP and PACAP. *Ann N Y Acad Sci*, 2000. **26**: 833–42.
112. Shioda, S., et al., Prevention of delayed neuronal cell death by PACAP and its molecular mechanism. *Nippon Yakurigaku Zasshi*, 2004. **123**(4): 243–52.
113. Grumolato, L., et al., Microarray and suppression subtractive hybridization analyses of gene expression in pheochromocytoma cells reveal pleiotropic effects of pituitary adenylate cyclase-activating polypeptide on cell proliferation, survival, and adhesion. *Endocrinology*, 2003. **144**(6): 2368–79.
114. Ait-Ali, D., et al., Neuropeptides, growth factors and cytokines: a cohort of informational molecules whose expression is up-regulated by the stress-associated slow transmitter PACAP in chromaffin cells. *Cell Mol Neurobiol*, 2010. **30**(8): 1441–9.
115. Eiden, L.E., et al., Discovery of PACAP-related genes through microarray analyses in cell culture and in vivo. *Ann N Y Acad Sci*, 2008. **1144**: 6–20.
116. Grumolato, L., et al., Selenoprotein T is a PACAP-regulated gene involved in intracellular Ca^{2+} mobilization and neuroendocrine secretion. *FASEB J*, 2008. **22**(6): 1726–68.
117. Stroth, N., et al., PACAP: a master regulator of neuroendocrine stress circuits and the cellular stress response. *Ann N Y Acad Sci*, 2011. **1220**: 49–59.

118. Mustafa, T., M. Grimaldi, and L.E. Eiden, The hop cassette of the PAC1 receptor confers coupling to Ca2+ elevation required for pituitary adenylate cyclase-activating polypeptide-evoked neurosecretion. *J Biol Chem*, 2007. **282**(11): 8079–91.
119. Laburthe, M., A. Couvineau, and V. Tan, Class II G protein-coupled receptors for VIP and PACAP: structure, models of activation and pharmacology. *Peptides*, 2007. **28**(9): 1631–9.
120. Mustafa, T. and L.E. Eiden, The secretin superfamily: PACAP, VIP and related peptides, in *Handbook of Neurochemistry and Molecular Neurobiology: XIII. Neuroactive Peptides and Proteins*, R. Lim, Editor. Heidelberg, Germany: Springer. 2006: 1–36.
121. Harmar, A.J., et al., International Union of Pharmacology. XVIII. Nomenclature of receptors for vasoactive intestinal peptide and pituitary adenylate cyclase-activating polypeptide. *Pharmacol Revs*, 1998. **50**: 265–70.
122. Cutler, D.J., et al., The mouse VPAC2 receptor confers suprachiasmatic nuclei cellular rhythmicity and responsiveness to vasoactive intestinal polypeptide in vitro. *Eur J Neurosci*, 2003. **17**(2): 197–204.
123. Harmar, A.J., et al., The VPAC2 receptor is essential for circadian function in the mouse suprachiasmatic nuclei. *Cell*, 2002. **109**: 497–508.
124. Colwell, C.S., et al., Disrupted Circadian Rhythms in VIP and PHI Deficient Mice. *Am J Physiol Regul Integr Comp Physiol*, 2003. **285**: R939–R949.
125. Aton, S.J., et al., Vasoactive intestinal polypeptide mediates circadian rhythmicity and synchrony in mammalian clock neurons. *Nat Neurosci*, 2005. **8**(4): 476–83.
126. Colwell, C.S. and J.A. Waschek, Role of PACAP in circadian function of the SCN. *Regul Peptides*, 2001. **102**: 49–68.
127. Hannibal, J., et al., Dissociation between light-induced phase shift of the circadian rhythm and clock gene expression in mice lacking the pituitary adenylate cyclase activating polypeptide type 1 receptor. *J Neurosci*, 2001. **21**: 4883–90.
128. Otto, C., et al., Altered emotional behavior in PACAP-type-I-receptor-deficient mice. *Brain Res Mol Brain Res*, 2001. **92**(1–2): 78–84.
129. Hashimoto, H., et al., Altered psychomotor behaviors in mice lacking pituitary adenylate cyclase-activating polypeptide (PACAP). *Proc Natl Acad Sci U S A*, 2001. **98**: 13355–60.
130. Winzell, M.S. and B. Ahren, Role of VIP and PACAP in islet function. *Peptides*, 2007. **28**(9): 1805–13.
131. Girard, B.A., et al., Noncompensation in peptide/receptor gene expression and distinct behavioral phenotypes in VIP- and PACAP-deficient mice. *J Neurochem*, 2006. **99**(2): 499–513.
132. Moody, T.W., et al., VIP and PACAP: recent insights into their functions/roles in physiology and disease from molecular and genetic studies. *Curr Opin Endocrinol Diabetes Obes*, 2011. **18**(1): 61–7.
133. Delgado, M., et al., Anti-inflammatory properties of the type 1 and type 2 vasoactive intestinal peptide receptors: role in lethal endotoxic shock. *Eur J Immunol*, 2000. **30**(11): 3236–46.
134. Journot, L., et al., Differential signal transduction by six splice variants of the pituitary adenylate cyclase-activating peptide (PACAP) receptor. *Biochem Soc T*, 1995. **23**: 133–137.
135. Pisegna, J.R. and S.A. Wank, Cloning and characterization of the signal transduction of four splice variants of the human pituitary adenylate cyclase activating polypeptide receptor. Evidence for dual coupling to adenylate cyclase and phospholipase C. *J Biol Chem*, 1996. **271**(29): 17267–74.
136. Spengler, D., et al., Differential signal transduction by five splice variants of the PACAP receptor. *Nature*, 1993. **365**: 170–5.
137. Banks, W.A., et al., Transport of pituitary adenylate cyclase-activating polypeptide across the blood-brain barrier and the prevention of ischemia-induced death of hippocampal neurons. *Ann N Y Acad Sci*, 1996. **805**: 270–7.

138. Lerner, E.A., A.O. Iuga, and V.B. Reddy, Maxadilan, a PAC1 receptor agonist from sand flies. *Peptides*, 2007. **28**(9): 1651–4.
139. Soares, M.B., et al., The vasoactive peptide maxadilan from sand fly saliva inhibits TNF-alpha and induces IL-6 by mouse macrophages through interaction with the pituitary adenylate cyclase-activating polypeptide (PACAP) receptor. *J Immunol*, 1998. **160**(4): 1811–6.
140. Doan, N.D., et al., Design and in vitro characterization of PAC1/VPAC1-selective agonists with potent neuroprotective effects. *Biochem Pharmacol*, 2011. **81**(4): 552–61.
141. Hammack, S.E., et al., Chronic stress increases pituitary adenylate cyclase-activating peptide (PACAP) and brain-derived neurotrophic factor (BDNF) mRNA expression in the bed nucleus of the stria terminalis (BNST): roles for PACAP in anxiety-like behavior. *Psychoneuroendocrinology*, 2009. **34**(6): 833–43.
142. Reglodi, D., et al., Effects of pretreatment with PACAP on the infarct size and functional outcome in rat permanent focal cerebral ischemia. *Peptides*, 2002. **23**(12): 2227–34.
143. Tan, Y.V., et al., Pituitary adenylyl cyclase-activating polypeptide is an intrinsic regulator of Treg abundance and protects against experimental autoimmune encephalomyelitis. *Proc Natl Acad Sci U S A*, 2009. **106**(6): 2012–7.
144. Xia, C.F., et al., Adrenomedullin gene delivery protects against cerebral ischemic injury by promoting astrocyte migration and survival. *Hum Gene Ther*, 2004. **15**(12): 1243–54.
145. Wang, X., et al., Discovery of adrenomedullin in rat ischemic cortex and evidence for its role in exacerbating focal brain ischemic damage. *Proc Natl Acad Sci U S A*, 1995. **92**(25): 11480–4.
146. Cserepes, B., et al., Cardioprotective action of urocortin in early pre- and postconditioning. *Ann N Y Acad Sci*, 2007. **1095**: 228–39.
147. Song, S., et al., The effect of pretreatment with calcitonin gene-related peptide on attenuation of liver ischemia and reperfusion injury due to oxygen free radicals and apoptosis. *Hepatogastroenterology*, 2009. **56**(96): 1724–9.
148. Delcourt, N., et al., PACAP type I receptor transactivation is essential for IGF-1 receptor signalling and antiapoptotic activity in neurons. *EMBO J*, 2007. **26**(6): 1542–51.

12 General Strategies of Peptide Synthesis

Aleksandra Misicka

CONTENTS

12.1 INTRODUCTION

The first synthesis of the free peptide (glycylglycine) was published by Fischer and Fourneau in 1901. However, over the next several decades, progress in peptide synthesis was slow because of the lack of an easily cleaved amino-protecting group in the presence of a peptide bond. The situation was changed by the development of the benzoxycarbonyl (Z) group by Bergmann and Zervas (1932). With the use of this protecting group, du Vigneaud et al. (1954) performed the first chemical synthesis of a biologically active peptide, oxytocin. This achievement was a crucial milestone in the field of peptide chemistry and du Vigneaud was awarded the Nobel Prize. Since then, the syntheses of many other biologically active peptides have been elaborated, but these types of peptide syntheses, performed in solution, are highly labor intensive and time consuming.

The revolution came about in 1963 when Merrifield presented his ingenious concept of peptide synthesis on a solid polymeric support (Merrifield 1963). This method, which is now known as the solid phase peptide synthesis (SPPS), has revolutionized peptide synthesis and Merrifield was acknowledged with the Nobel Prize in 1984 for this discovery (Figure 12.1).

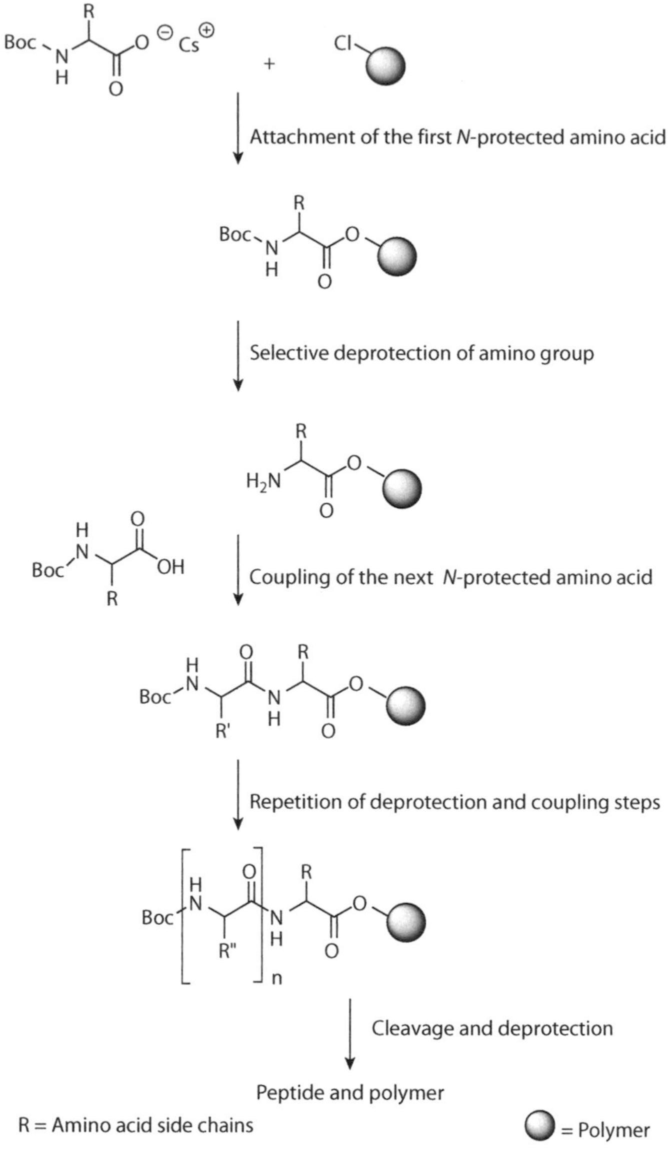

FIGURE 12.1 General scheme of SPPS introduced by Merrifield, R.B., *J. Am. Chem. Soc.,* 85, 2149–2152, 1963.

In SPPS, the peptide chain is assembled in the usual manner, starting from the *C*-terminus of the synthesized peptide. The first amino acid of the peptide is connected via its carboxyl group to an insoluble polymer (resin) that can be easily separated from either reagents or dissolved products through the use of filtration. After deprotection of the N^{α}-amino group of the first amino acid, the peptide chain is built by the attachment of consecutive N^{α}-protected amino acids with the use of coupling reagents. The synthesizing peptide chain remains attached to the polymeric support until the desired sequence of amino acids has been obtained. After finishing the synthesis on the polymeric support, the complete peptide can be removed from the polymeric support with the simultaneous cleavage of all protecting groups.

SPPS is the convenient methodology of choice for peptide synthesis and requires the proper combination of protecting groups, coupling reagents, and solid supports, including linkers. The purpose of this chapter is to present the basic mechanisms of peptide bond formation and the principles of the commonly employed strategies for the SPPS.

12.2 PROTECTING GROUP STRATEGIES USED IN SOLID PHASE PEPTIDE SYNTHESIS

Protection of the α-amino functionality of amino acids is one of the most important issues in peptide chemistry. This protection is essential to avoid the polymerization of activated amino acids. Besides the α-amino group, proteinogenic amino acids can have other functional groups in the side chain, for example, hydroxyl, amino, thiol, or guanidinium groups. A growing number of available amino acids not found in nature make the set of the possible side chain functions even bigger. In SPPS, the protection of all functional groups, including indole ring in tryptophan and the imidazole ring in histidine, is recommended. Groups used for side chain protection must be orthogonal to the α-amino-protecting group, which means that they can be separately removed in any order and in the presence of the remaining ones.

The α-amino-protecting group employed by Merrifield in SPPS was the *tert*-butyloxycarbonyl (Boc) group (Anderson and McGregor 1957; McKay and Albertson 1957). Boc protection is removed in acidic condition; therefore, the application of Boc protection for α-amino group excludes any acid-sensitive group as a permanent protection for the side chains and must employ groups that resist the conditions necessary for iterative Boc cleavage in SPPS. The best suitable groups for this purpose are benzyl-type (Bzl)-protecting groups and this strategy is called Boc/Bzl. The final cleavage from the polymeric support and deprotection of all protecting groups require the use of strong acids, such as hydrogen fluoride (HF) or trifluoromethanesulfonic acid (TFMSA). HF is a dangerous chemical, which reacts with glass, and an expensive polytetrafluoroethylene-lined apparatus is needed to perform it. The Boc/Bzl strategy is therefore currently used only in specialized laboratories.

The most common α-amino-protecting group currently used in SPPS is the 9-fluorenylmethoxy-carbonyl (Fmoc) group, the other urethane-type protected group introduced by Carpino, cleaved by a secondary amine (Carpino and Han 1972). Usually, piperidine in dimethylformamide is used in a standard deprotection, an alternative can be 1,8-diazabicyclo[5.4.0]undec-7-ene (DBU), efficient for difficult Fmoc cleavages. In the case of base-sensitive glycopeptides, a weaker base, such as morpholine, is recommended. Most suitable for the orthogonal protection of side chain functionalities are acid-removable groups, such as *t*Bu, and this strategy is called Fmoc/*t*Bu. The final cleavage and deprotection is achieved under mildly acid conditions, mostly in trifluoroacetic acid (TFA) (Figure 12.2). Therefore, the Fmoc-based method is the method of choice for the routine synthesis of peptides. A very useful review of methods and protocols recommended for Fmoc SPPS is available (Amblard et al. 2006) (Figure 12.3 and Table 12.1).

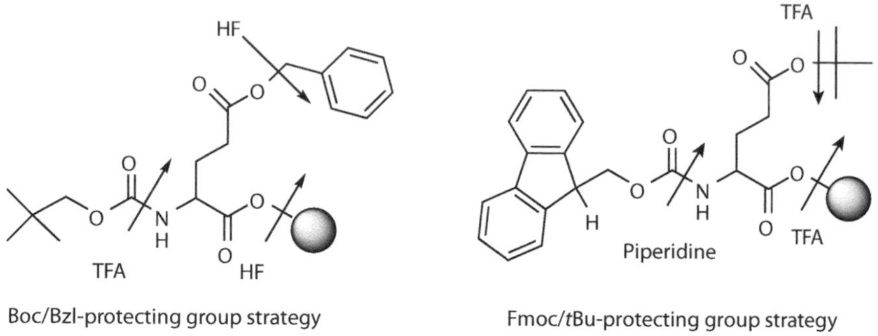

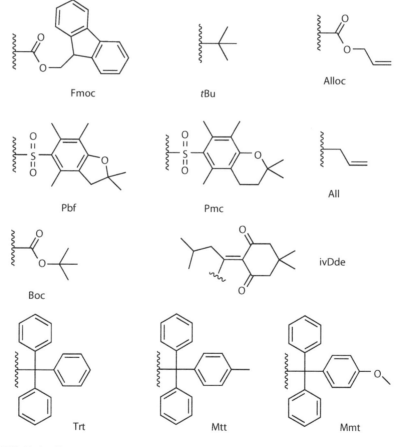

HF
TFA
HF
TFA
Piperidine
TFA
Boc/Bzl-protecting group strategy
Fmoc/tBu-protecting group strategy

FIGURE 12.2 Examples of strategies used in SPPS.

Fmoc
tBu
Alloc
Pbf
Pmc
All
Boc
ivDde
Trt
Mtt
Mmt

FIGURE 12.3 Protection groups used in Fmoc SPPS.

TABLE 12.1

The Most Popular Side-Protecting Groups Used in Fmoc-Based SPPS

Amino Acid	Side Chain	Protecting Group	Removal Condition
Arginine	NH / ∿HN—NH₂	Pmc / Pbf	95% TFA
Aspartic acid	O / ∿OH	tBu	95% TFA
Glutamic acid		All	Pd(Ph₃P)₄/PhSiH₃
Asparagine	O / ∿NH₂	Trt	95% TFA
Glutamine			
Cysteine	∿ SH	Trt	95% TFA
Serine	∿ OH	tBu	TFA
Threonine		Trt	1% TFA
Tyrosine			
Lysine	∿ NH₂	Boc	TFA
		Alloc	Pd(Ph₃P)₄/PhSiH₃
		Mtt	1% TFA in DCM
		Mmt	1% TFA in DCM
		ivDde	hydrazine
Tryptophan	∿H₂C (indole)	Boc(NH_ind)	TFA
Histidine	∿H₂C (imidazole)	Trt(NHτ)	TFA
		Mtt(NHτ)	1% TFA in DCM

However, for the solid-phase preparation of complex peptides, such as cyclic peptides and bioconjugated peptides, as well as for the construction of libraries of peptides using a combinatorial approach, a range of orthogonal protecting groups is needed (Albericio 2000). A comprehensive review article summarizing the great efforts undertaken to develop amino acid-protecting groups, including an arsenal of suitable orthogonal protecting groups for other functional groups occurring in amino acids, has also recently become available (Isidro-Llobet, Alvarez, and Albericio 2009). New protecting groups have been developed to (1) increase the number of available orthogonal protectants allowing the synthesis of cyclic and bioconjugated peptides; (2) diminish epimerization during activation; and (3) avoid side reactions, which can occur during peptide synthesis, such as aspartimide formation, dehydration of the carboxamide side chain of asparagines/glutamine, δ-lactam formation, and deguanidination of arginine.

12.3 COUPLING REAGENTS FOR PEPTIDE BOND FORMATION

Formation of the peptide bond needs to be highly efficient to yield high purity peptides using SPSS. Generally, peptide bond formation is a nucleophilic substitution reaction of an amino group at the activated carboxyl group involving a tetrahedral intermediate. Activation is achieved by the introduction of electron-accepting moieties that allow the dissociation of a tetrahedral complex with the formation of a peptide bond. The variations of the removed group provide a broad-spectrum of methods for peptide bond formation. The most simple and rapid procedure applied in SPPS is the *in situ* activation of a carboxylic group by the use of carbodiimides or onium salts, which are called coupling reagents.

12.3.1 CARBODIIMIDES

Carbodiimides, primarily dicyclohexylcarbodiimide (DCC) and diisopropylcarbodiimide (DIC), were coupling reagents of choice for many years (Sheehan and Hess 1955). The mechanism for coupling carboxylic acids to amines is shown in Figure 12.4. In the case of DCC, dicyclohexylurea (DCU), the by-product formed, is insoluble in the reaction solvents and can be easily removed by filtration. This is a great advantage in peptide synthesis performed in solution, but for SPPS, other dicarbodiimides are recommended. One of the undesirable side reactions associated with the use of carbodimides is the base-catalyzed acyl migration from isourea oxygen to nitrogen, which results in the formation of *N*-acylurea that does not undergo further aminolysis (Figure 12.4). The other danger is the risk of epimerization.

The systematic investigations to suppress *N*-acylurea formation and suppress racemization resulted in the development of the DCC-additive protocol. The chief preferred additive was *N*-hydroxybenzotriazole (HOBt), as it reacts rapidly with *O*-acylurea forming the corresponding active benzotriazolyl esters, which then undergo fast aminolysis (König and Geiger 1970) (Figure 12.5).

The other benzotriazole-based additive, *N*-hydroxy-7-azabenzotriazole (HOAt), developed later by Carpino (1993), has been proven in many comparative studies

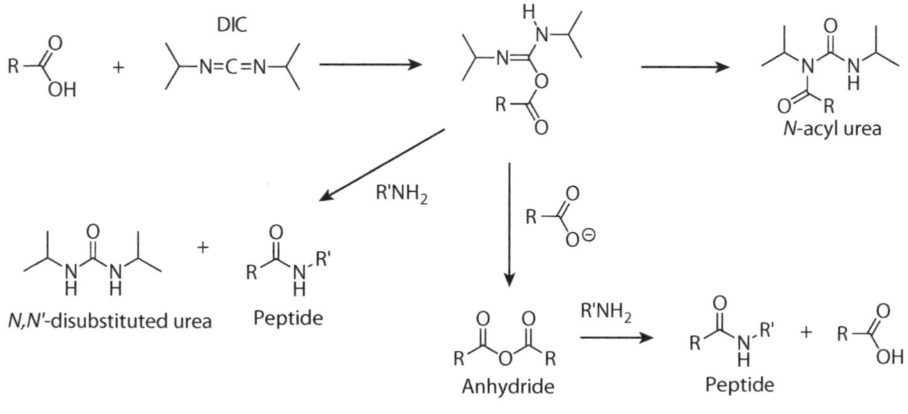

FIGURE 12.4 Mechanism of DIC coupling.

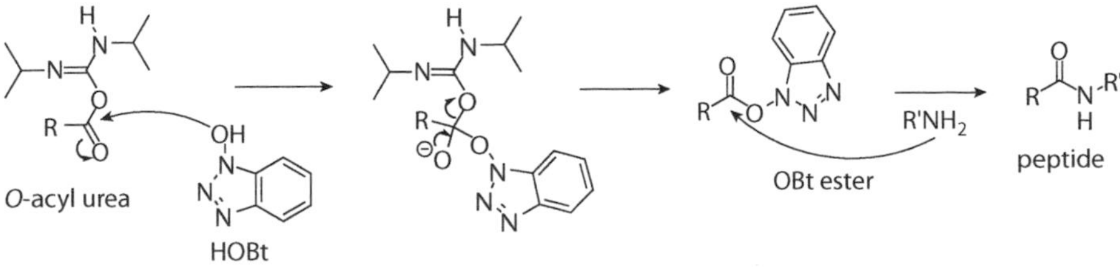

FIGURE 12.5 Mechanism of activation by HOBt when used as an additive with DIC.

to be the best additive in terms of yield, kinetics, and reduced epimerization levels. HOAt has been widely used along with DIC in order to suppress racemization and other side reactions, especially in difficult couplings (Carpino and El-Faham 1999).

Recent studies revealed the explosive properties of benzotriazoles, which has led to increasing transport restrictions (Wehrstedt, Wandrey, and Heitkamp 2005; Malow, Wehrstedt, and Neuenfeld 2007). From this perspective, ethyl 2-hydroxyimino-2-cyanoacetate (called Oxyma), which had been previously reported (Itoh 1973; Izdebski 1979), fulfills the requirement not only in terms of racemization control, and efficiency in difficult couplings, but also in terms of its safety profile as well. Oxyma decomposes at a slower rate and exerts only one-third of the pressure observed in the experiments with HOBt and HOAt. Oxyma was recently carefully tested as an additive in combination with carbodiimide, showing its superiority to HOBt in terms of the suppression of racemization and of coupling efficiency (Figure 12.6). Interestingly, in some demanding assemblies of MeAla or Aib analogs of enkephalin, its performance was even superior to that of HOAt (Subiros-Funosas, Prohens, et al. 2009).

12.3.2 Phosphonium Salts

Phoshonium salts based on benzotriazoles are very efficient coupling reagents. The first reagent of this type was benzotriazol-1-yloxy-tris(dimethylamino)-phosphonium hexafluorophosphate (BOP), developed by Castro et al. (1975). It is a very efficient coupling reagent for amide bond formation, often successfully applied in cases where other reagents had failed (Brunel, Salmi, and Letourneux 2005). The disadvantages of this reagent are the formation of the highly toxic and carcinogenic compound, hexamethylphosphoramide (HMPA), as its by-product and a high risk of racemization during the coupling reaction. Its use has, therefore, been limited. Instead, other reagents of this type were introduced: benzotriazol-1-yloxy-tris(pyrrolidino)-phosphonium hexafluorophosphate (PyBOP) (Coste, Lenguyen, and Castro 1990) and the HOAt analog 7-azabenzotriazol-1-yloxy-tris(pyrrolidino)-phosphonium hexafluorophosphate (PyAOP), where the diaminomethyl groups were replaced with pyrrolidine substituents that do not produce the hazardous by-products (Albericio et al. 1997). The latest compound shows one of the best reactivities from the proposed phoshonium salts in coupling reactions (Figure 12.7).

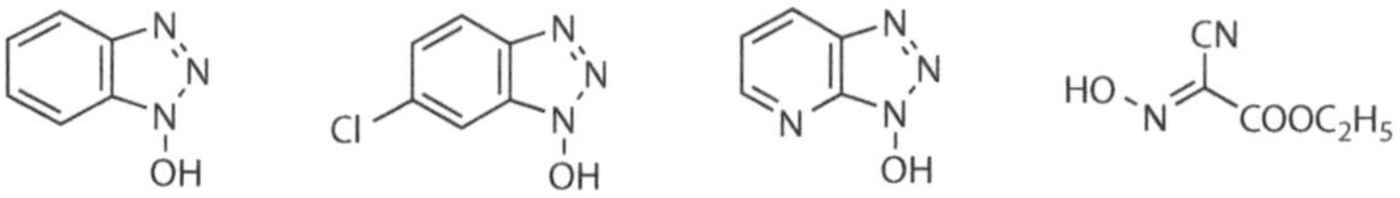

FIGURE 12.6 Structures of additives used in carbodimides activation.

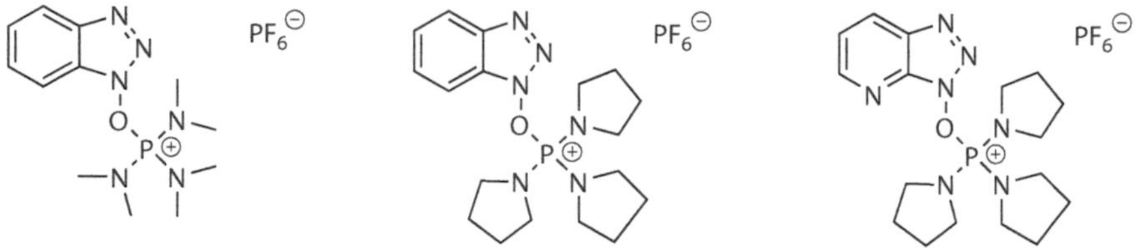

FIGURE 12.7 Structures of phosphonium-type coupling reagents.

12.3.3 URONIUM/AMINIUM SALTS

Many coupling reagents presently used are based on HOBt or HOAt system and uronium/aminium salts developed by Knorr et al. (1989). These compounds become popular because of their high efficiency and low tendency to induce racemization of the activated amino acid or peptide residue. The most efficient uronium salts are shown in Figure. 12.8.

Depending on solvent, isolation method, and counter ion, the uronium or aminium isomers have been structurally identified (Figure 12.9). Coupling reagents based on uronium salts were first reported as the O-isomer, but X-ray crystallography of some of these salts showed that these compounds were in fact N-isomers (Carpino et al. 2002).

Uronium salts react with a carboxylic group to form the corresponding O-hydroxybenzotriazolyl (OBt)- or 7-aza-1-hydroxy benzotriazolyl (OAt)-active esters (Figure 12.10). A tertiary amine (generally diisopropylamine) is required to produce the carboxylate of N-protected amino acids, which reacts with coupling reagents.

Uronium salts can irreversibly block the free N-terminal amino functionality of the peptide-resin by forming guanidinium by-products (Figure 12.11), thus the order of addition and timing is crucial (Story and Aldrich 1994).

The most popular uronium/aminium salts used as *in situ* activation reagents in peptide synthesis are O-(1H-benzotriazole-1-yl)-N,N,N',N'-tetramethyluronium hexafluoro-phosphate (HBTU) and O-(1H-benzotriazole-1-yl)-N,N,N',N'-tetramethyluronium tetrafluoroborate (TBTU), which differ from each other only by a counterion. Studies using HBTU and TBTU have shown that the counterion had no practical effect on coupling reactions. Both reagents, in addition to having a high reactivity, have also been shown to limit enantiomerization during fragment condensation. As in the case of phosphonium salts, the other uronium salts based on HOAt (compared to HOBt), O-(7-azabenzotriazol-1-yl)-N,N,N',N'-tetramethyluronium hexafluorophosphate (HATU) and

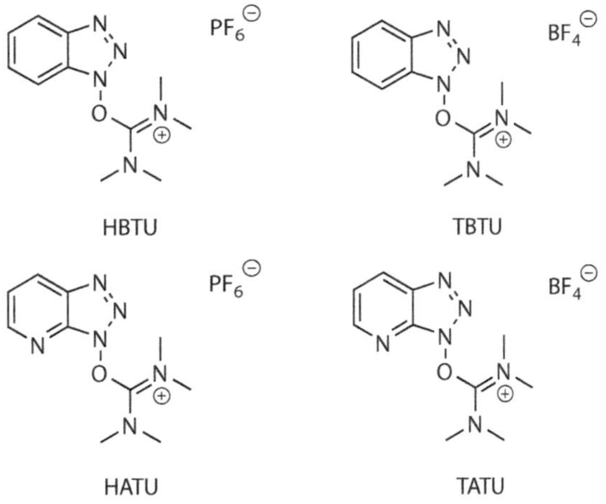

FIGURE 12.8 Structures of the most efficient uronium salts.

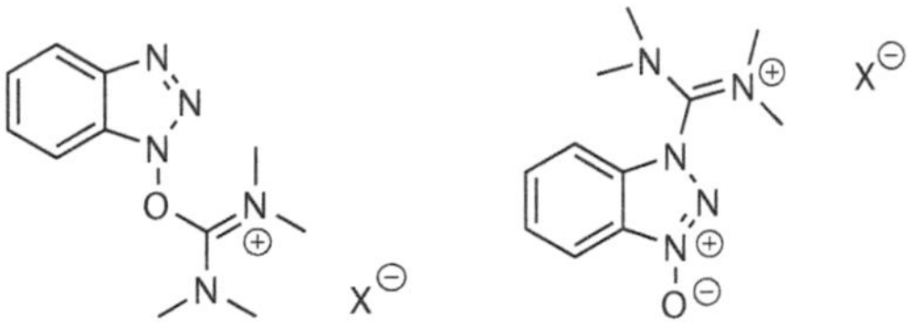

FIGURE 12.9 Uronium and aminium isomers.

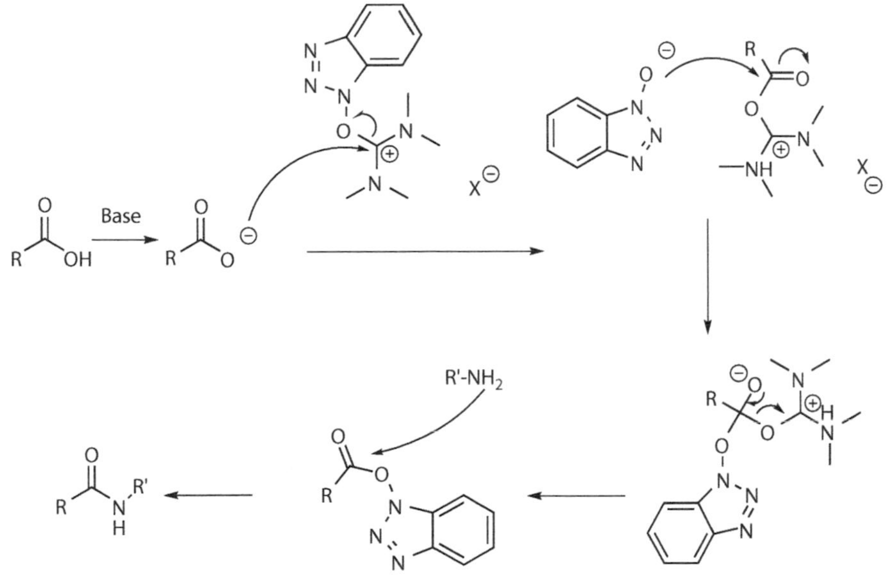

FIGURE 12.10 Activation process using uronium/aminium type reagents. (From Valeur, E., *Chem. Soc. Rev.*, 38, 606–631, 2009. Reproduced by permission of The Royal Society of Chemistry.)

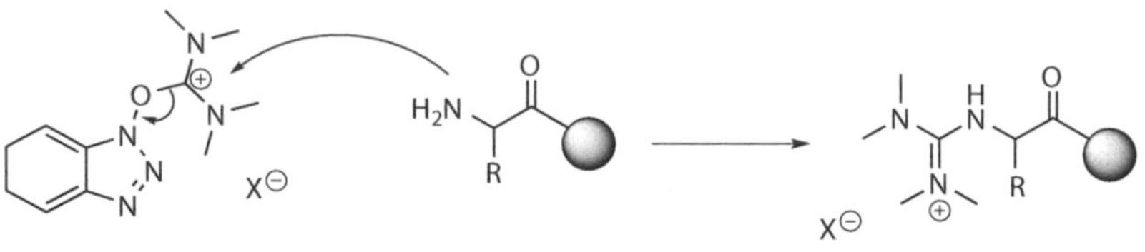

FIGURE 12.11   Guanidinium formation with uronium-type coupling reaction.

O-(7-azabenzotriazol-1-yl)-N,N,N',N'-tetra-methyluronium   tetrafluoroborate
(TATU), are more efficient with regard to yielding coupling reactions and giving
products with less epimerization (Carpino 1993). Much work has been carried
out in the last decade on the variation of the substituents and modifications in the
ring of the benzotriazole-based reagents, but no evidence that suggests that any of
the new reagents reported were more beneficial than reagents routinely used like
HBTU (for routine couplings) or HATU (for hindered couplings), as summarized in
a recent critical review (Valeur and Bradley 2008), has been presented. The authors
discussed in detail many of the benzotriazole-type coupling reagents proposed
for amide formation and emphasized that many of the newly introduced coupling
reagents have not been compared to the classic reagents of that type, such as HATU,
which currently is the gold standard for peptide synthesis.

The potentially explosive properties of benzotriazole-based coupling reagents
were up to now almost always disregarded, but recently safety concerns have led to a
strong demand for improvements, because shipping and storage of the benzotriazole-
based reagents turned out to be difficult. One of the recently developed uronium-type
coupling reagents, not benzotriazole based, is 1-[(1-(cyano-2-ethoxy-2-oxoethylide-
neaminooxy)dimethylamino-morpholino)] uronium hexafluorophosphate (COMU)
(El-Faham et al. 2009). This compound is based on Oxyma, which shows a better
safety profile. A similar compound, N,N,N',N'-tetramethyluronium hexafluorophosphate (HOTU), was previously developed (Breipohl and Koenig 1991) (Figure 12.12).
The presence of the morpholine group in COMU has a marked influence on the
polarity of the carbon skeleton in comparison with the tetramethyl derivatives. The
effectiveness of COMU was proved in the synthesis of a highly demanding Aib[2,3]
enkephalin in comparison to the best coupling reagents (El-Faham et al. 2009).

12.3.4   TRIAZINE-BASED COUPLING REAGENTS

Triazine-type reagents represent another type of efficient coupling reagents. Recently,
a series of triazine-based coupling reagents (TBCRs) designed according to the concept
of "superactive esters" was developed (Kaminski et al. 2005). The concept of the
acceleration of the coupling process in the case of "superactive esters" is based on the
facile departure of the removed group by its rearrangement in an energetically favored,
consecutive process to a stable, chemically inert, and neutral side product. The best of
this type of triazine-coupling reagents, 4-(4,6-dimethoxy-1,3,5-triazin-2-yl)-4-methyl-
morpholinium tetrafluoroborate (DMT-NMM/BF$_4$), was proved to be highly versatile in
SPPS of difficult peptide sequences and head-to-tail constrained cyclopeptide analogs.
This coupling reagent was also successfully applied in the synthesis of N-glycosyl

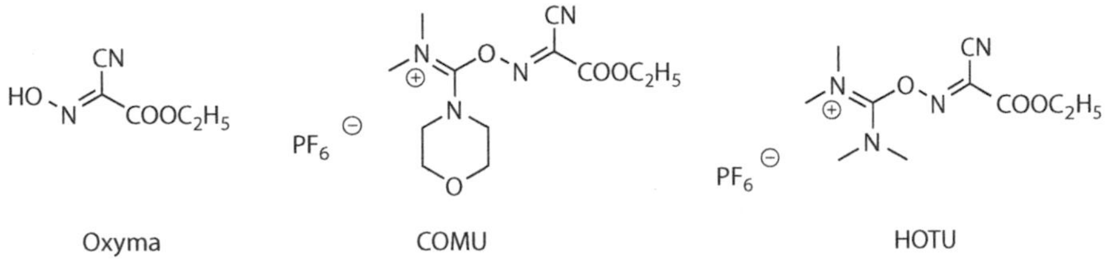

Oxyma   COMU   HOTU

FIGURE 12.12   Structures of Oxyma and uronium salts coupling reagents based on Oxyma.

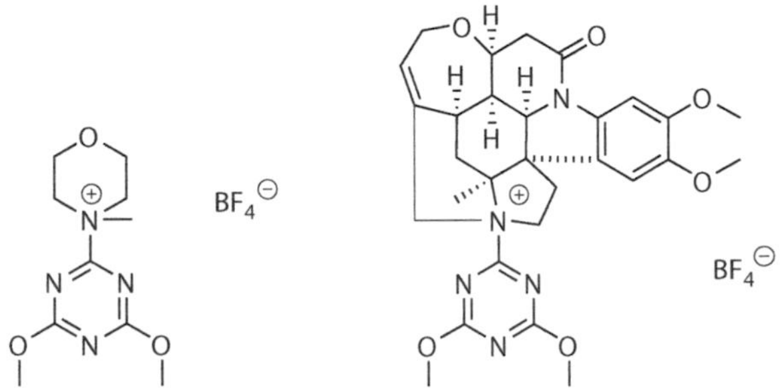

FIGURE 12.13   Structures of triazine-based coupling reagents: (a) DMT-NMM/BF$_4$ and (b) Chiral *N*-triazinylbrucinium tetrafluoroborate.

amino acids (Paolini et al. 2007). Recently, a chiral TBCR, *N*-triazinylbrucinium tetrafluoroborate, was developed as an enantioselective coupling reagent. The usefulness of this compound was shown in the activations of racemic carboxylic acids yielding enantiomerically enriched amides, esters, and dipeptides, with an enantiomeric ratio ranging from 8:92 to 0.5:99.5 (Kolesinska and Kaminski 2009) (Figure 12.13).

12.4   POLYMERIC SUPPORTS SUITABLE FOR SOLID PHASE PEPTIDE SYNTHESIS

The polymeric support suitable for SPPS must be chemically inert, mechanically stable, and completely insoluble in the solvent used and easily separated by filtration. It must contain a sufficient number of reactive sites where the first amino acid of the peptide chain to be synthesized can be attached. The bond connecting the peptide chain to the resin should be stable during the synthesis of the peptide chain and must be cleavable at the end of the procedure.

Polystyrene (PS) containing 1% of a divinylbenzene-crosslinked polymer was the choice of Merrifield for his pioneering study. This type of polymer is still the common matrix used for routine synthesis in SPPS. The alternative matrixes, crosslinked polyamide-based resin (PA) and composite PS-polyethylene glycol (PEG)-based resins, are much more hydrophilic than PS resins. The swelling is superior to those of polystyrene-based resins and, therefore, they are more suitable for the synthesis of difficult sequences and large peptides. The next-generation polymer supports consist solely of a polyethylene glycol matrix. Its amphiphilic nature helps to achieve the synthesis of particular sequences that were not easily accessible previously. A striking example is the syntheses of β-amyloid (1-42) performed on ChemMatrix, a totally PEG-based resin (Garcia-Martin, Quintanar-Audelo, and Garcia-Ramos 2006).

Currently, no general rule exists for selecting the most convenient solid support. However, it is important to consider the type of chemistry to be carried out during the synthesis, the swelling–solvent ratio, and the length and sequence of the desired product. It is also relevant to take into account the nature of the resin bead size uniformity, cross-linking, and loading. In the case of difficult peptide sequences, it is crucial to select an appropriate resin composition. A review of currently available polystyrene (PS) resins grafted with PEG, or copolymerized with PEG, and totally based PEG-resin was recently published (Garcia-Martin and Albericio 2008).

The introduction of anchoring groups (linkers) on polymeric supports is the precondition for their application in peptide synthesis. Many different linker systems are now available for peptide synthesis, and some examples of linkers for Boc/Bzl (Figure 12.14) and Fmoc/tBu (Figure 12.15) strategy are presented in the following.

The cleavage of the final peptide from resins, shown in Figure 12.14, can be achieved in strong acid conditions, such as HF or TFMSA, providing a peptide acid or a peptide amide.

Depending on the chosen linker compatible with the Fmoc/tBu strategy, the cleavage of the final peptide can be performed in 95% TFA (Wang and Rink amide resin) or in a milder condition, such as 1% of TFA in dichloromethane (SASRIN, HMPB resin), providing peptide acids (Wang, SASRIN, HMPB) or peptide amides (Rink amide resin).

One of the most significant developments in linker technology of the last decade proved to be 2-chlorotrityl chloride resin (Barlos et al. 1989). Attachment of the first amino acid to the O-chlorotrityl chloride resin is accomplished using the reaction with the Fmoc amino acid derivative in the presence of a base in anhydrous conditions. This reaction does not involve an activated species, so it is free of racemization. Therefore, this resin is recommended for the immobilization of sensitive residues, such as Cys and His, prone to enantiomerization. As a trityl-based resin, it is highly acid labile (0.5% TFA/DCM, AcOH/DCM) allowing the cleavage of protected peptide segments from the resin. The steric hindrance of the bulk of the trityl linker prevents diketopiperazine formation and therefore these resins are recommended for the synthesis of Pro and Gly C-terminal peptides.

Other types of resins are available for special applications, for example, Hycram resin, which allows the cleavage of the final product from the resin in neutral

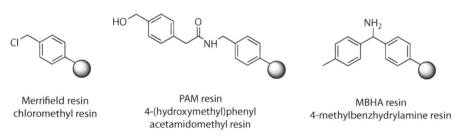

Merrifield resin	PAM resin	MBHA resin
chloromethyl resin	4-(hydroxymethyl)phenyl	4-methylbenzhydrylamine resin
	acetamidomethyl resin	

FIGURE 12.14 Examples of resins suitable for Boc/Bzl strategy.

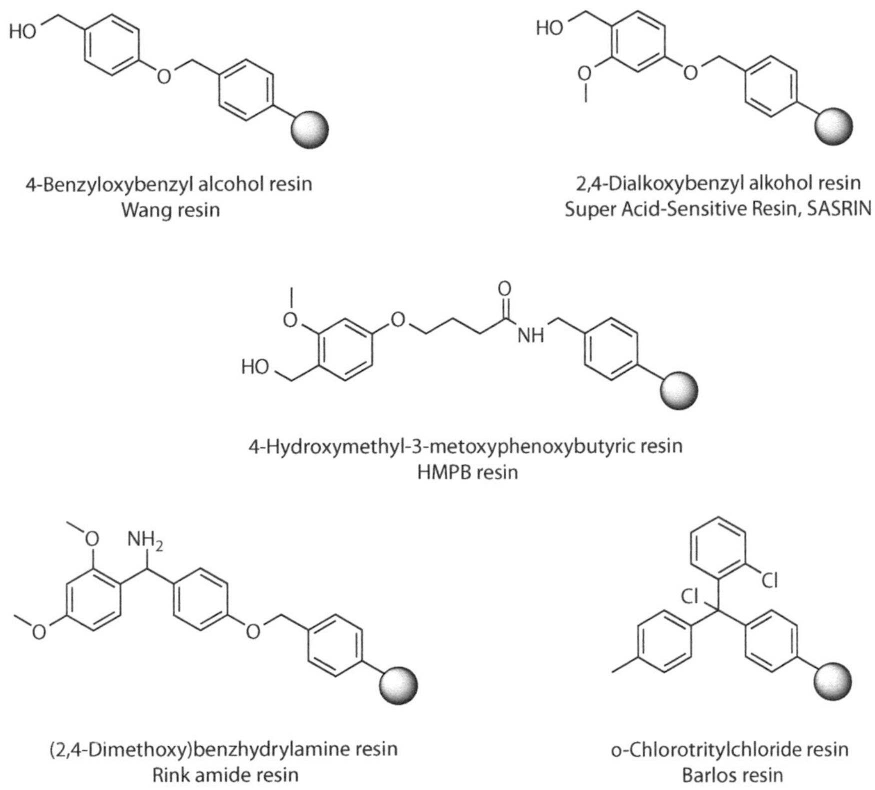

4-Benzyloxybenzyl alcohol resin
Wang resin

2,4-Dialkoxybenzyl alkohol resin
Super Acid-Sensitive Resin, SASRIN

4-Hydroxymethyl-3-metoxyphenoxybutyric resin
HMPB resin

(2,4-Dimethoxy)benzhydrylamine resin
Rink amide resin

o-Chlorotritylchloride resin
Barlos resin

FIGURE 12.15 Examples of resin suitable for Fmoc/*t*Bu strategy.

conditions by treatment with a Pd(0) catalyst that affects the allyl transfer of suitable nucleophile or photolabile linkers, which then allows cleavage by photolysis. The field of solid phase technique has grown enormously in the last decade and the number of specialized resins increased dramatically, enabling the recent developments in peptide chemistry.

12.5 MICROWAVE-ASSISTED SOLID PHASE PEPTIDE SYNTHESIS

The first attempts to use microwave (MW) irradiation in organic chemistry were conducted by Gedye et al. (1986) and since then this methodology has been successfully applied in solution-phase synthesis. The first attempts to apply MWs in SPPS were undertaken by Chen who prepared a dipeptide in a kitchen oven (Chen, Chiou, and Wang 1991), next followed by the synthesis of acyl carrier protein (ACP) peptide fragment 65–74 with good yields (Yu, Chen, and Wang 1992). For a while, the field did not advance much because of the general belief that undesired side reactions would be accelerated by MW. However, in the last few years, it has been demonstrated that typical side reactions, including racemizations and aspartimide formation during MW-assisted SPPS, can be controlled. Lowering the MW coupling temperature from 80°C to 50°C limits racemization of amino acids prone to this reaction and the use of piperazine instead of the typically used pyridine for cleavage sequences prone to aspartimide formation eliminates this problem (Palasek, Cox, and Collins 2007). Along with the progress of the technology, manual and automated peptide synthesizers equipped with MW capability, with temperature and pressure control of the reaction mixture, were constructed and are now commercially available.

The main advantages of MW-assisted chemistry are shorter reaction times and higher yields. These advantages originate in part from the unique heat profiles of MW reactions, which cannot be reproduced by classical heating. One important mechanism in MW-assisted chemical reactions is the dipolar polarization mechanism, which is of particular importance for peptide synthesis (de la Hoz, Diaz-Ortiz, and Moreno 2005). This mechanism is thus useful in improving the coupling efficiency of SPPS, which often faces many problems in the assembly of long and difficult sequences due to incomplete reactions possibly caused by aggregation and steric hindrance. These problems have only partly been solved by new coupling reagents and solid supports, and the use of MWs gives another possibility to meet the requirement of fast and efficient solid-phase synthetic strategy.

In the last few years, many articles have been published demonstrating the utility of MW-assisted SPPS. Different types of resins, loading rates, coupling reagents, solvents, and cleavage protocols were investigated (Coantic, Subra, and Martinez 2008; Galanis, Albericio, and Grotli 2009; Pedersen, Sorensen, and Jensen 2009; Subiros-Funosas, Acosta, et al. 2009). MW energy has been recently described to be valuable for the synthesis of difficult sequences, which were previously shown to be difficult to obtain by any conventional strategy. These include gramicidin A, glyco-analog of CSF (Rizzolo et al. 2007), or easily aggregating peptides like amylin (Muthusamy et al. 2010) or β-amyloid (1-42) (Bacsa, Bosze, and Kappe 2010). The labeling of peptides with a variety of fluorophores and quenchers (Fara, Diaz-Mochon, and Bradley 2006), coupling of fatty acids to the peptide side chain (Ni et al. 2010), or the synthesis of glycopeptides (Heggemann et al. 2010) are listed among other developed applications of MW-assisted SPPS. In every one of the presented studies, a shortened time and increased yield of the desired compound was observed, proving that the MW-assisted solid phase synthesis is another valuable tool for peptide chemists.

12.6   DISULFIDE BOND FORMATION

Disulfides in natural peptides and proteins play an important role in the maintenance of biologically active conformation(s). Natural peptides may contain one (e.g., oxytocin, somatostatin, and calcitonin) or more (e.g., endothelin, defensins, and conotoxins) disulfide bonds. In living organisms, disulfide bond formation is one of the enzymatic posttranslational processes. Interest in disulfide formation has not been limited to the reproduction of the patterns of natural structures, but very often peptide analogs containing disulfide bridge are used for structure-activity studies of natural linear peptides. In addition, some of the new peptide-based drugs contain one or more disulfide bridges (e.g., Ziconide, a novel drug for severe chronic pain contains three disulfide bridges).

The synthesis of peptides with only one disulfide bond is usually not difficult. The precursor-peptide synthesis is achieved with both thiols being protected by the same protecting group. After cleavage, the cyclization of the cysteine side chain protecting groups is performed under a high dilution to avoid intermolecular disulfide bridge formation. Air (du Vigneaud et al. 1954), potassium ferricyanide (Hope, Murti, and du Vigneaud 1962), iodine (Flouret et al. 1979), thalium trifluoroacetate (Fuji et al. 1987), or dimethylsulfoxide (DMSO; Tam et al. 1991) is used as oxidizing agent. The other possibility is performing the oxidation step while the peptide is still anchored on resin (Albericio et al. 1997); however, disulfide bond formation in solution provides a compound of higher purity when compared with solid-phase methods.

Some protecting groups for the side chain thiol functionality of cysteine allow simultaneous cleavage and disulfide formation. Treatment of peptides containing an acetamidomethyl (Acm) or trityl (Trt) group with iodine simultaneously cleaves the protecting group and oxidizes the free thiols to yield the disulfide. Similarly, the treatment of S-tBu protectants with mixture of alkyltrichlorosilanes and sulfoxides (MeSiCl$_3$/Ph$_2$SO) produces disulfide in one step (Table 12.2). The useful reviews of thiol protecting groups and methods used for disulfide bond formation together with detailed practical protocols are available (Andreu et al. 1994; Andreu and Nicolas 2000).

Synthesis of peptides with two or more disulfide bonds is much more complicated. The air oxidation method is a frequent approach for synthesizing the native structure of a natural peptide with multiple disulfide bonds from a linear polythiol precursor. In the case when this method is not satisfactory, a sequential approach to disulfide formation is used. Cysteine residues are paired, selectively protected, deprotected, and oxidized. One elegant example of the successful application of a two-step procedure for disulfide bond formation is the synthesis of Hirudin variant 1 (HV1), a small protein consisting of 65 amino acids and 3 disulfide bonds. The linear sequence was assembled on 2-chlorotrityl resin by the sequential condensation of protected fragments, with the use of monomethoxy trityl (Mmt) and Acm groups for cysteine side chain protection. Disulfide bridges were next successfully formed in solution. After cleavage of Mmt group, an oxidative folding step was applied to form two of the three disulfide bonds, followed by Acm iodine oxidation to form the third disulfide bond (Goulas, Gatos, and Barlos 2006).

Another application of this type of approach is the synthesis of 68-mer peptides (proteinase inhibitor LEKTI domain 6) containing two disulfide bonds, which has been reported recently (Vasileiou et al. 2009). An optimized, site-directed disulfide bond formation employing the Mmt and Acm groups for the protection of cysteine residues and DMSO and iodine as oxidants was used.

The use of MW-assisted heating to form a disulfide loop on the peptide bound to a resin during the synthesis of α-conotoxin MII, a 16-residue bicyclic peptide, was also recently reported (Galanis, Albericio, and Grotli 2008). The first disulfide bridge was prepared on a solid support using an MW pulse strategy with a good yield during a reduced time; the second disulfide bond was formed in solution by the oxidative removal of Acm by iodine. However, despite extensive research, the controlled formation of intramolecular disulfide bridges still remains one of the main challenges in the field of peptide chemistry.

TABLE 12.2
Groups Used for Side Chain Protection of Cysteine

Protecting Group	Abbreviation	Structure	Compatibility	Removal	Removal with Oxidation
S-benzyl	Bzl		Boc	HF, TFMSA, Na/liq. NH$_3$	
S-4-methylbenzyl	Meb		Boc	HF, TFMSA, Na/liq. NH$_3$	
S-*tert*-butyl	tBu		Boc, Fmoc	HF, TFMSA, Hg(II)	MeSiCl$_3$/Ph$_2$SO
S-acetamidomethyl	Acm		Boc, Fmoc	Ag(I), Hg(II)	I$_2$, Tl(III)
S-monomethoxy trityl	Mmt		Fmoc	1% TFA in DCM, Ag(I), Hg(II)	I$_2$ Tl(III)
S-trityl	Trt		Fmoc	TFA, Hg(II), Ag(I)	I$_2$ Tl(III)

12.7 CONCLUSIONS

Since the introduction of the SPPS method by Merrifield, tremendous progress has been made in the field of peptide synthesis. With the ever growing inventory of available resins/linkers, protecting groups, coupling reagents, and the implementation of MW-assisted heating, the synthesis of increasingly complex peptides is now possible. In addition, during the last few years, many new methodologies have been developed for the synthesis of peptide analogs with posttranslational modifications (glyco-, phospho-, sulfo-, or disulfide-peptides), fluorescent- or biotin-labeled peptides, or peptide–protein conjugates. Such modified peptides may provide tools to study the functions and mechanisms of action of novel biologically active peptides.

REFERENCES

Albericio, F. 2000. Orthogonal protecting groups for Nα-amino and C-terminal carboxyl functions in solid-phase. *Biopolymers* (*Pept. Sci.*). 55: 123–139.

Albericio, F., Cases, M., Alsina, J., et al. 1997. On the use of PyAOP, a phosphonium salt derived from HOAt, in solid-phase peptide synthesis. *Tetrahedron Lett.* 38: 4853–4856.

Albericio, F., Hammer, R.P., Garcia-Echeverria, C., et al. 1991. Cyclization of disulfide-containing peptides in solid-phase synthesis. *Int. J. Pept. Prot. Res.* 37: 402–413.

Amblard, M., Fehrentz, J-A., Martinez, J., et al. 2006. Methods and protocols of modern solid phase peptide synthesis. *Mol. Biotechnol.* 22; 239–254.

Anderson, G.W., McGregor, A.C. 1957. t-Butyloxycarbonylamino acids and their use in peptide synthesis. *J. Am. Chem. Soc.* 79: 6180–6183.

Andreu, A., Albericio, F., Sole, N.A., et al. 1994. Formation of disulfide bond in synthetic peptides and proteins. In *Methods in Molecular Biology*, Vol. 35. Peptide Synthesis Protocols. Ed. M.W. Pennington and B.W. Dunn. Humana Press, Totowa, NJ.

Andreu D., Nicolas, E. 2000. Disulfide bond formation in synthetic peptides and proteins: the state of the art. In *Solid Phase Synthesis*, Ed. S.A. Kates and F. Albericio, 365–375. Marcel Dekker, NY.

Bacsa, B., Bosze, S., Kappe, C.O. 2010. Direct solid-phase synthesis of the β-amyloid (1-42) peptide using controlled microwave heating. *J. Org. Chem.* 75: 2103–2106.

Barlos, K., Gatos, D., Kallitsis, J., et al. 1989. Darstellung geschützter peptid-fragmente unter einsatz substituierter triphenylmethyl-harze, *Tetrahedron Lett.* 30: 3943–3946.

Bergmann, M., Zervas, L. 1932. Über ein allgemeines Verfahren der Peptid-Synthese. *Ber. Deut. Chem. Ges.* 65B: 1192–1201.

Breipohl, G., Koenig, W. 1991. A new coupling agent for peptide synthesis. *Ger. Offen.* DE 90–4016596.

Brunel, J.M., Salmi, C., Letourneux, Y. 2005. Efficient peptide coupling method of conjugated carboxylic acids with methyl ester amino acids hydrochloride. Application to the synthesis of Fa-Met, an important enzymatic substrate. *Tetrahedron Lett.* 46: 217–220.

Carpino, L.A. 1993. 1-Hydroxy-7-azabenzotriazole. An efficient peptide coupling additive. *J. Am. Chem. Soc.* 115: 4397–4398.

Carpino, L.A., El-Faham, Y. 1999. The diisopropylcarbodiimide/1-hydroxy-7-azabenzotriazole system: segment coupling and stepwise peptide assembly. *Tetrahedron.* 55: 6813–6830.

Carpino, L.A., Han, G.Y. 1972. The 9-fluorenylmethoxycarbonyl amino-protecting group. *J. Org. Chem.* 37: 3404–3409.

Carpino, L.A., Imazumi, H., El-Faham, A., et al. 2002. The uronium/guanidinium peptide coupling reagents: finally the true uronium salts. *Angew. Chem. Int. Ed.* 41: 442–445.

Castro, B., Dormoy, J.R., Evin, G., et al. 1975. Reactifs de couplage peptidique I (1)—l'hexafluorophosphate de benzotriazolyl N-oxytrisdimethylamino phosphonium (B.O.P.). *Tetrahedron Lett.* 16: 1219–1222.

Chen, S.T., Chiou, S.H., Wang, K.T. 1991. Enhancement of chemical-reactions by microwave irradiation. *J. Chin. Chem. Soc.* 38: 85–91.

Coantic, S., Subra, G., Martinez, J. 2008. Microwave-assisted solid phase peptide synthesis on high loaded resins. *Int. J. Pept. Res. Ther.* 14: 143–147.

Coste, J., Lenguyen, D., Castro, B. 1990. PyBOP: a new peptide coupling reagent devoid of toxic by-product. *Tetrahedron Lett.* 31: 205–208.

de la Hoz, A., Diaz-Ortiz, A., Moreno, A. 2005. Microwaves in organic synthesis. Thermal and non-thermal microwave effects. *Chem. Soc. Rev.* 34: 164–178.

du Vigneaud, V., Ressler, C., Swan, J.M., et al. 1954. The synthesis of oxytocin. *J. Am. Chem. Soc.* 76: 3115–3121.

El-Faham, A., Subirós-Funosas, R., Prohens, R., et al. 2009. COMU: a safer and more effective replacement for benzotriazole-based uronium coupling reagents. *Chem. Eur. J. 15*: 9404–9416.

Fara, M.A., Diaz-Mochon, J.J., Bradley, M. 2006. Microwave-assisted coupling with DIC/HOBt for the synthesis of difficult peptoids and fluorescently labeled peptides—gentle heat goes a long way. *Tetrahedron Lett.* 47: 1011–1014.

Fischer, E., Fourneau, E. 1901. Ueber einige derivate des glykocolls. *Ber. Dtsch. Chem. Ges.* 34: 2868–2877.

Flouret, G., Terada, S., Kato, T., et al. 1979. Synthesis of oxytocin using iodine for oxidative cyclization and silica gel adsorption chromatography for purification. *Int. J. Pept. Prot. Res.* 13: 137–141.

Fuji, N., Otaka, A., Funakoshi, S., et al. 1987. Sulphoxide-directed disulphide bond-forming reaction for the synthesis of cystine peptides. *J. Chem. Soc., Chem. Commun.* 21: 1676–1678.

Galanis, A.S., Albericio, F., Grotli, M. 2008. Enhanced microwave-assisted method for on-bead disulfide bond formation: synthesis of α-conotoxin MII. *Pept. Sci.* 92: 23–34.

Galanis, A.S., Albericio, F., Grotli, M. 2009. Solid-phase peptide synthesis in water using microwave-assisted heating. *Org. Lett.* 11: 4488–4491.

Garcia-Martin, F.G., Albericio, F. 2008. Solid supports for the synthesis of peptides. From the first resin used to the most sophisticated in the market. *Chimica Oggi* 26: 29–36.

Garcia-Martin, F., Quintanar-Audelo, M., Garcia-Ramos, Y. 2006. ChemMatrix, a poly(ethyleneglycol)-based support for solid-phase synthesis of complex peptides. *J. Comb. Chem.* 8: 213–220.

Gedye, R., Smith, F., Westaway, K., et al. 1986. The use of microwave ovens for rapid organic synthesis. *Tetrahedron Lett.* 27: 279–282.

Goulas, S., Gatos, D., Barlos, K. 2006. Convergent solid-phase synthesis of hirudin. *J. Pept. Sci.* 12: 116–123.

Heggemann, C., Budka, C., Schomburg, B., et al. 2010. Antifreeze glycopeptide analogues: microwave-enhanced synthesis and functional studies. *Amino Acids.* 38: 213–222.

Hope, D.B., Murti, V.V.S., du Vigneaud, V. 1962. A highly potent analogue of oxytocin, desamino-oxytocin. *J. Biol. Chem.* 237: 1563–1566.

Isidro-Llobet, A., Alvarez, M., Albericio, F. 2009. Amino acid-protecting groups. *Chem. Rev.* 109: 2455–2504.

Itoh, M. 1973. Peptides. IV. Racemization suppression by the use of ethyl 2-hydroximino-2-cyanoacetate and its amide. *Bull. Chem. Soc. Jpn.* 46: 2219–2221.

Izdebski, J. 1979. New reagent suppressing racemizations in peptide synthesis by the DCC metod. *Pol. J. Chem.* 53: 1049–1057.

Kaminski, Z.J., Kolesinska, B., Kolesinska, J., et al. 2005. *N*-Triazinylammonium tetrafluoroborates. A new generation of efficient coupling reagents useful for peptide synthesis. *J. Am. Chem. Soc.* 127: 16912–16920.

Knorr, R., Trzeciak, A., Bannwarth, W., et al. 1989. New coupling reagents in peptide chemistry. *Tetrahedron Lett.* 30: 1927–1930.

Kolesinska, B., Kaminski, Z.J. 2009. Design, synthesis, and application of enantioselective coupling reagent with a traceless chiral auxiliary. *J. Org. Lett.* 11: 765–768.

König, W., Geiger, R. 1970. A new method for synthesis of peptides: activation of the carboxyl group with dicyclohexylcarbodiimide using 1-hydroxybenzotriazoles as additives. *Chem. Ber.* 103: 788–798.

Malow, M., Wehrstedt, K.D., Neuenfeld, S. 2007. On the explosive properties of 1H-benzotriazole and 1H-1,2,3-triazole. *Tetrahedron Lett.* 48: 1233–1235.

McKay, F., Albertson, N.F. 1957. New amine-masking groups for peptide synthesis. *J. Am. Chem. Soc.* 79: 4686–4690.

Merrifield, R.B. 1963. Solid phase peptide synthesis, I. The synthesis of a tetrapeptide. *J. Am. Chem. Soc.* 85: 2149–2152.

Muthusamy, K., Albericio, F., Arvidsson, P.I., et al. 2010. Microwave assisted SPPS of amylin and its toxicity of the pure product to RIN-5F cells. *Pept. Sci.* 94: 339–349.

Ni, S.J., Zhang, H.B., Huang, W.L., et al. 2010. Solid phase synthesis of fatty acid modified glucagon-like peptide-1(7-36) amide under thermal and controlled microwave irradiation. *Chin. Chem. Lett.* 21: 27–30.

Palasek, S.A., Cox, Z.J., Collins, J.M. 2007. Limiting racemization and aspartimide formation in microwave-enhanced Fmoc solid phase peptide synthesis. *J. Pept. Sci.* 13: 143–148.

Pedersen, S., Sorensen, K.K., Jensen, K.J. 2009. Semi-automated microwave- assisted SPPS: optimization of protocols and synthesis of difficult sequences. *Pept. Sci.* 94: 206–212.

Rizzolo, F., Sabatino, G., Chelli, M., et al. 2007. A convenient microwave-enhanced solid-phase synthesis of difficult peptide sequences: case study of gramicidin A and CSF114(Glc). *Int. J. Pep. Res. Ther.* 13: 203–208.

Sheehan, J.C., Hess, G.P. 1955. A new method of forming peptide bonds. *J. Am. Chem. Soc.* 77: 1067–1068.

Story, S.C., Aldrich J.V. 1994. Side-product formation during cyclization with HBTU on a solid support. *Int. J. Pept. Protein Res.* 43: 292–296.

Subiros-Funosas, R., Acosta, G.A., El-Faham, A., et al. 2009. Microwave irradiation and COMU: a potent combination for solid-phase peptide synthesis. *Tetrahedron Lett.* 50: 6200–6202.

Subiros-Funosas, R., Prohens, R., Barbas, R., et al. 2009. Oxyma: an efficient additive for peptide synthesis to replace the benzotriazole-based HOBt and HOAt with lower risk of explosion. *Chem. Eur. J.* 15: 9394–9403.

Tam, J.P., Wu, C.R., Liu, W., et al. 1991. Disulfide bond formation in peptides by dimethyl sulfoxide. Scope and applications. *J. Am. Chem. Soc.* 113: 6657–6662.

Valeur, E., Bradley, M. 2009. Amide bond formation: Beyond the myth of coupling reagents. *Chem. Soc. Rev.* 38: 606–631.

Vasileiou, Z., Barlos, K.K., Gatos D., et al. 2009. Synthesis of the proteinase inhibitor LEKTI domain 6 by the fragment condensation method and regioselective disulfide bond formation. *Pept. Sci.* 94: 339–349.

Wehrstedt, K.D., Wandrey, P.A., Heitkamp, D. 2005. Explosive properties of 1-hydroxybenzotriazoles. *J. Hazard. Mater.* 126: 1–7.

Yu H.-M., Chen, S.T., Wang, K.-M. 1992. Enhanced coupling efficiency in solid-phase peptide synthesis by microwave irradiation. *J. Org. Chem.* 57: 4781–4784.

13 Addressing the Interactions between Opioid Ligands and Their Receptors
Classic and Nonclassic Cases

Luca Gentilucci

CONTENTS

13.1 INTRODUCTION

Opioids are widely utilized as analgesics for sedating acute pain (Heitz, Witkowski, and Viscusi 2009) or for prolonged treatment of chronic pain caused by cancer, nerve damage, arthritis, and other diseases (Przewlocki and Przewlocka 2001; Bountra, Munglani, and Schmidt 2003; Gentilucci 2004; Manchikanti et al. 2010). The mechanisms involved in the transition from acute to chronic pain are complex, with the involvement of interacting receptor systems and intracellular ion flux, second messenger systems, new synaptic connections, and apoptosis (Vadivelu and Sinatra 2005).

Unfortunately, a prolonged administration often produces several undesired side effects, in particular, tolerance, the decrease of drug efficacy with repeated administration, causing patients to require escalating doses of opioid to maintain the same level of analgesia, and dependence, a physical, behavioral, and cognitive syndrome in which the use of the drug takes on a high priority. Many other adverse symptoms are also connected to opioids: constipation, dry mouth, edema, headache, insomnia, itching and skin rash, nausea and vomiting, urinary retention, weight gain, and so on.

In certain cases, chronic pain becomes resistant to single-agent therapy over time, necessitating a multimodal approach to therapy. The combination of standard analgesics with NSAIDs or adjuvant drugs can improve the efficacy of the treatment (Raffa et al. 2003; McDaid et al. 2010), particularly for cancer pain. Despite the widespread availability of analgesic therapies, cancer pain remains undertreated (Montagnini and Zaleon 2009). Besides, the degree to which these therapies are successful in different patients varies, and several factors have been identified as or hypothesized to be the cause for this large variability. Among these factors, receptor mutations cause interindividual variability of the clinical effects of opioids. With the recent advances in genetic research, inherited causes of the variability of opioid therapy can be investigated (Loetsch and Geisslinger 2005).

As a consequence, it appears that there is an urgent need for novel therapeutic protocols, as well as for innovative compounds, that can inhibit pain transmission based on alternative action mechanisms, not involved in the arising of harmful side effects. The rationale design of new, safer analgesics requires comprehending in detail the modality of receptor activation triggered by the interaction with the agonists. This chapter discusses the current opinions on the mechanisms of ligand–receptor interaction and activation and stresses the recently emerged evidences on the uncommon mechanisms by which some ligands deprived of relevant pharmacophores interact and eventually activate the opioid receptor (OR).

13.2 STRUCTURE AND FUNCTIONS OF THE OPIOID RECEPTORS

The opioid receptors are members of the seven-helix transmembrane G-protein-coupled receptors (GPCR) family. Three different types were cloned in the early

1990s, beginning with the mouse δ-opioid receptor (DOR) and followed by μ-opioid receptor (MOR) and κ-opioid receptor (KOR). These three classes have been further divided into several subclasses: MOR1 and MOR2, DOR1 and DOR 2, KOR1, KOR 2, and KOR 3; however, they all share extensive structural homologies (Chaturvedi et al. 2000). In addition, the orphan opioid-receptor-like (ORL1) receptor shares around 60% of the sequence homology with ORs, but it produces different, and even opposite, pharmacological effects (Meunier 1997).

The receptors bind to cytoplasmic G-proteins, which consist of the three subunits α, β, and γ. When the appropriate ligand binds the receptor at the extracellular side of the membrane, guanosine diphosphate (GDP) dissociates from the α-subunit of the G-protein; guanosine triphosphate (GTP) then binds to the empty guanine nucleotide-binding pocket of the G-protein, causing the dissociation of the α-subunit and the βγ-subunits. The GTP-bound α-subunit and the βγ-subunit then activate a number of signaling pathways: activation of inwardly rectifying K-channels, inhibition of voltage-operated K-channels, and inhibition of adenylyl cyclase. Other responses of unknown intermediate mechanisms are the activation of phospholipase A (PLA), phospholipase C (PLC), MAP Kinase, and some Ca-channels.

The precise 3D structures of ORs have not yet been discovered. The first crystal structure of a GPCR that appeared in the literature was that of *Bos taurus* Rhodopsin (Bt_Rho) (Palczewski et al. 2000), whereas the first crystal structures of a nonrhodopsin GPCR, the *Homo sapiens* β2-adrenergic receptor (Hs_Adrb2) bound to partial inverse agonist carazolol 14-16, were obtained at the end of 2007. There are currently 22 crystal structures of inverse agonist-bound class A GPCRs described in the literature. Modeling based on the rhodopsin crystal structure gave the opportunity to investigate the structural features of ORs, both in the active and in the nonactive states (Ananthan, Zhang, and Hobrath 2009; Mobarec, Sanchez, and Filizola 2009).

ORs exist in homooligomeric or heterooligomeric complexes (Levac, O'Dowd, and George 2002), as is the case with many other members of the GPCR family (Filizola and Weinstein 2002), so the pharmacological responses of the receptors are cross-modulated and cross-regulated. ORs are quite promiscuous and can form heterodimers also with nonopioid receptors, for example, MOR with α2a-adrenoceptors (Milligan 2004). In addition, opioid receptors also interact with a myriad of other proteins, which alter ligand binding or functional responses as well as receptor localization and processing (Presland 2004).

The MORs produce the strongest analgesia, but they can also cause euphoria, respiratory depression, physical dependence, bradycardia, and so on. Therefore, pure MOR agonists, such as morphine, provide the best analgesics, but also the maximum side effects. Their use is best limited to short-term "rescue" analgesia, though chronic pain, especially in the case of cancer, may require continual use in the later stages of the disease. KORs trigger a lesser analgesic response and may cause miosis, sedation, and dysphoria. DORs possess a lower antinociceptive efficacy, but they might have a reduced addictive potential. Mixed agonist/antagonists, such as butorphanol, are not considered useful in the management of chronic pain.

13.3 OPIOID LIGANDS

13.3.1 MOR Agonists Derived from Morphine and Fentanyl

The majority of opioid analgesics currently used in hospital practise are MOR agonists derived from morphine (Bountra et al. 2003; Breivik 2005; Johnson, Fudala, and Payne 2005). The progressive simplification of morphine structure leads to the morphinans, such as butorphanol, to the benzomorphans, such as, pentazocine, ketocycloazocine, and bremazocine, to the piperidines, such as fentanyl, pethidine, and so on, to the phenylpropylamines, including methadone (Figure 13.1).

The structures of the MOR-selective compounds, codeine, buprenorphine, methadone, fentanyl, and sufentanyl, are shown in Figure 13.2.

13.3.2 DOR and KOR Agonists

DOR agonists are attractive potential analgesics, as these compounds exhibit strong antinociceptive activity with relatively few side effects (Varga et al. 2004). A few DOR-selective analgesics have been identified through the modification of morphine, including the DOR-agonists, TAN-67 and SB 213698, or the piperazine derivatives, BW 373U86 and SNC 80 (Figure 13.3) (Calderon and Coop 2004). In the past decade, several novel classes of δ-opioid agonists have been synthesized.

The many unwanted side effects associated to the MOR agonists stimulated considerable interest in developing ligands for KOR as potential analgesics and for the treatment of a variety of other disorders (Aldrich and McLaughlin 2009). Among the KOR-selective analgesics, it is worth mentioning the arylacetamide derivative U50,488 (Vonvoigtlander, Lahti, and Ludens 1983) and its derivatives spiradoline (U62,066) and enadoline (CI-977) (Figure 13.3), which have entered

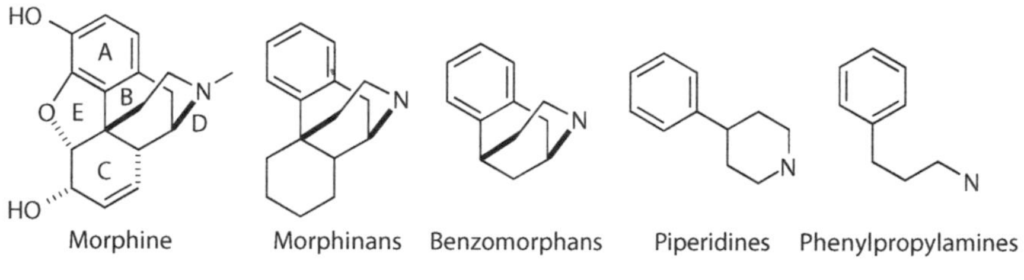

FIGURE 13.1 Structures of morphine-derived agonists.

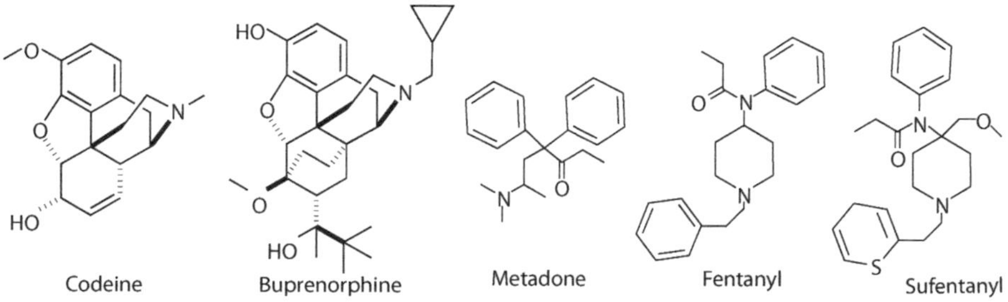

FIGURE 13.2 Structures of some MOR-active compounds utilized in clinical practise.

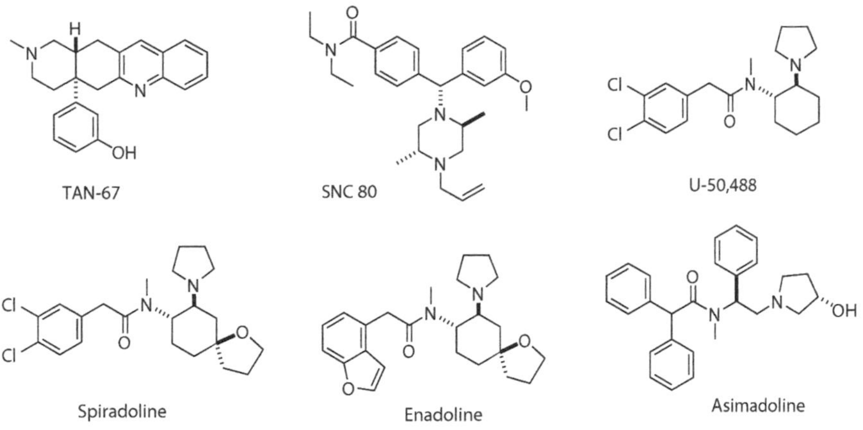

FIGURE 13.3   Structures of some DOR- and KOR-active compounds.

clinical trials as centrally acting analgesics: Finally, the risk of CNS-mediated, mechanism-related side effects of sedation and dysphoria stimulated the discovery of analogs with limited brain penetration to produce a peripherally mediated analgesic effect in inflammatory conditions, such as asimadoline, EMD-61753 (Figure 13.3).

13.3.3 Endogenous Opioid Peptides

The main groups of endogenous opioid peptides (Gentilucci 2004; Janecka et al. 2010) found in the mammals are enkephalins (YGGFM, YGGFL), dynorphins (YGGFLRRIRPKLKWDNQ,   YGGFLRRQFKVVT),   and   β-endorphins (YGGFMTSEKSQTPLVTLFKNAIIKNAYKKGE), deriving from the propeptides, proenkephalin, prodynorphin, and proopiomelanocortin, respectively. These peptides were isolated during the last 20–30 years; more recently, a couple of novel peptides having an unusual and short sequence have been discovered in the brain and named endomorphin-1 (YPWF-NH$_2$) and endomorphin-2 (YPFF-NH$_2$) (Zadina et al. 1997) and are currently regarded as the real endogenous analgesics in mammals, being released in response to pain stimuli with a strong antinociceptive effect against neuropathic pain and acute pain (Horvath 2000; Fichna et al. 2007). Another neuropeptide that joined the family of endogenous opioid peptides in the last few years is nociceptin (FGGFTGARKSARKLANQ), or orphanin FQ (noc/oFQ). This peptide has a nociceptive potential; however, recent data seem to indicate that nociceptin can act also as a spinal analgesic (Meunier 1997).

The lengths of the endogenous opioid peptides vary from 4 to 31 amino acids, nevertheless their structures are somewhat correlated. They all show a common fragment at the N-terminus, the "message," while the C-terminus is known as the "address," and is responsible for moderate receptor selectivity. Indeed, enkephalins show a slight selectivity for DOR over MOR; β-endorphin equally binds to both DOR and MOR; dynorphin A and its truncated fragments show a certain preference for KOR, whereas nociceptin has a good selectivity for ORL1 receptors. On the other hand, EM1 and EM2 show a noteworthy selectivity for MOR.

The endogenous opioid peptides have been extensively studied as potentially useful analgesics, as they do not produce undesired immunomodulatory, cardiovascular, and respiratory side effects (Wilson et al. 2000; Przewlocki and Przewlocka 2001; Fichna et al. 2007). Unfortunately, they cannot be used for clinical practise for treating pain in humans, as they are readily degraded *in vivo* by several peptidases (Witt et al. 2001; Gentilucci, De Marco, and Cerisoli 2010).

Besides their analgesic effects, opioid peptides are involved in regulating a number of nonopioid physiological functions (Vaccarino and Kastin 2001; Wollemann and Benyhe 2004; Fichna et al. 2007; Bodnar 2009), such as locomotion, gastrointestinal, renal, and hepatic functions, cardiovascular responses, respiration and thermoregulation, immunological responses, sexual activity, pregnancy, and also behavioral and emotional effects, stress response, learning, memory, eating and drinking, alcohol and drug consumption, mood, and seizures.

13.3.4 Nonopioid Peptides

In addition, some classes of nonopioid peptides display some activity toward the opioid receptors (Teschemacher 1993). For instance, food protein fragments have been found to elicit opioid effects (Teschemacher 2003). Many of these "food hormones" are fragments of the milk proteins α-, β-, or κ-casein; α-lactalbumin; β-lactoglobulin; or lactotransferrin (e.g., β-casomorphin-7, YPFPGPI, and β-casomorphin-5, YPFPG). Other opioid-active proteins can be obtained from plants, wheat gluten, rice albumin, spinachs, or from bovine serum albumin, meat proteins, and so on.

Amphibian skin contains the peptides dermorphins (YaFGYPS-NH$_2$, etc.), deltorphins (YmFHLMD-NH$_2$, etc.) (Negri, Melchiorri, and Lattanzi 2000), and others that are often homologous or even identical to the gastrointestinal hormones and neurotransmitters of the mammals. Finally, several other kinds of nonopioid or antiopioid peptides have demonstrated analgesic activity. Examples are hitogranin and related peptides (Ruan, Prasad, and Lemaire 2000), neuropeptide FF (Roumy and Zajac 1999), cholecystokinin (Noble, Smadja, and Roques 1994), interleukin-2 (Jiang et al. 2000), spinorphin (Honda et al. 2001), rubiscolins (Yang et al. 2001), and so on.

13.3.5 Peptidomimetics

As discussed in the previous sections, native opioid peptides generally suffer with scarce receptor selectivity. A second important limitation to the use of native peptides as pharmacological tools is their rapid enzymatic *in vivo* degradation by many endo- and exopeptidases. Finally, the peptides display a poor ability to cross the blood–brain barrier (BBB) and this causes scarce bioavailability (Witt et al. 2001). These inherent limitations prompted the development of a huge number of opioid peptidomimetic (Hruby and Balse 2000; Gentilucci 2004; Eguchi 2004; Janecka and Kruszynski 2005; Janecka et al. 2010) and nonpeptidomimetic ligands (Kaczor and Matosiuk 2002; Lipkowski et al. 2004), aiming to improve receptor affinity, selectivity, and bioavailability (Gentilucci et al. 2010).

Due to the extraordinary number of peptidomimetics so far described in the literature, a detailed discussion is beyond the scope of this review. To appreciate the different strategies adopted to design the peptide analogs, the modifications introduced in the structure of endomorphins can be utilized as prototypic examples (Keresztes, Borics, and Tóth 2010): introduction of D-amino acids (Okada et al. 2000; Paterlini et al. 2000; Fichna et al. 2005); introduction of β-Pro or homo-Pro in place of Pro2, giving Y-(R)-βPro-WF-NH$_2$ (Cardillo et al. 2000) and Y-(S)-homoPro-WF-NH$_2$ (Cardillo et al. 2002), respectively; the palindromic dimerization of EM2 [YPFF-NH-CH$_2$-CH$_2$-]$_2$ (Gao et al. 2005), which was inspired from the palyndromic peptide biphalin, YaGF-NH-NH-FGaY (Lipkowski, Konecka, and Sroczynska 1982); the introduction of isosteric or isoelectronic peptide bonds, for example, a reduced (CH$_2$NH) amide bond between Tyr1 and Pro2 (Zhao et al. 2005); introduction of modified groups in position 3 (Fichna et al. 2005); introduction of conformational constraints in different positions, such as a β-turn backbone mimetic (Tömböly et al. 2008), the *cis*-Tyr1-pseudoPro2 (Keller et al. 2001), or the *cis*-Dmt1-Pro2 bond (Okada et al. 2003), or the *trans* constrained Tyr1-Chx2 (Doi et al. 2002); *N* to *C* cyclization (Gentilucci 2004; Gentilucci et al. 2008) or side chain to *C* cyclization (Perlikowska, do-Rego, and Cravezic 2010).

Some selected examples of outstanding opioid peptidomimetics are shown in Figure 13.4: the MOR-selective DAMGO, YaGMeF-Glyol (Handa et al. 1981), the cyclopeptides JOM-6, Y-c(S-Et-S) [cF-D-Pen]NH$_2$, and the DOR-selective JOM-13, Y-c[cF-D-Pen]OH) (Mosberg et al. 1988) and DPDPE, Y-c[D-Pen-GF-D-Pen]-OH (Akiyama et al. 1985).

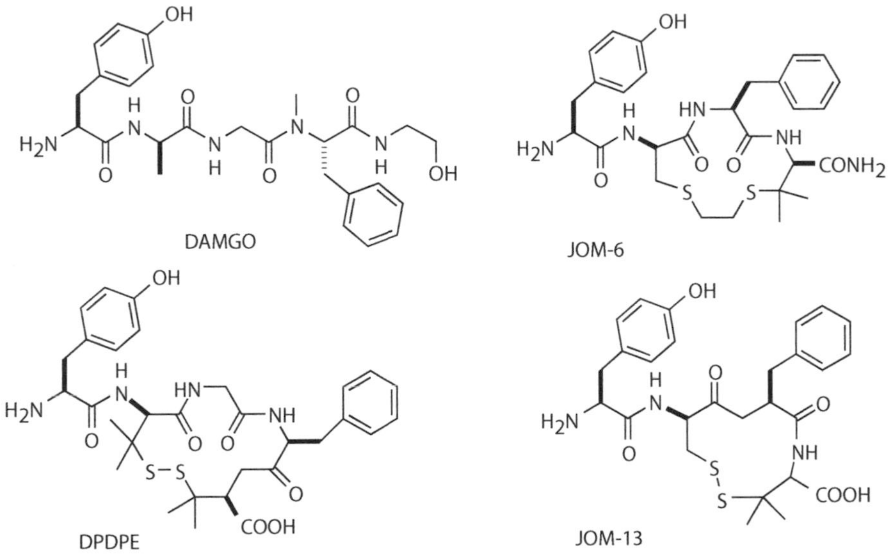

FIGURE 13.4 Selected examples of opioid peptidomimetics.

13.4   OPIOID LIGAND–RECEPTOR INTERACTION MODELS

13.4.1   INVESTIGATION OF THE INTERACTIONS BETWEEN OPIOID LIGANDS AND RECEPTORS

The binding pockets of the receptors in the active and the inactive states are not identical, which implies distinct interactions for the agonist and the antagonist, and the precise differences between the active and the inactive forms of the receptors have not been fully elucidated (Fowler et al. 2004). The many efforts dedicated to investigating the agonist—receptor interaction did not converge toward a unique model. Rather, different classes of agonists seem to adopt specific interaction modes (e.g., opioid peptides, morphine, and fentanyls). Over the years, for these categories, a number of somewhat differing models have been proposed, mostly for the lack of direct information on the structures of ligand–receptor complexes. Indeed, the X-ray diffraction of membrane-bound GPCRs met with considerable experimental obstacles. As a consequence, most of the present knowledge about ligand interactions with the receptors derives from structure–activity relationship studies (SAR) (Hruby and Agnes 1999; Hruby and Balse 2000; Kaczor and Matosiuk 2002; Bernard, Coop, and MacKerell 2003; Eguchi 2004).

Collecting all the information independently available for the ligands and the receptors, plausible models of reciprocal contacts and interactions have been simulated by computational methods using homology models constructed based on the 3D structure of rhodopsin-like GPCRs as a template (Sagara et al. 1996; Pogozheva, Lomize, and Mosberg 1998; Subramanian et al. 2000; Mosberg and Fowler 2002; Zhang et al. 2005). For most opioid ligands, the construction of structural models began with placing the protonated amine close to the conserved Asp in TM-III. Docking procedures allowed calculation of the selective ligands' best orientation within the binding site and determined which receptor residues were likely to be involved in the interaction.

13.4.2   MORPHINE

The models proposed to explain morphine binding postulated an ionic interaction between the cationic tertiary amino group and Asp147 in TM-III. Figure 13.5 also shows the hydrophobic interaction between the phenol group of morphine and the phenol group of Tyr299 in TM-VI and the H-bond between the phenolic-OH of the A ring and both the amino group of Lys303 at TM-VI and the phenolic-OH of Tyr148 at TM-III (Sagara et al. 1996). An alternative model (Pogozheva et al. 1998) proposes a different pose of the rigid framework of morphine within the receptor pocket, so that the phenolic-OH interacts with His297 at TM-VI in a similar way as the phenolic-OH of opioid peptides.

13.4.3   FENTANYLS

Different fentanyl derivatives have been utilized for molecular-docking computations (Dosen-Micovic, Ivanovic, and Micovic 2006). The protonated tertiary nitrogen of ohmefentanyl electrostatically interacts with Asp149 (mouse). Other relevant

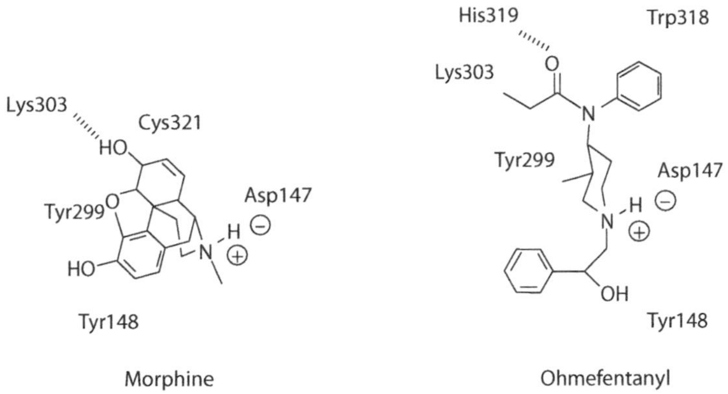

FIGURE 13.5 Sketches of the relevant ligand–receptor interactions for morphine (left) and ohmefentanyl (right).

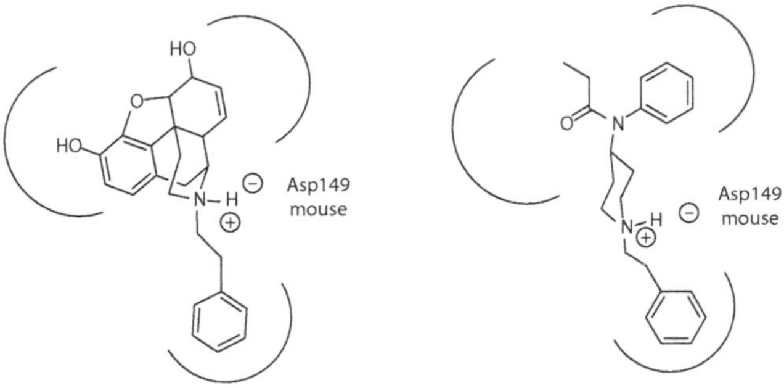

FIGURE 13.6 Different binding poses of *N*-phenethylnormorphine (left) and *cis*-3-methylfentanyl (right).

interactions are an H-bond between amide carbonyl of ohmefentanyl and imidazole of His319, and π–π interactions between two aryl groups of ligand and of Trp318, His319, and Tyr148 (Figure 13.5).

Another suggestive study compared the bounded conformations of *cis*-3-methylfentanyl and *N*-phenethylnormorphine (Figure 13.6); the cationic amines of both ligands bind the same Asp and the *N*-phenethyl groups are situated into equivalent binding domains, but the remaining parts of the ligands occupy different receptor regions (Subramanian et al. 2000).

13.4.4 MOR-SELECTIVE PEPTIDE AND PEPTIDOMIMETIC AGONISTS

Intensive work has been done on the SAR analysis of short peptides, such as enkephalins, endomorphins (EMs), morphiceptin (YPFP), and their constrained analogs. These studies lead to the general opinion that Tyr1 is the primary pharmacophore, responsible for receptor binding and activation, whereas the spatial disposition of Phe3, situated 11–12 Å far apart from Tyr1, seems to be more important for receptor specificity (Podlogar et al. 1998; Paterlini et al. 2000; Janecka et al.

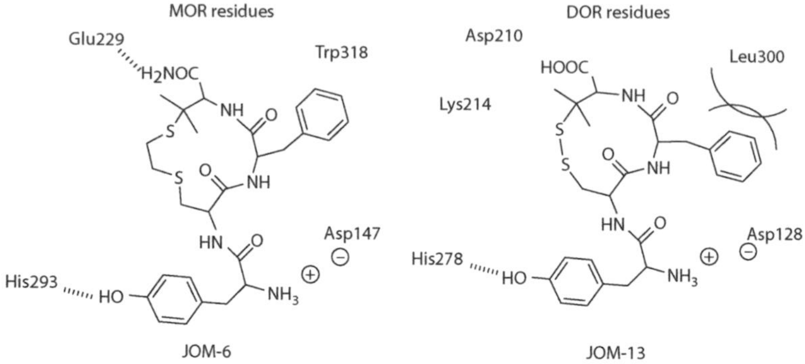

FIGURE 13.7 The different display of the Phe aromatic side chain in the MOR agonist JOM-6 and in the DOR agonist JOM-13 is the main selectivity probe responsible for receptor selectivity.

2002; Eguchi 2004; Gentilucci 2004; In et al. 2005; Shao et al. 2007; Liu et al. 2009; Borics and Toth 2010).

The cyclic peptidomimetic JOM-6 (Mosberg and Fowler 2002; Fowler et al. 2004) is by far the most effective molecular probe utilized to understand the interactions with the MOR in the bottom of the receptor pocket (Figure 13.7). Exhaustive studies revealed the fundamental ionic bond between the protonated amine of Tyr1 and the carboxylate side chain of Asp147 of TM-III, whereas the phenolic-OH group forms ionic interactions or H-bond with His297 of TM-VI. The π–π interaction of the phenolic ring with the indole of Trp293 of TMH VI seems fundamental for triggering the activation of the G-protein. The remaining portion of the ligand occupies the outer part of the pocket; Phe is responsible for the subtype specificity of the ligands (Figure 13.7). The binding pocket of the active and the inactive states is not identical, thus determining distinct interactions for agonists and antagonists (Fowler et al. 2004).

13.4.5 DOR Agonists

It is commonly believed that the tyramine moiety of DOR-agonists interact with Asp128 and His278 (Pogozheva et al. 1998). The comparison of the DOR-selective JOM-13 (Figure 13.7) and the MOR-selective JOM-6 docked into the respective receptors revealed that the Phe residues of the two ligands must accommodate into specific pockets (Mosberg and Fowler 2002; Eguchi 2004). These have different sizes and shapes, originated by the presence of Trp318 in the MOR, and of Leu300 in DOR, and this would be responsible for the preference shown by the two peptides for the different receptors.

13.4.6 KOR Agonists

The precise mechanism by which KOR agonists, such as U69,593 and other nonopiates, interact with KOR is not known in detail. The model reported in Figure 13.8 (Pogozheva

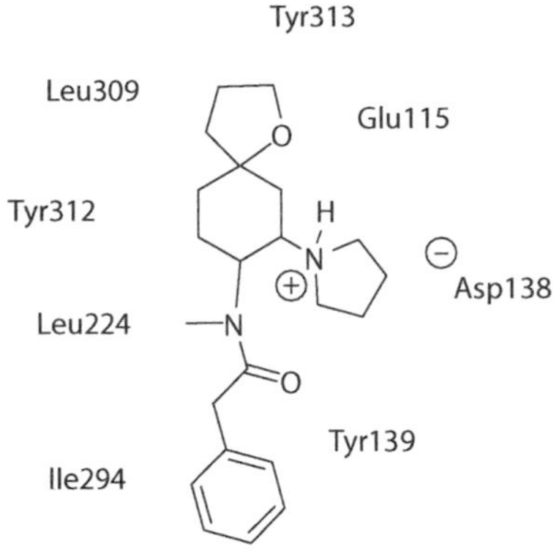

FIGURE 13.8   Sketches of the relevant ligand–receptor interactions for U69,593.

et al. 1998) shows the ionic interaction of the carboxyl side chain of Asp138 in TM-III with cationic pyrrolidine nitrogen, and the carbonyl of arylacetamide portion hydrogen bonded to Tyr139. It was suggested that the selectivity of U69,593 for the KOR is connected to the fitting of a KOR-specificity pocket.

13.5   OPIOID LIGANDS DEPRIVED OF SOME PHARMACOPHORES

13.5.1   DERIVATIZATION OR DELETION OF TYR1

The large majority of the agonists possess a tyramine moiety, generally believed fundamental for ligand–receptor interaction. Although the phenolic OH of Tyr1 does play an important role in the interaction of peptides with the ORs, this role is not critical. For instance, the transposition of the Tyr1 with Phe3 in YkFA gave an analog with lower affinity but with higher selectivity than YkFA (Burden et al. 1999). The same kind of residue exchange in the cyclic peptide JOM-6, rendered the MOR agonist JH-54 with only fourfold reduced affinity (Mosberg, Ho, and Sobczyk-Kojiro 1998). In a similar way, Tyr can be substituted in the DPDPE sequence with a variety of aromatic residues lacking hydroxy group without drastic loss of potency and affinity (Arnold and Schiller 2000; McFadyen et al. 2000). Peptidomimetics containing (*S*)-4-(carboxamido) phenylalanine (Cpa) as a surrogate of Tyr1 displayed comparable binding affinities and potencies with respect to their parent peptides (Dolle et al. 2004), as the carboxamido residue also acts as an H-bond donor.

On the other hand, the removal or the derivatization of the charged *N*-terminal amino group generally transformed the opioid agonists into inactive compounds or into antagonists (Schiller et al. 2003; Weltrowska et al. 2004). For instance, cyclic hexapeptides derived from β-casomorphin, having the terminal amino group deleted or formylated, displayed both MOR and DOR antagonist properties (Schiller et al. 2000). The substitution of Tyr1 in the structure of EM-1 with a benzyl carbamate group gave a peptide, which maintained affinity for MOR (Cardillo et al. 2003). The enkephalin analog (2S)-Mdp-aGFL-NH$_2$, Mdp being

(2S)-2-methyl-3-(2',6'-dimethyl-4'-hydroxyphenyl)-propionic acid, turned out to be a quite potent DOR antagonist and a somewhat less potent MOR antagonist (Lu, Weltrowska, et al. 2001). Similarly, the enkephalin and the Dyn A analogs obtained by replacement of Tyr1 with 3-(2',6'-dimethyl-4'-hydroxyphenyl)propanoic acid (Dhp), or Mdp (Lu, Nguyen, et al. 2001), are examples of Dyn A-derived-antagonists with high KOR selectivity, dynantin. Finally, the dyn A analog, cyclo(N,5) [Trp3,Trp4,Glu5] dynorphin A-(1-11)NH2, is an *N*-terminal-to-side chain cyclic peptide lacking the basic *N*-terminus that showed good KOR affinity (Vig, Murray, and Aldrich 2003).

From the inspection of the above reported examples, it is evident that the removal or derivatization of the tyramine is generally incompatible with receptor activation. The very few compounds deprived of the protonable amino group, which maintained an agonist behavior (Gentilucci, Squassabia, and Artali 2007), are enlisted here by the year of publication (Figure 13.9): the MOR-selective bicyclic enkephalin mimetic (**1**) (Eguchi et al. 2002); the KOR-selective neoclerodane diterpene salvinorin A (**2**) (Roth et al. 2002); the MOR active cyclic analog of EM-1 c[YpwFG], also referred to as CycloEM (**3**) (Cardillo et al. 2004); and the "Carba"-analogues of fentanyl equipped with a guanidino group (**4**) (Weltrowska et al. 2010).

The highly constrained 6,6-bicyclic system **1** (Figure 13.9), which has no terminal amino group, showed an initial level of analgesic activity similar to that of morphine, although the *in vivo* half-life was shorter than that of morphine (Eguchi et al. 2002). This compound belongs to a series of constrained peptidomimetics designed to mimic

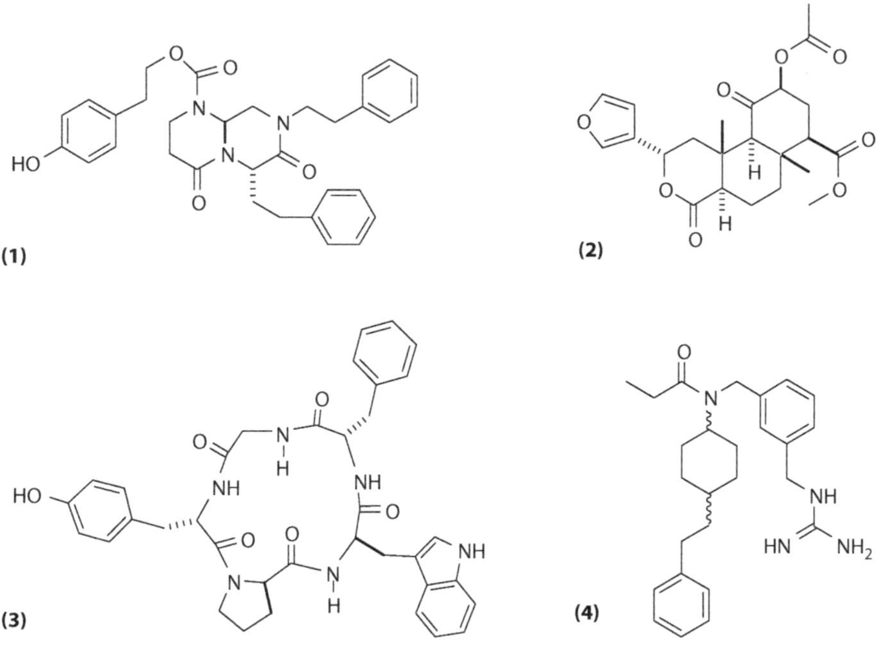

(1)   **(2)**

(3)   **(4)**

FIGURE 13.9 Structures of opioid ligands lacking the protonable amino group: the MOR agonists (**1**), cycloendomorphin-1 (**3**), "Carba"-analogues of fentanyl (**4**), and the KOR-agonist, salvinorin A (**2**).

enkephalin or EM β-turn models. On the basis of 2D NMR analysis and molecular mechanics, the authors noticed a partial superimposition of its structure with a *trans*-EM1 type III β-turn backbone structure. Based on this assumption and the MOR-selectivity profile of **1**, the authors implicitly suggested that the interaction with the receptor might mimic that of EM or enkephalins, even in the absence of an ionic interaction.

In the fentanyl derivative **4** (Figure 13.9) having a guanidine appendage, the piperidine ring nitrogen was replaced by a carbon, giving the *cis*- and the *trans*-isomeric carba-analogs (Weltrowska et al. 2010). Both the isomers showed reduced MOR- and DOR-binding affinities as compared to the parent fentanyl, but retained opioid full agonist activity with about half the potency of Leu-enkephalin. The binding of compound **4** was simulated by molecular docking. The compounds likely retain the same binding mode of the "normal" fentanyls; however, it was observed that the ionic bridge between the guanidine moiety of the ligands and an Asp in the second extracellular loop significantly contributes to the receptor binding.

On the other hand, the compounds **2** and **3** (Figure 13.9) are truly deprived of any ionic interactions and are agonists of the respective MOR and KOR. It can be postulated that these compounds must have a receptor-binding mode quite different from that of the classical nitrogen-containing agonists (Gentilucci et al. 2008).

13.5.2 THE CASE OF SALVINORIN

Salvinorin A (**2**) is a neoclerodane diterpene hallucinogen (Roth et al. 2002; Yan et al. 2005) isolated from *Salvia divinorum* and unique among the other opioid agonists, because it is the first nonnitrogen-containing selective KOR agonist. This compound has the potential to treat disorders characterized by alterations in perception, including schizophrenia, Alzheimer's disease, and bipolar disorder.

The first docking model proposed for salvinorin was based on nonopioid KOR-selective agonists, such as U69,593 (Roth et al. 2002), and showed four hydrogen-bonding interactions as the main interaction forces, involving the amide-NH of Gln115 in TM-II and furanoic oxygen, the phenolic-OH of Tyr139 in TM-III and lactone-CO, the phenolic-OH of Tyr312 in TM-VII and methoxycarbonyl group, and the phenolic-OH of Tyr313 in TM-VII and acetyl group (Figure 13.10a). Further, the screening of salvinorin derivatives for binding affinity and functional activity suggested that the methyl ester and furan rings are required for activity, but that the lactone and ketone functionalities are not (Munro et al. 2005).

More recently, the same authors utilized improved data of the KOR and calculated a new model, in which the molecule seems to be stabilized in the binding pocket by interactions with tyrosine residues in TM-VII (Tyr313 and Tyr320) and TM-II (Tyr119), but the interaction of Tyr313 with 2-acetoxy group of salvinorin A is hydrophobic (Figure 13.10b).

Also, this second model was subsequently revisited. More recently acquired structure–function data of salvinorin analogs (Beguin et al. 2005; Harding et al. 2005) led to a third model (Figure 13.10c). The potential binding-site interactions of salvinorin A with the KOR were studied using a combination of wild-type, chimeric, and single-point mutant opioid receptors (Kane et al. 2006). A unique binding epitope for salvinorin A interacted with the receptor in Gln115 and Tyr119 from

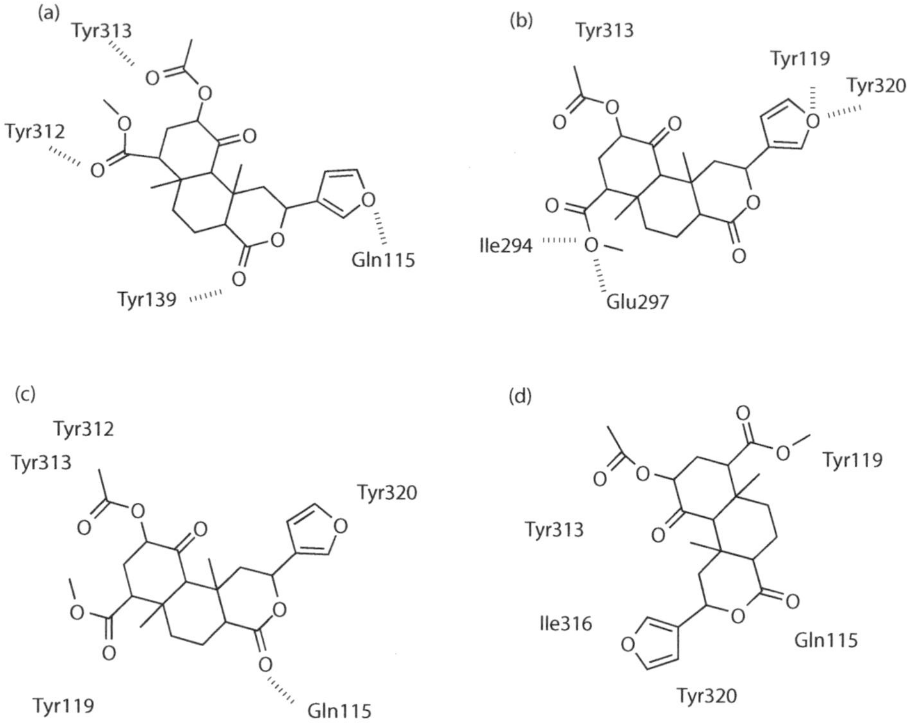

FIGURE 13.10 The different receptor-bound structures of salvinorin A appeared in the literature (2002–2008, see text).

TM-II and, Tyr312, Tyr313, and Tyr320, from TM-VII. In particular, Gln115 most likely serves as a hydrogen bond donor. No recognition elements were identified in TM-VI that have been previously considered important both in the transition from an inactive to an active state of the receptor or in modulating selectivity. No support was found for the involvement of Asp138 in TM-III or Glu297 at the rim of TM-VI.

An ulterior refinement required the use of site-directed mutagenesis and molecular-modeling techniques (Kane, McCurdy, and Ferguson 2008). This model (Figure 13.10d) shows salvinorin A vertically posed within a pocket spanning transmembrane II and VII, with the 2' substituent directed toward the extracellular domains, and relevant contacts with the residues Q115, Y119, Y313, I316, and Y320. In particular, the mutation of I316 was found to completely abolish binding. The model explains the role that hydrophobic contacts play in binding and gives insight into the MOR selectivity of 2'-benzoyl salvinorin (herkinorin).

13.5.3 THE CASE OF CYCLOENDOMORPHINS

The cyclopentapetide CycloEM (**3**) is the most active member of a mini-library of diastereomeric, 3D distinct analogs of EM-1, designed to be more stable *in vivo* and more lipophilic (Cardillo et al. 2004). The library was originated by introducing all of the amino acids of the sequence in L or D configuration, and the cyclic structure was obtained by connecting the *N*- and *C*-terminus with a Gly bridge.

The diastereomeric CycloEMs were tested as MOR ligands in binding experiments, using tritiated DAMGO as the specific radioligand. Apparently, the diverse stereochemistry arrays of the compounds confer them alternative 3D displays and different pharmacological features (Gentilucci, Tolomelli, and Squassabia 2007), as only the diastereoisomer **3** revealed a good receptor affinity in the 10^{-8} M range. The agonism of CycloEM was assessed by the inhibition of the forskolin-stimulated production of cAMP in intact human neuroblastoma cells. Besides, CycloEM induced receptor internalization, as observed also for DAMGO. Further studies proved that cycloEM is more lipophilic and resistant toward enzymatic hydrolysis than EM-1, due to cyclic structure and the presence of D-amino acids (Bedini et al. 2010).

In vivo antinociception was evaluated by a visceral pain model in mice that allows detecting both central and peripheral analgesia. Given after acetic acid, CycloEM reduced the number of writhes in a dose-dependent manner; the administration of a KOR antagonist (norbinaltorphimine) or of a DOR (naltrindole) antagonist before CycloEM did not affect antinociception. On the other hand, the effect was antagonized by the MOR antagonists, β-funaltrexamine or naloxone methiodide, which does not readily cross the BBB, and only by a high dose of the MOR1-selective antagonist naloxonazine. CycloEM was also more effective in preempitive antinociception. Indeed, i.p. given before acetic acid produced a stronger decrease in the number of stretches. The cyclopeptide maintained a good antinociceptive effect when delivered by subcutaneous injection. The antinociceptivity of CycloEM was also assessed by the tail-flick test, a protocol utilized for testing central analgesia against acute pain. The cyclopeptide i.p. administered gave a moderate response only at higher doses, suggesting a partial penetration of the BBB. Taken all together, data indicate that CycloEM is among the first EM-1 analogs showing referential visceral analgesic activity after systemic administration, mainly by a peripheral mechanism, although it has some central effects at high doses (Bedini et al. 2010). The tests with selective opioid antagonists confirmed the specificity of CycloEM toward MOR.

As CycloEM represents a lipophilic opioid agonist lacking the protonable amino group, investigations were performed to provide insights into how the compound might interact and activate the MOR (Gentilucci et al. 2008). Conformation analysis of the peptide in solution was performed by NMR spectroscopy, and VT-NMR was utilized to infer the presence of intramolecular H-bonds. 2D ROESY was utilized to obtain interproton distances, so that plausible 3D structures were obtained by molecular dynamics performed in a box of explicit water molecules, using the ROESY distances as constraints. Clustering of the structures gave a single significant structure, representing the average of slightly different conformations in fast equilibrium. Two main conformations were observed by unrestrained molecular dynamics in explicit water, each showing well-defined secondary elements and the H-bonds expected on the basis of VT-NMR (Figure 13.11).

The two structures were introduced within a homology model of the MOR for molecular docking computations. After the recognition of the possible binding sites, energy minimization, and molecular dynamics, the results were further refined by

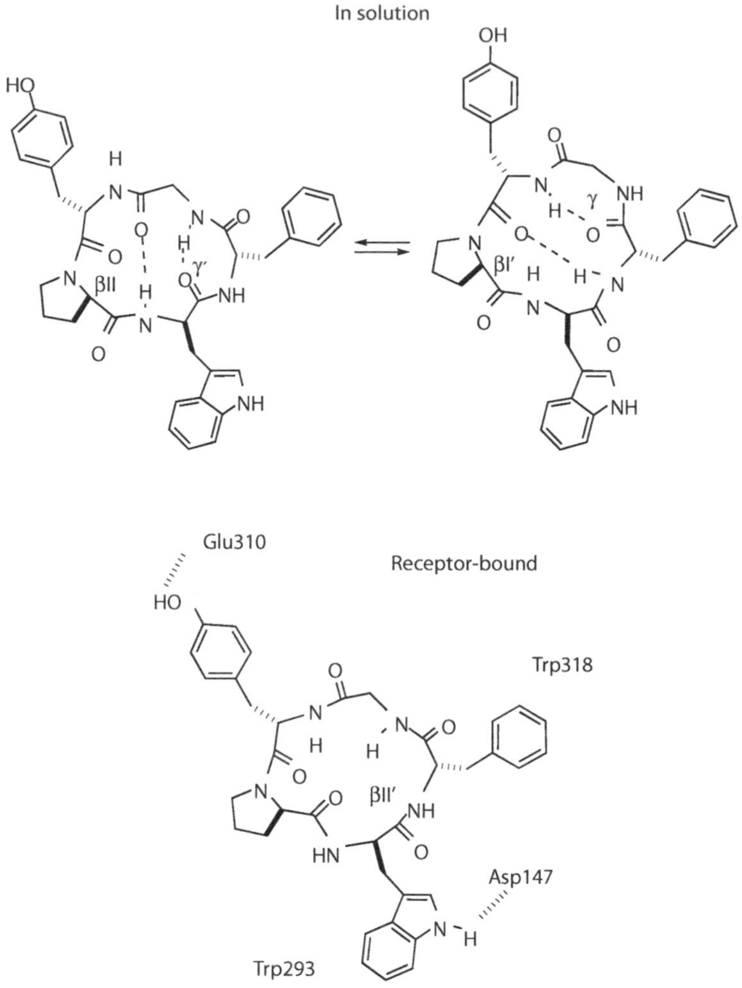

FIGURE 13.11 The preferred, interconverting structures of cycloEM in solution are different from the receptor-bound one.

hybrid QM/MM computations, to give a unprecedented orientation, with the peptide deeply inserted among the TMH III, V, VI, and VII, featuring the indole NH of D-Trp3 H-bonded to Asp147 of TMH III, plus other interactions with key receptor residues (Gentilucci et al. 2008). This H-bond partially compensates the absence of the ionic interaction; also ligand polarization represents a strong contribution to the overall binding energy, as revealed by the QM/MM computations. The interaction between D-Trp of the ligand and Trp293 of the TMH VI is of particular interest, as the induced fit in this region of the receptor might produce a conformational change of the entire TMH VI, responsible in turn for receptor activation (Fowler et al. 2004).

To corroborate this unprecedented model, a second generation of cycloEMs was prepared by introducing different residues in place of Gly5 with higher or conversely reduced conformational freedom. The new analogs displayed a variety of affinities for the MOR, lower than that of the first generation CycloEM. In any case, their structures were analyzed by conformational analysis and docking. All of the derivatives adopted at the receptor the same binding fashion of CycloEM, but with different calculated docking scores and energies. The correlation between the calculated docking energies and the free energies derived from the experimental affinity data showed a nice linear regression with excellent statistic parameters, demonstrating the reliability of the model.

The docking protocol highlighted an overall bioactive conformation with D-Trp-Phe embedded in a inverse type II β-turn. However, the cyclopeptide in solution adopts alternative conformations with different kinds of β- and γ- turn. In theory, it is possible to improve the efficacy of such a flexible molecule, taking into consideration the concept, "the more the structure of the free ligand in solution resembles the receptor-bound structure, the stronger the binding and the affinity" (Kessler 1982). Despite the many limitations, this assumption sometimes works very well and can be utilized as an indicative guide for structure optimization. Accordingly, the fostering of the β-turn motif on D-Trp-Phe should give cyclopeptides with improved affinity. Neither D-Pro2 nor Gly5 in CycloEM really interact the receptor, but they are fundamental in stabilizing the cyclopentapeptide geometries observed in solution. In particular, D-Pro manifests specific conformational features with respect to the other amino acids.

Based on these assumptions, a third generation of CycloEMs was designed by retaining the amino acids of the pharmacophores, D-Trp, Phe, and Tyr, and introducing L-Ala and/or Gly in the positions 2 and 5. Gratifyingly, c[YGwFG] had a receptor affinity in the nanomolar range, almost as good as DAMGO, and about 10-fold lower compared to the parent CycloEM, a very good result for a compound lacking the cationic amino group (Gentilucci et al. 2011). Also, this compound induced antinociception activity in the visceral pain model in mice and a lesser but significant antinociception in the tail-flick test.

This cyclopeptide has a greater conformational freedom in solution; however, the docking analysis revealed a preferential, plausible orientation at the receptor with the same 3D display of the pharmacophoric side chains and the most relevant contacts discussed for CycloEM, including the H-bond between Trp indolyl NH and Asp[147] of TMH III.

More recent results indicate that the D-Trp-Phe motive retains its ability to bind MOR and analgesic *in vivo* activity in short linear sequences. Interestingly, the D-Trp-Phe sequence shows an intrinsic tendency to adopt an inverse type II β-turn (De Marco et al. 2011). These results lead to the conclusion that peptides or cyclopeptides comprising a D-Trp-Phe motif may represent leads for the development of novel lipophilic opioids, characterized by atypical mechanism of receptor activation.

The cycloEMs represent the first MOR-agonists, described in the literature (since 2004), in which bioactivity resides in the D-Trp-Phe sequence. Interestingly, they share a structure similar to that of the KOR ligand CJ-15,208 c[pFWF] (Saito et al. 2002), isolated as a metabolite of a fungus, and its derivatives (Dolle et al. 2009; Aldrich et al. 2011), in which agonist activity has been discovered only recently (Aldrich et al. 2011). These results strongly support the hypothesis proposed for the cycloEMs that the Trp-Phe dipeptide might represent an unusual kind of message–address motif. The stereochemistry of Trp, the relative positioning of Trp and Phe, and the secondary conformation strongly influence affinity and selectivity and, likely, agonism versus antagonism.

13.6 CONCLUSIONS

This chapter reviews the classic models reported in the literature to describe ligand–receptor interactions of opioid agonists, such as morphine, fentalyls, as well as of some endogenous peptides and their mimetics. From the comparison of the diverse poses of the ligands within the receptor pockets, it is evident that "great care must be taken in evaluating the structural and conformational properties of all kinds of ligand derivatives, regardless of the close structural relationships displayed across a given series" (Subramanian et al. 2000). Attention has also been paid to emerging data on nonclassical opioid ligands with uncommon structure. These data support the opinion that the typical pharmacophores generally considered necessary for receptor interaction and activation are not a *condito sine qua non* (Gentilucci 2004; Gentilucci et al. 2008; Weltrowska et al. 2010). The observation that a few unusual molecules can activate the receptors, though they lack the protonated amine led to the re-discussion of the importance of the classic ionic interaction between this amine and a conserved Asp residue from TMH VI.

Compounds, such as the MOR agonists **1**, *c*[YpwFG], cycloendomorphin-1 (**3**), and the KOR agonist salvinorin A (**2**), can be considered nonclassical opioids. Also the fentanyl analog **4** binds the MOR in an unconventional way, albeit an ionic bond between a cationic group of the ligand and a different Asp on EL II still exists.

For their higher liphophilicity and the atypical mode of receptor activation, these compounds might be promising candidates for the development of novel, highly bioavailable analgesics or antipsychotics, useful toward those forms of pain resistant to therapies with common opiates, not accompanied by the same adverse symptoms.

ACKNOWLEDGMENTS

The author thanks Fondazione Umberto Veronesi, Milano, Roma, Italy, for financial support (Project no. 11 2012-3, Tryptoids: development of new therapeutic agents of natural origins for pain management and sedation in patients with advanced stages of cancer); Prof. F. Nyberg; and Prof. J. E. Zadina (this manuscript was thought in occasion of the INRC2010 Meeting in Malmo, and was concluded the same day of the fifth anniversary of the flood of New Orleans by Katrina); special thanks also to G. Cardillo, S. Spampinato, A. Tolomelli, R. De Marco, F. Squassabia, R. Spinosa, and A. Bedini.

REFERENCES

Akiyama, K., Gee, K. W., Mosberg, H. I., Hruby, V. J., Yamamura, H. I. 1985. Characterization of [3H][2-D-penicillamine, 5-D-penicillamine]-enkephalin binding to delta opiate receptors in the rat brain and neuroblastoma-glioma hybrid cell line (NG 108-15). *Proc. Natl. Acad. Sci. USA.* 82: 2543–7.

Aldrich, J. V., McLaughlin, J. P. 2009. Peptide kappa opioid receptor ligands: potential for drug development. *AAPS J.* 11: 312–22.

Aldrich, J. V., Kulkarni, S. S., Senadheera, S. N., Ross, N. C., Reilley, K. J., Eans, S. O., Ganno, M. L., Murray, T. F., McLaughlin, J. P. 2011. Unexpected opioid activity profiles of analogues of the novel peptide kappa opioid receptor ligand CJ-15,208. *Chem. Med. Chem.* 6: 1739–45.

Ananthan, S., Zhang, W., Hobrath, J. V. 2009. Recent advances in structure-based virtual screening of G-protein coupled receptors. *AAPS J.* 11: 178–85.

Arnold, Z. S., Schiller, P. W. 2000. Optically active aromatic amino acids. Part VI. Synthesis and properties of (Leu5)-enkephalin analogues containing O-methyl-L-tyrosine1 with ring substitution at position 3'. *J. Pept. Sci.* 6: 280–9.

Bedini, A., Baiula, M., Gentilucci, L., Tolomelli, A., De Marco, R., Spampinato, S. 2010. Peripheral antinociceptive effects of the cyclic endomorphin-1 analogue c[YpwFG] in a mouse visceral pain model. *Peptides.* 31: 2135–40.

Beguin, C., Richards, M. R., Wang, Y., et al. 2005. Synthesis and in vitro pharmacological studies of C(4) modified salvinorin A analogues. *Bioorg. Med. Chem. Lett.* 15: 2761–5.

Bernard, D., Coop, A., MacKerell, A. D. Jr. 2003. 2D conformationally sampled pharmacophore: a ligand-based pharmacophore to differentiate δ opioid agonists from antagonists. *J. Am. Chem. Soc.* 125: 3101–7.

Bodnar, R. J. 2009. Endogenous opiates and behaviour: 2008. *Peptides.* 30: 2432–79.

Borics, A., Toth, G. 2010. Structural comparison of mu-opioid receptor selective peptides confirmed four parameters of bioactivity. *J. Mol. Graph. Model.* 28: 495–505.

Bountra, C., Munglani, R., Schmidt, W. K. 2003. *Pain: Current Understanding, Emerging Therapies, and Novel Approaches to Drug Discovery.* New York: Marcel Dekker.

Breivik, H. 2005. Opioids in chronic non-cancer pain, indications and controversies. *Eur. J. Pain.* 9: 127–30.

Burden, J. E., Davis, P., Porreca, F., Spatola, A. F. 1999. Synthesis and biological activities of position one and three transposed analogs of the opioid peptide YKFA. *Bioorg. Med. Chem. Lett.* 9: 3441–6.

Calderon, S. N., Coop, A. 2004. SNC 80 and related delta opioid agonists. *Curr. Phar. Des.* 10: 733–42.

Cardillo, G., Gentilucci, L., Melchiorre, P., Spampinato, S. 2000. Synthesis and binding activity of endomorphin-1 analogues containing β-amino acids. *Bioorg. Med. Chem. Lett.* 10: 2755–8.

Cardillo, G., Gentilucci, L., Qasem, A. R., Sgarzi, F., Spampinato, S. 2002. Endomorphin-1 analogues containing beta-proline are mu-opioid receptor agonists and display enhance enzymatic hydrolysis resistance. *J. Med. Chem.* 45: 2571–8.

Cardillo, G., Gentilucci, L., Tolomelli, A., Qasem, A. R., Spampinato, S., Calienni, M. 2003. Conformational analysis and μ-opioid receptor affinity of short peptides, endomorphin models in a low polarity solvent. *Org. Biomol. Chem.* 1: 3010–14.

Cardillo, G., Gentilucci, L., Tolomelli, A., et al. 2004. Synthesis and evaluation of the affinity toward l-opioid receptors of atypical, lipophilic ligands based on the sequence c[-Tyr-Pro-Trp-Phe-Gly-]. *J. Med. Chem.* 47: 5198–203.

Chaturvedi, K., Christoffers, K. H., Singh, K., Howells, R. D. 2000. Structure and regulation of opioid receptors. *Biopolymers.* 55: 334–46.

De Marco, R., Gentilucci, L., Tolomelli, A., Feddersen, S., Spampinato, S., Bedini, A., Artali, R. 2011. Novel opioid peptide agonists and antagonists. *Pharm. Rep.* 63: 226.

Doi, M., Asano, A., Komura, E., Ueda, Y. 2002. The structure of an endomorphin analogue incorporating 1-aminocyclohexane-1-carboxlylic acid for proline is similar to the β-turn of Leu-enkephalin *Biochem. Biophys. Res. Commun.* 297: 138–42.

Dolle, R. E., Machaut, M., Martinez-Teipel, B., et al. 2004. (4-Carboxamido)phenylalanine is a surrogate for tyrosine in opioid receptor peptide ligands. *Bioorg. Med. Chem. Lett.* 14: 3545–8.

Dolle, R. E., Michaut, M., Martinez-Teipel, B., Seida, P. R., Ajello, C. W., Muller, A. L., De Haven, R. N., Carroll, P. J. 2009. Nascent structure–activity relationship study of a diastereomeric series of kappa opioid receptor antagonists derived from CJ-15,208. *Bioorg. Med. Chem. Lett.* 19: 3647–50.

Dosen-Micovic, L., Ivanovic, M., Micovic, V. 2006. Steric interactions and the activity of fentanyl analogs at the l-opioid receptor. *Bioorg. Med. Chem.* 14: 2887–95.

Eguchi, M. 2004. Advances in selective opioid receptoragonists and antagonists. *Med. Res. Rev.* 24: 182–212.

Eguchi, M., Shen, R. Y., Shea, J. P., Lee, M. S., Kahn, M. 2002. Design, synthesis, and evaluation of opioid analogues with non-peptidic beta-turn scaffold: enkephalin and endomorphin mimetics. *J. Med. Chem.* 45: 1395–8.

Fichna, J., do-Rego, J.-C., Kosson, P., Costentin, J., Janecka, A. 2005. Characterization of antinociceptive activity of novel endomorphin-2 and morphiceptin analogs modified in the third position. *Biochem. Pharmacol.* 69: 179–85.

Fichna, J., Janecka, A., Costentin, J., do-Rego. J.-C. 2007. The endomorphin system and its evolving neurophysiological role pharmacological reviews. *Pharmacol. Rev.* 59: 88–123.

Filizola, M. Weinstein, H. 2002. Structural models for dimerization of G-protein coupled receptors: the opioid receptor homodimers. *Biopolymers.* 66: 317–25.

Fowler, C. B., Pogozheva, I. D., Lomize, A. L., LeVine, H., III, Mosberg, H. I. 2004. Complex of an active mu-opioid receptor with a cyclic peptide agonist modeled from experimental constraints *Biochemistry.* 43: 15796–810.

Gao, Y., Liu, X., Wei, J., Zhu, B., Chena, Q., Wang, R. 2005. Structure-activity relationship of the novel bivalent and C-terminal modified analogues of endomorphin-2. *Bioorg. Med. Chem. Lett.* 15: 1847–50.

Gentilucci, L. 2004. New trends in the development of opioid peptide analogues as advanced remedies for pain relief. *Curr. Topics Med. Chem.* 4: 19–38.

Gentilucci, L., De Marco, R., Cerisoli, L. 2010. Chemical modifications designed to improve peptide stability: incorporation of non-natural amino acids, pseudo-peptide bonds, and cyclization. *Curr. Pharm. Des.* 16: 3185–203.

Gentilucci, L., Squassabia, F., Artali, R. 2007. Re-discussion of the importance of ionic interactions in stabilizing ligand-opioid receptor complex and in activating signal transduction. *Curr Drug Targets.* 8: 185–96.

Gentilucci, L., Squassabia, F., De Marco, R., et al. 2008. Investigation of the interaction between the atypical agonist c[YpwFG] and MOR. *FEBS J.* 275: 2315–37.

Gentilucci, L., Tolomelli, A., Squassabia, F. 2007. Topological exploration of cyclic endomorphin-1 analogues, structurally defined models for investigating the bioactive conformation of MOR agonists. *Prot. Pept. Lett.* 14: 51–6.

Gentilucci, L., Tolomelli, A., De Marco, R., Spampinato, S., Bedini, A., Artali, R. 2011. The inverse type II β-turn on D-Trp-Phe, a pharmacophoric motif for MOR agonists. *Chem. Med. Chem.* 6: 1640–53.

Handa, B. K., Land, A. C., Lord, J.A., Morgan, B. A., Rance, M. J., Smith, C. F. 1981. Analogues of beta-LPH61-64 possessing selective agonist activity at mu-opiate receptors. *Eur. J. Pharm.* 70: 531–40.

Harding, W. W., Tidgewell, K., Byrd, N., et al. 2005. Neoclerodane diterpenes as a novel scaffold for μ opioid receptor ligands. *J. Med. Chem.* 48: 4765–71.

Heitz, J. W., Witkowski, T. A., Viscusi, E. R. 2009. New and emerging analgesics and analgesic technologies for acute pain management. *Curr. Opin. Anaesthesiol.* 22: 608–17.

Honda, M., Okutsu, H., Matsuura, T., Miyagi, T., Yamamoto, Y., Hazato, T., Ono, H. 2001. Spinorphin, an endogenous inhibitor of enkephalin-degrading enzymes, potentiates Leu-enkephalin-induced anti-allodynic and antinociceptive effects in mice. *Jpn. J. Pharmacol.* 87: 261–7.

Horvath, G. 2000. Endomorphin-1 and endomorphin-2: pharmacology of the selective endogenous mu-opioid receptor agonists. *Pharmacol. Ther.* 88: 437–63.

Hruby, V. J., Agnes, R. S. 1999. Conformation–activity relationships of opioid peptides with selective activities at opioid receptors. *Biopolymers.* 51: 391–410.

Hruby, V. J., Balse, P. M. 2000. Conformational and Topographical considerations in designing agonist peptidomimetics from peptide leads. *Curr. Med. Chem.* 7: 945–70.

In, Y., Minoura, K., Tomoo, K., et al. 2005. Structural function of C-terminal amidation of endomorphin. Conformational comparison of mu-selective endomorphin-2 with its C-terminal free acid, studied by 1H-NMR spectroscopy, molecular calculation, and X-ray crystallography. *FEBS J.* 272: 5079–97.

Janecka, A., Fichna, J., Mirowski, M., Janecki, T. 2002. Structure-activity relationship, conformation and pharmacology studies of morphiceptin analogues-selective mu-opioid receptor ligands. *Mini Rev. Med. Chem.* 2: 565–72.

Janecka, A., Kruszynski, R. 2005. Conformationally restricted peptides as tools in opioid receptor studies. *Curr. Med. Chem.* 12: 471–81.

Janecka, A. Perlikowska, R., Gach, K., Wyrebska, A., Fichna, J. 2010. Development of opioid peptide analogs for pain relief. *Curr. Pharm. Des.* 16: 1126–35.

Jiang, C. L., Xu, D., Lu, C. L., Wang, Y. X., You, Z. D., Liu, X. Y. 2000. Interleukin-2: structural and biological relatedness to opioid peptides. *Neuroimmunomodulation.* 8: 20–4.

Johnson, R. E., Fudala, P. J., Payne, R. 2005. Buprenorphine: considerations for pain management. *J. Pain Sympt. Manag.* 29: 297–326.

Kaczor, A., Matosiuk, D. 2002. Non-peptide opioid receptor ligands—recent advances. Part I—agonists. *Curr. Med. Chem.* 9: 1567–89.

Kane, B. E., Nieto, M. J., McCurdy, C. R., Ferguson, D. M. 2006. A unique binding epitope for salvinorin A, a non-nitrogenous kappa opioid receptor agonist. *FEBS J.* 273: 1966–74.

Kane, B. E., McCurdy, C. R., Ferguson, D. M. 2008. Toward a structure based model of salvinorin A recognition of the κ-opioid receptor. *J. Med. Chem.* 51: 1824–30.

Keller, M., Boissard, C., Patiny, L., et al. 2001. Pseudoproline-containing analogues of morphiceptin and endomorphin-2: evidence for a cis Tyr-Pro amide bond in the bioactive conformation. *J. Med. Chem.* 44: 3896–903.

Keresztes, A., Borics, A., Tóth, G. 2010. Recent advances in endomorphin engineering. *Chem. Med. Chem.* 5: 1176–96.

Kessler, H. 1982. Conformation and biological activity of cyclic peptide. *Angew. Chem. Int. Ed. Engl.* 21: 512–23.

Levac, B. A., O'Dowd, B. F., George, S. R. 2002. Oligomerization of opioid receptors: generation of novel signaling units. *Curr. Opin. Pharmacol.* 2: 76–81.

Lipkowski, A. W., Konecka, A. M., Sroczynska, I. 1982. Double enkephalins—synthesis, activity on guinea pig ileum and analgesic effect. *Peptides.* 3: 697–700.

Lipkowski, A. W., Misicka, A., Carr, D. B., Ronsisvalle, G., Kosson, D., Bonney, I. M. 2004. Neuropeptide mimetics for pain management. *Pure Appl. Chem.* 76: 941–50.

Liu, X., Kai, M., Jin, L., Wang, R. 2009. Molecular modeling studies to predict the possible binding modes of endomorphin analogs in mu opioid receptor. *Bioorg. Med. Chem. Lett.* 19: 5387–91.

Loetsch, J., Geisslinger, G. 2005. Are mu-opioid receptor polymorphisms important for clinical opioid therapy? *Trends Mol. Med.* 11: 82–9.

Lu, Y., Nguyen, T. M., Weltrowska, G., Berezowska, I., Lemieux, C., Chung, N. N., Schiller, P. W. 2001. [2,6-Dimethyltyrosine1]-dynorphin A(1-11)-NH2 analogues lacking an N-terminal amino group: potent and selective κ opioid antagonists. *J. Med. Chem.* 44: 3048–53.

Lu, Y., Weltrowska, G., Lemieux, C., Chung, N. N., Schiller, P. W. 2001. Stereospecific synthesis of (2S)-2-methyl-3-(2',6'-dimethyl-4'-hydroxyphenyl)-propionic acid (Mdp) and its incorporation into an opioid peptide. *Bioorg. Med. Chem. Lett.* 11: 323–5.

McDaid, C., Maund, E., Rice, S., Wright, K., Jenkins, B., Woolacott N. 2010. Paracetamol and selective and non-selective non-steroidal anti-inflammatory drugs (NSAIDs) for the reduction of morphine-related side effects after major surgery: a systematic review. *Health Technol. Assess.* 14: 1–153.

McFadyen, I. J., Sobczyk-Kojiro, K., Schaefer, M. J., et al. 2000. Tetrapeptide derivatives of [D-Pen(2),D-Pen(5)]-enkephalin (DPDPE) lacking an N-terminal tyrosine residue are agonists at the mu-opioid receptor. *J. Pharmacol. Exp. Ther.* 295: 960–6.

Manchikanti, L., Benyamin, R., Datta, S., Vallejo, R., Smith, H. 2010. Opioids in chronic noncancer pain. *Expert Rev. Neurother.* 10: 775–89.

Meunier, J. C. 1997. Nociceptin/orphanin FQ and the opioid receptor-like ORL1 receptor. *Eur. J. Pharmacol.* 340: 1–15.

Milligan, G. 2004. G protein-coupled receptor dimerization: function and ligand pharmacology. *Mol. Pharmacol.* 66: 1–7.

Mobarec, J. C., Sanchez, R., Filizola, M. 2009. Modern homology modeling of G-protein coupled receptors: which structural template to use? *J. Med. Chem.* 52: 5207–16.

Montagnini, M. L., Zaleon, C. R. 2009. Pharmacological management of cancer pain. *J. Opioid Manag.* 5: 89–96.

Mosberg, H. I., Fowler, C. B. 2002. Development and validation of opioid ligand–receptor interaction models: the structural basis of mu vs. delta selectivity. *J. Pept. Res.* 60: 329–35.

Mosberg, H. I., Ho, J. C., Sobczyk-Kojiro, K. 1998. A high affinity, mu-opioid receptor-selective enkephalin analogue lacking an N-terminal tyrosine. *Bioorg. Med. Chem. Lett.* 8: 2681–4.

Mosberg, H. I., Omnaas, J. R., Medzihradsky, F., Smith, C. B. 1988. Cyclic disulfide- and dithioether-containing opioid tetrapeptides: development of a ligand with enhanced delta opioid receptor selectivity and potency. *Life Sci.* 43: 1013–20.

Munro, T. A., Rizzacasa, M. A., Roth, B. L., Toth, B. A., Yan, F. 2005. Studies toward the pharmacophore of salvinorin A, a potent kappa opioid receptor agonist. *J. Med. Chem.* 48: 345–8.

Negri, L., Melchiorri, P., Lattanzi, R. 2000. Pharmacology of amphibian opiate peptides. *Peptides.* 21: 1639–47.

Noble, F., Smadja, C., Roques, B. P. 1994. Role of endogenous cholecystokinin in the facilitation of mu-mediated antinociception by delta-opioid agonists. *J. Pharmacol. Exp. Ther.* 271: 1127–34.

Okada. Y., Fujita, Y., Motoyama, T., et al. 2003. Structural studies of [2′,6′-dimethyl-L-tyrosine¹]endomorphin-2 analogues: enhanced activity and *cis* orientation of the Dmt-Pro amide bond. *Bioorg. Med. Chem. Lett.* 11: 1983–94.

Okada, Y., Fukumizu, A., Takahashi, M., et al. 2000. Synthesis of stereoisomeric analogues of endomorphin-2, H-Tyr-Pro-Phe-Phe-NH$_2$, and Examination of their opioid receptor binding activities and solution conformation. *Biochem. Biophys. Res. Commun.* 276: 7–11.

Palczewski, K., Kumasaka, T., Hori, T., et al. 2000. Crystal structure of rhodopsin: a G protein-coupled receptor. *Science.* 289: 739–45.

Paterlini, M. G., Avitabile, F., Ostrowski, B. G., Ferguson, D. M., Portoghese, P. S. 2000. Stereochemical requirements for receptor recognition of the μ-opioid peptide endomorphin-1. *Biophys. J.* 78: 590–9.

Perlikowska, R., do-Rego, J. C., Cravezic, A. 2010. Synthesis and biological evaluation of cyclic endomorphin-2 analogs. *Peptides.* 31: 339–45.

Podlogar, B. L., Paterlini, M. G., Ferguson, D. M., et al. 1998. Conformational analysis of the endogenous mu-opioid agonist endomorphin-1 using NMR spectroscopy and molecular modelling. *FEBS Lett.* 439: 13–20.

Pogozheva, I. D., Lomize, A. L., Mosberg, H. I. 1998. Opioid receptor three-dimensional structures from distance geometry calculations with hydrogen bonding constraints. *Biophys. J.* 75: 612–34.

Presland, J. 2004. G-protein-coupled receptor accessory proteins: their potential role in future drug discovery. *Biochem. Soc. Trans.* 32: 888–91.

Przewlocki, R., Przewlocka, B. 2001. Opioids in chronic pain. *Eur. J. Pharmacol.* 429: 79–91.

Raffa, R. B., Clark-Vetri, R., Tallarida, R. J., Wertheimer, A. I. 2003. Combination strategies for pain management. *Exp. Opin. Pharm.* 4: 1697–708.

Roth, B. L., Baner, K., Westkaemper, R., et al. 2002. Salvinorin A: a potent naturally occurring nonnitrogenous κ-opioid selective agonist. *Proc. Natl. Acad. Sci. USA.* 99: 11934–9.

Roumy, M., Zajac, J. 1999. Europeptide FF selectively attenuates the effects of nociceptin on acutely dissociated neurons of the rat dorsal raphe nucleus. *Brain Res.* 845: 208–14.

Ruan, H., Prasad, J. A., Lemaire, S. 2000. Non-opioid antinociceptive effects of supraspinal histogranin and related peptides: possible involvement of central dopamine D(2) receptor. *Pharmacol. Biochem. Behav.* 67: 83–91.

Sagara, T., Egashira, H., Okamura, M., Fujii, I., Shimohigashi, Y., Kanematsu, K. 1996. Ligand recognition in μ opioid receptor: experimentally based modeling of μ opioid receptor binding sites and their testing by ligand docking. *Bioorg. Med. Chem.* 4: 2151–66.

Saito, T., Hirai, H., Kim, Y. J., Kojima, Y., Matsunaga, Y., Nishida, H., Sakakibara, T., Suga, O., Sujaku, T., Kojima, N. 2002. The tryptophan isomers of the cyclic tetrapeptide CJ-15,208, reported to be a kappa opioid receptor (KOR) antagonist. *J. Antibiot.* 55: 847–54.

Schiller, P. W., Berezowska, I., Nguyen, T. M.-D., et al. 2000. Novel ligands lacking a positive charge for the delta-and mu-opioid receptors. *J. Med. Chem.* 43: 551–9.

Schiller, P. W., Weltrowska, G., Nguyen, T. M.-D., Lemieux, C., Chung, N. N., Lu, Y. 2003. Conversion of δ-, κ-, and μ-receptor selective opioid peptide agonists into δ-, κ- and μ-selective antagonists. *Life Sci.* 73: 691–8.

Shao, X., Gao, Y., Zhu, C., et al. 2007. Conformational analysis of endomorphin-2 analogs with phenylalanine mimics by NMR and molecular modeling. *Bioorg. Med. Chem.* 15: 3539–47.

Subramanian, G., Paterlini, M. G., Portoghese, P. S., Ferguson, D. M. 2000. Molecular docking reveals a novel binding site model for fentanyl at the mu-opioid receptor. *J. Med. Chem.* 43: 381–91.

Teschemacher, H. 1993. Atypical opioid peptides. In *Handbook of Experimental Pharmacology, 104/1*, ed. A. Herz, 499–528. Berlin: Springer-Verlag.

Teschemacher, H. 2003. Receptor ligands derived from food proteins. *Curr. Pharm. Design.* 9: 1331–44.

Tömböly, C., Ballet, S., Feytens, D., et al. 2008. Endomorphin-2 with a beta-turn backbone constraint retains the potent mu-opioid receptor agonist properties. *J. Med. Chem.* 51: 173–7.

Vaccarino, A. L., Kastin, A. J. 2001. Endogenous opiates: 2000. *Peptides.* 22: 2257–328.

Vadivelu, N., Sinatra, R. 2005. Recent advances in elucidating pain mechanisms. *Curr. Opin. Anaesthesiol.* 18: 540–7.

Varga, E. V., Navratilova, E., Stropova, D., Jambrosic, J., Roeske, W. R., Yamamura H. I. 2004. Agonist-specific regulation of the delta-opioid receptor. *Life Sci.* 76: 599–612.

Vig, B. S., Murray, T. F., Aldrich, J. V. A. 2003. Novel N-terminal cyclic dynorphin A analogue cycloN,5[Trp3,Trp4,Glu5] dynorphin A-(1-11)NH2 that lacks the basic N-terminus. *J. Med. Chem.* 46: 1279–82.

Vonvoigtlander, P. F., Lahti, R. A., Ludens, J. H. 1983. U50,488: a kappa-selective agent with poor affinity for mu₁ opiate binding sites. *J. Pharmacol. Exp. Ther.* 224: 7–12.

Weltrowska, G., Chung, N. N., Lemieux, C., et al. 2010. "Carba"-analogues of fentanyl are opioid receptor agonists. *J. Med. Chem.* 53: 2875–81.

Weltrowska, G., Lu, Y., Lemieux, C., Chung, N. N., Schiller, P. W. 2004. A Novel cyclic enkephalin analogue with potent opioid antagonist activity. *Bioorg. Med. Chem. Lett.* 14: 4731–3.

Wilson, A. M., Soignier, R. D., Zadina, J. E., et al. 2000. Dissociation of analgesic and rewarding effects of endomorphin-1 in rats. *Peptides.* 21: 1871–4.

Witt, K. A., Gillespie, T. J., Huber, J. D., Egleton, R. D., Davis, T. P. 2001. Peptide drug modifications to enhance bioavailability and blood-brain barrier permeability. *Peptides.* 22: 2329–43.

Wollemann, M., Benyhe, S. 2004. Non-opioid actions of opioid peptides. *Life Sci.* 75: 257–70.

Yan, F., Mosier, P. D., Westkaemper, R. B., et al. 2005. Identification of the molecular mechanisms by which the diterpenoid salvinorin A binds to κ-opioid receptors. *Biochemistry.* 44: 8643–51.

Yang, S., Yunden, J., Sonoda, S., et al. 2001. Rubiscolin, a δ-selective opioid peptide derived from plant Rubisco. *FEBS Lett.* 509: 213–17.

Zadina, J. E., Hackler, L., Ge, L.-J., Kastin, A. J. 1997. A potent and selective endogenous agonist for the mu-opiate receptor. *Nature.* 386: 499–502.

Zhang, Y., Sham, Y. Y., Rajamani, R., Gao, J., Portoghese, P. S. 2005. Homology Modeling and molecular dynamics simulations of the mu opioid receptor in a membrane–aqueous system. *Chem. Bio. Chem.* 6: 853–9.

Zhao, Q.-Y., Chen, Q., Yang, D.-J., et al. 2005. Endomorphin 1[ψ] and endomorphin 2[ψ], endomorphins analogues containing a reduced (CH_2NH) amide bond between Tyr^1 and Pro^2, display partial agonist potency but significant antinociception. *Life Sci.* 77: 1155–65.

14 Use of Peptides as Drug Leads—A Case Study on the Development of Dipeptides Corresponding to the Heptapeptide Substance P(1-7), with Intriguing Effects on Neuropathic Pain

Rebecca Fransson and Anja Sandström

CONTENTS

14.1   INTRODUCTION

14.1.1   PEPTIDES AS DRUG LEADS

Peptides are major players in a variety of physiological processes in mammals and microorganisms. Being neurotransmitters in the nervous system, the neuropeptides constitute a large and significant group of peptides, which often coexist with other neurotransmitters in the neurons (Hökfelt et al. 2000; Hökfelt, Bartfai, and Bloom 2003). As a result of their effect on living organisms, biologically active peptides are interesting starting points in drug discovery. Although peptides can serve as valuable research tools in the first phase of drug discovery projects and for the initial study of biological mechanisms of various diseases, they are not suitable as pharmaceuticals intended for oral administration. This is due to the inherent drawbacks of the peptide structure, which is associated with rapid degradation by proteolytic enzymes and low bioavailability.

Peptides are polyfunctional molecules usually composed of a sequential arrangement of 2–50 amino acid residues linked by amide (peptide) bonds. Besides the complexity arising from the presence of basic, acidic, and hydrophobic amino acids, peptides also have a large degree of conformational freedom due to the presence of numerous rotatable bonds. Hence, peptides can fold into complex tertiary structures that are crucial for their molecular recognition ability and function. The design and development of low-molecular-weight and bioavailable drug-like molecules that mimic the action of peptides strive to overcome the problems with peptides. The term "peptidomimetics" is often used for such compounds. Peptidomimetics are molecules with significantly reduced peptidic character that mimic the three-dimensional bioactive conformation of peptides and subsequently retain the ability to interact with the biological target and cause the same biological effect (Ripka and Rich 1998; Vagner, Qu, and Hruby 2008).

Rational design of peptidomimetics from biologically active peptides is a complementary alternative to the high throughput screening (HTS) procedures that have come to dominate industrial drug discovery strategies for hit and lead identification. In rational peptide lead optimization, a stepwise procedure for the transformation of biologically significant peptides into small drug-like pseudopeptides or peptidomimetics is often performed. The process normally begins with elucidation of structure–activity relationships and the minimal active sequence of the peptide. This is achieved by evaluation of binding affinities of peptide analogs to the target protein. Key residues for the biological activity, that is, the pharmacophores, are also identified during the initial phase. In practise, such information is normally gathered through amino acid scans, truncations, and carboxy-(C-) and amino-(N-) terminal modifications (Vlieghe et al. 2010). Amino acid scans determine the importance of amino acid side chains by evaluation of peptide analogs, in which the amino acids have been replaced one at a time with alanine or glycine. Alanine and glycine are the smallest amino acids available, having a methyl and a hydrogen side chain, respectively. Amino acid truncations from both ends of a peptide provide information about the minimal sequence needed for retained activity. The importance of a basic *N*-terminal or an acidic *C*-terminal is determined by the introduction of capping groups. Based on the information gained from these studies, further structural modifications are undertaken to improve stability, potency, and selectivity. Conformational restrictions are frequently used to explore the bioactive conformation and this can be achieved by cyclization, *N*-methylation, isosteric substitution, or by secondary structure replacement. Some of these general strategies have been thoroughly reviewed in the literature previously and a detailed description of these is beyond the scope of this chapter (Olson et al. 1993; Hanessian et al. 1997).

The development of angiotensin-converting enzyme (ACE) inhibitors and HIV protease inhibitors constitute two successful examples of rational design, leading to approved drugs. For these enzyme inhibitors, the design was also substantially aided by X-ray crystallography and molecular modeling of the target proteins (Leung, Abbenante, and Fairlie 2000). There are peptidomimetic ligands with both a rational design and a good affinity to peptide receptors, for example, bioactive peptides, somastostatin (Hirschmann et al. 1992, 1993), melanocyte-stimulating hormone (MSH) (Fotsch et al. 2003), and enkephalins (Belanger and Dufresne 1986). Herein, the process of using peptides as a starting point for drug discovery will be exemplified in a case study of the heptapeptide, substance P(1-7) [SP(1-7)]; a peptide of interest in our study. Furthermore, results from *in vivo* studies of SP(1-7) and the produced SP(1-7) analogs will be reviewed. In particular, promising results related to neuropathic pain and opioid withdrawal symptoms will be discussed.

14.1.2 SUBSTANCE P(1-7)

SP(1-7) (H-Arg-Pro-Lys-Pro-Gln-Gln-Phe-OH) is the major metabolite of the well-known neuropeptide, substance P (Arg-Pro-Lys-Pro-Gln-Gln-Phe-Phe-Gly-Leu-Met-NH$_2$) (Lee et al. 1981; Sakurada et al. 1985). SP(1-7) is formed after enzymatic processing of the undecapeptidic mother peptide, involving at least three different enzymes, including substance P endopeptidase (SPE), ACE, and neutral endopeptidase (NEP) (Figure 14.1) (Hallberg et al. 2003). In the brain, SP and the corresponding neurokinin 1 receptor (NK1 receptor) are expressed in areas related to depression (Kramer et al. 1998), anxiety (De Araujo, Huston, and Brandao 2001), and stress (Culman and Unger 1995; Ebner et al. 2008), as well as in areas involved in motivational properties and reward (Hasenohrl et al. 1991; Huston et al. 1993). In the spinal cord, SP is expressed in pain processing pathways (Zubrzycka and Janecka 2000). Several of the aforementioned diseases and mental conditions have been addressed in various drug discovery programs. The metabolite SP(1-7) has also been identified in the central nervous system (CNS) (Rimon et al. 1984; Sakurada et al. 1985). Interestingly, SP(1-7) has been shown to modulate and, in certain cases, oppose the effects of the parent peptide (Hallberg et al. 2003). Thus, in contrast to the mother peptide, SP(1-7) has an antinociceptive instead of nociceptive effect at the mouse spinal cord level (Sakurada, Watanabe, and Sakurada 2004). Moreover, SP(1-7) diminishes

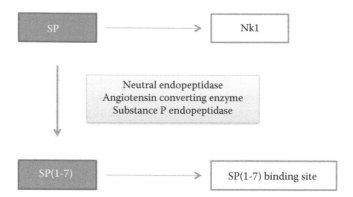

FIGURE 14.1 Schematic illustration of the enzymatic processing of the NK-1 receptor agonist substance P to the bioactive fragment SP(1-7), which acts on a distinct and specific binding site.

the vasodilatory response induced by SP in a blister model of inflammation (Wiktelius, Khalil, and Nyberg 2006) and the SP-dependent enhancement of opioid withdrawal and tolerance (Kreeger and Larson 1993; Zhou et al. 2003). Counteracting behavior is observed for several other metabolites derived from neuropeptides, such as for opioid peptides, calcitonin gene-related peptide, bradykinin and nociceptin (Hallberg et al. 2003). For example, the selective κ-opioid ligand dynorphin produces dysphoria in the reward system in the brain but after conversion into its N-terminal bioactive fragment, Leu-enkephalin, it becomes an δ-opioid agonist with euphoric properties (Hallberg et al. 2003).

The underlying physiological mechanisms of SP(1-7) at a molecular level, including receptor recognition, are still unclear. However, specific binding sites for SP(1-7) in the rat and mouse spinal cords have been identified (Igwe et al. 1990; Botros et al. 2006). Based on current understanding, the binding site is distinct from the NK1 receptors and the μ-opioid receptors, an issue that will be addressed in more detail in section 14.4. Even though the intriguing effects of SP(1-7) have been known for quite some time, SP(1-7) has not, until now, been addressed in a medicinal chemistry program.

14.2 PEPTIDE LEAD OPTIMIZATION OF SP(1-7)

14.2.1 STARTING POINTS FOR DRUG DISCOVERY

For a successful peptide lead optimization, access to a reliable *in vitro* assay system is of utmost importance. This allows for precise SAR evaluations based on the binding affinity of peptide analogs to the target protein. Binding affinities are often expressed as inhibitor/ligand concentrations achieving 50% displacement of natural ligands (IC_{50}-values) or dissociation constants for inhibitor/ligand binding (K_i-values).

In 2006, Nyberg and coworkers demonstrated the presence of specific binding sites for SP(1-7) in rat spinal cord, which is in accordance with the results from the mouse spinal cord previously reported by Igwe and coworkers (Igwe et al. 1990; Botros et al. 2006). A receptor binding assay using spinal cord tissue homogenate and measurement of binding affinity for various compounds by displacement of tritiated SP(1-7) was developed. To study the mechanism of action of SP(1-7), ligands for μ- and NK-receptors, as well as various C- and N-terminal SP fragments were screened for their potential binding to the specific binding site of SP(1-7) (Table 14.1). In comparison to SP(1-7), all tested ligands showed significantly weaker binding to this site, indicative of a specific target protein, yet unknown but neither identical to the tachykinin receptor nor to the μ receptor, for SP(1-7). Interestingly, the strong μ-receptor agonists, endomorphin-2 (EM-2) and endomorphin-1 (EM-1), were shown to differentially interact with the binding site of SP(1-7). Thus, EM-2 had a 10-fold lower affinity as compared to SP(1-7), whereas EM-1 had a 1400-fold lower affinity (for relative affinities, see Table 14.1). This is in accordance with other differences shown for these two μ-receptor agonists in various studies. For instance, the binding characteristics and their distribution in CNS have been reported to vary (Horvath 2000; Zadina 2002) and different binding affinities of EM-1 and EM-2 to the different subtypes of μ-receptors (μ1- and μ2-receptors) have been noted (Pasternak and Wood 1986; Sakurada et al. 2001). Besides the exciting mechanistic aspects of the SP(1-7)/EM-2 correlation, the study by Botros et al. also led to the identification of EM-2 as a lead compound in the development of low-molecular-weight ligands to the SP(1-7) binding site. Although the affinity of EM-2 was 10-fold lower, the smaller size motivated the use of this tetrapeptide for further development.

TABLE 14.1
Affinity of Different Peptides to the Specific Binding Site of SP(1-7), Expressed in Relative Affinity to that of SP(1-7)

Peptide	Ligand Type	Relative Affinity
SP(1-7)	SP fragment/SP(1-7) binding site agonist	100
SP(1-6)	SP fragment	0.04
SP(1-8)	SP fragment	1.0
SP	NK1 receptor agonist	0.5
[D-Pro^{2,4}]SP(1-7)	SP(1-7) binding site antagonist	42
[Sar^9, Met(O$_2$)^{11}]SP	NK1 receptor agonist	0.09
R-396	NK2 receptor antagonist	<0.007
Senktide	NK3 receptor agonist	<0.007
DAMGO	μ-receptor agonist	5.7
Endomorphin-1	μ-receptor agonist	0.07
Endomorphin-2	μ-receptor agonist	10
Naloxone	Nonselective opioid receptor antagonist	<0.007
Naloxonazine	Selective μ-opioid receptor antagonist	<0.007
Tyr-MIF-1	Opioid receptor ligand	6.8
Tyr-W-MIF-1	Opioid receptor ligand	<0.007

14.3 STRUCTURE–ACTIVITY RELATIONSHIP STUDY AND TRUNCATION OF SP(1-7) AND EM-2

A thorough SAR study of the binding of SP(1-7) and EM-2 toward the SP(1-7) binding site was performed to determine the key amino acids responsible for binding to the target protein and the importance of the *C*- and *N*-terminal ends of the peptides. Classical strategies such as Ala-scans and *C*- and *N*-terminal modifications of the two target peptides were performed (Figures 14.2 and 14.3) (Fransson et al. 2008; Fransson, Botros et al. 2010). Thus, a series of peptide analogs where each amino acid residue of the two target peptides was replaced by an alanine were produced and evaluated. Both *C*-terminal carboxylic acids and carboxamides were also included. The results were remarkably similar between SP(1-7) and EM-2 peptides (Figures 14.2 and 14.3), thus the *C*-terminal parts of SP(1-7) and EM-2 were essential for binding, whereas the *N*-terminal regions were not. The *C*-terminal phenylalanine was absolutely crucial for strong affinity in both cases. Moreover, the potency of SP(1-7) showed a fivefold increase upon amidation of the terminal carboxyl group (Figure 14.2). Similarly, the binding affinity of EM-2 was reduced by a factor of four upon removal of the amide, as apparent from evaluation of the EM-2 analog with a *C*-terminal carboxylic acid (Figure 14.3).

As suggested from the two Ala-scans, *N*-terminal truncations of the two peptides were allowed (Table 14.2). For SP(1-7), removal of the *N*-terminal arginine resulted in a 20-fold decrease in affinity. However, truncation down to the tripeptide level could be done without further loss in affinity and *C*-terminal amidation of all the truncated SP(1-7) analogs increased affinity 5- to 10-fold. Hence, H-Gln-Gln-Phe-NH$_2$

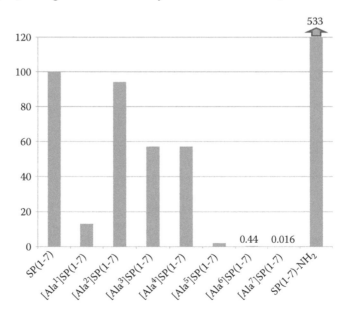

FIGURE 14.2 Relative affinity of alanine substituted SP(1-7) analogs and SP(1-7)-amide to the SP(1-7) binding site, as compared to SP(1-7) (H-Arg-Pro-Lys-Pro-Gln-Gln-Phe-OH) itself.

100
90
80
70
60
50
40
30
20
10
0
0.60
EM-2
[Ala1] EM-2
[Ala2] EM-2
[Ala3] EM-2
[Ala4] EM-2
EM-2-OH

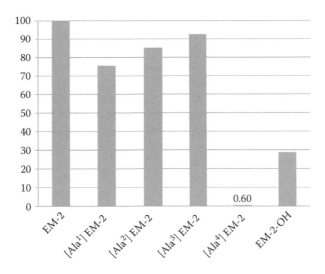

FIGURE 14.3 Relative affinity of alanine substituted EM-2 analogs and a deamidated EM-2 derivate to the SP(1-7) binding site, as compared to EM-2 (H-Tyr-Pro-Phe-Phe-NH_2) itself.

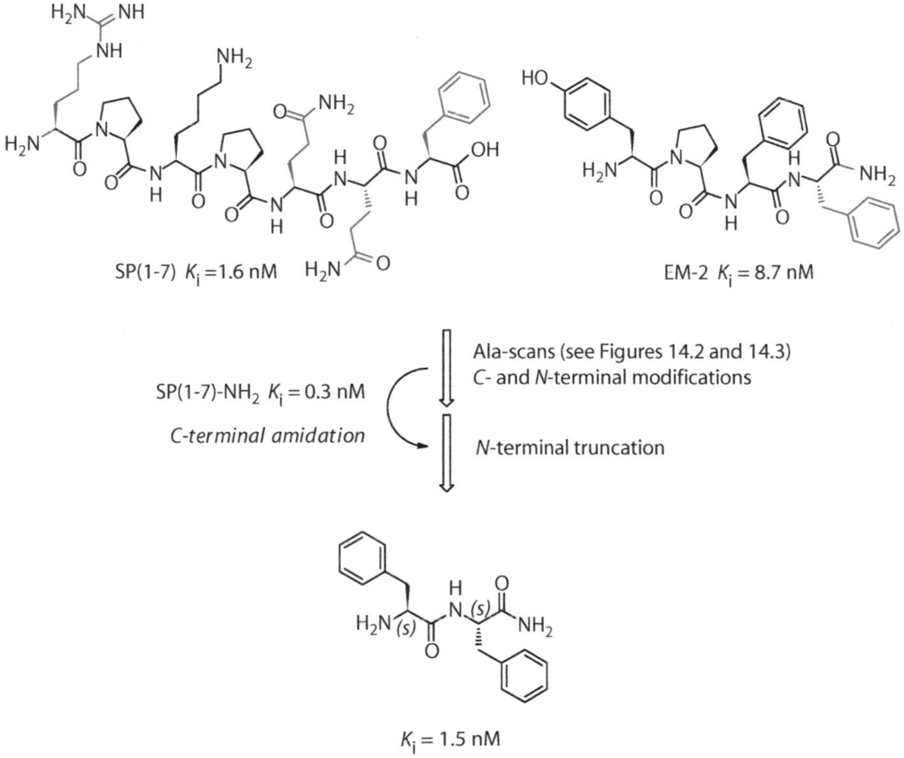

SP(1-7) Ki =1.6 nM
EM-2 Ki = 8.7 nM
Ala-scans (see Figures 14.2 and 14.3)
C- and N-terminal modifications
SP(1-7)-NH2 Ki = 0.3 nM
C-terminal amidation
N-terminal truncation
Ki = 1.5 nM

FIGURE 14.4 (See color insert) Summary of the optimization process starting from the mother peptides SP(1-7) and EM-2 and leading to a dipeptide with equal binding affinity as SP(1-7) itself.

exhibited a K_i of 1.9 nM. For EM-2, removal of the two N-terminal amino acids Tyr-Pro improved the binding affinity six times resulting in the notable discovery of the dipeptide H-Phe-Phe-NH$_2$ with a K_i-value similar to that of SP(1-7) itself. It should be emphasized that the Tyr-Pro sequence is the critical fragment for binding to the μ-receptor, a fact that highlights the double nature of EM-2 (Okada et al. 2003; Kruszynski et al. 2005; Fichna et al. 2007). The Ala-scan of EM-2 led us to believe that further truncation down to a single phenylalanine could be possible, but evaluation of H-Phe-NH$_2$ revealed a complete loss in binding affinity (K_i of 5028 nM).

The binding features of our new dipeptide lead compound H-Phe-Phe-NH$_2$ were further explored via a stereochemical scan (Table 14.2). Thus, all four stereoisomers of H-Phe-Phe-NH$_2$ were synthesized and evaluated and the natural L-Phe-L-Phe isomer was found to be preferred. In addition, a few closely related analogs to H-Phe-Phe-NH$_2$ encompassing noncoded amino acids were evaluated (Table 14.2), but none were as potent as the original dipeptide. Taken together, these observations indicate the presence of a discrete binding pocket in the SP(1-7) binding site matching the H-Phe-Phe-NH$_2$ compound very well.

Altogether, the optimization process described above and which is summarized in Figure 14.4 started with the heptapeptide SP(1-7) and resulted in the remarkable discovery of the dipeptide H-Phe-Phe-NH$_2$ possessing equal binding affinity as the parent peptide.

14.4 THERAPEUTIC POTENTIAL OF SP(1-7) AND ITS ANALOGS (*IN VIVO* STUDIES)

14.4.1 OPIOID DEPENDENCE AND NEUROPATHIC PAIN

Because of their superior role in the treatment of pain, the opiates constitute a very important class of therapeutics. However, opiate usage is known to be associated with addiction, due to the euphoric effects caused by the action of opiates in the reward system in the brain. At present, methadone and buprenorphine are used in the treatment of opiate addiction (Kakko et al. 2007; Kreek et al. 2010). Despite differences in their agonistic and Absorption, Distribution, Metabolism, Elimination (ADME) profiles, they are both synthetic opioid analogs with high affinities to the μ-receptors. The μ-receptors are believed to be responsible for most of the opioid analgesic effects in addition to some of the major side effects, for example, respiratory depression, euphoria, and dependence. Consequently, there is a great need for both an antinociceptive agent and drugs for opioid addiction that utilize alternative signal pathways (Contet, Kieffer, and Befort 2004). Opioid dependence and addiction have also become a pressing issue in the treatment of chronic pain (Ballantyne and LaForge 2007).

Although chronic neuropathic pain constitutes a major public health problem and a vast economic burden to society, it is still a disregarded and undertreated diagnosis (Haanpaa et al. 2009). International Association for the Study of Pain defines neuropathic pain as *pain initiated or caused by primary lesion or dysfunction in the nervous system* but the more distinct definition *pain arising as a direct consequence of lesion or disease affecting the somatosensory system* is under consideration among clinicians and researchers (Dworkin et al. 2003; Treede et al. 2008).

TABLE 14.2
K_i Values of SP(1-7) and EM-2 Peptide Analogs for Inhibition of SP(1-7) Binding to Rat Spinal Cord Membrane

Sequence	$K_i \pm$ SEM (nM)
H-Arg-Pro-Lys-Pro-Gln-Gln-Phe-OH (SP(1-7))	1.6 ± 0.1
H-Tyr-Pro-Phe-Phe-NH$_2$ (EM-2)	8.7 ± 0.1
Terminally modified SP(1-7) peptides	
Ac-Arg-Pro-Lys-Pro-Gln-Gln-Phe-OH	7.1 ± 0.04
H-Arg-Pro-Lys-Pro-Gln-Gln-Phe-NH$_2$	0.3 ± 0.02
Truncated SP(1-7) Peptides	
H-Pro-Lys-Pro-Gln-Gln-Phe-OH	29.6 ± 0.8
H-Pro-Lys-Pro-Gln-Gln-Phe-NH$_2$	2.8 ± 0.25
H-Lys-Pro-Gln-Gln-Phe-OH	30.9 ± 0.4
H-Lys-Pro-Gln-Gln-Phe-NH$_2$	4.4 ± 0.1
H-Pro-Gln-Gln-Phe-OH	26.2 ± 0.7
H-Pro-Gln-Gln-Phe-NH$_2$	4.5 ± 0.3
H-Gln-Gln-Phe-OH	20.4 ± 0.8
H-Gln-Gln-Phe-NH$_2$	1.9 ± 0.05
Truncated EM-2 peptides	
H-Pro-Phe-Phe-NH$_2$	10.9 ± 0.7
H-Phe-Phe-NH$_2$	1.5 ± 0.1
H-Phe-NH$_2$	$5,028 \pm 31$
Terminally modified Phe-Phe peptides	
H-Phe-Phe-OH	$>10,000$
Ac-Phe-Phe-NH$_2$	18.5 ± 1.7
Phe-Phe Analogs	
(L)-Phe-(D)-Phe-NH$_2$	540 ± 20
(D)-Phe-(D)-Phe-NH$_2$	64 ± 2
(D)-Phe-(L)-Phe-NH$_2$	175 ± 13
Phe-Phe analogs having noncoded amino acids	
H-Phe-Phg-NH$_2$	$>10,000$
H-Leu-Phe-NH$_2$	10.2 ± 1.0
H-Tyr(OMe)-Phe(2-Me)-NH$_2$	$>10,000$
H-Phe-Thi-NH$_2$	251 ± 4
H-(D)-Phg-Phe-NH$_2$	$2,247 \pm 115$
H-(L)-Phg-Phe-NH$_2$	182 ± 7
H-Cha-Phe(3-F)-NH$_2$	$>10,000$

Frequent pain due to neuropathy affects the quality of life for numerous people with various underlying diagnoses or causes, such as diabetes, polyneuropathy, human immunodeficiency virus (HIV) infection, sensory neuropathy, poststroke syndromes, postherpetic neuralgia, inherited neurodegeneration, multiple scleroses, chemotherapy, inflammations, and breast cancer surgery (Dworkin et al. 2003; Haanpaa et al.

2009). A history of diabetes is frequently (in up to 60% of cases) associated with the development of neuropathy (Feldman et al. 1999). Neuropathy is further related to *hyperalgesia*, defined as hypersensitivity to noxious stimuli, and chronic painful diabetes (Dyck et al. 2000; Davies et al. 2006). At present, neuropathic pain is insufficiently treated with opioids as well as drugs originating from antiepileptic and tricyclic antidepressives (Dworkin et al. 2003).

14.4.2 IN VIVO ASSAYS

The *in vivo* effect of SP(1-7) and its analogs were evaluated in two different animal models. In the first model, the attenuated effect on morphine withdrawal signs was investigated (Zhou et al. 2009). Adult male Wistar rats were injected with morphine until tolerance to the drug was achieved. This period was followed by a single dose of naloxone to precipitate somatic sign of withdrawal. The test compounds were administrated 30 minutes before the naloxone challenge and the withdrawal signs were observed directly after the naloxone injection and continuously monitored over a 30-minute period.

   In the second animal model, the therapeutic role of SP(1-7) and its analogs in neuropathic pain was studied. The antinociceptive response was evaluated in nondiabetic and diabetic mice. The mice became diabetic after administration of streptozotocin and the antihyperalgesic response of the test compounds was evaluated by the tailflick test after intrathecal administration (Carlsson et al. 2010).

14.5 ABILITY OF SP(1-7) AND ITS SYNTHETIC ANALOGS TO ATTENUATE THE RESPONSE TO OPIOID WITHDRAWAL

Both SP(1-7) and its amidated *C*-terminal analog, SP(1-7)-amide, attenuated the expression of naloxone-precipitated withdrawal in morphine-dependent rats when administrated intracerebroventricularly (Zhou et al. 2003, 2009). Furthermore, the SP(1-7)-amide was shown to reduce withdrawal signs, such as teeth chattering, ptosis, and writhing in a dose-dependent fashion. In agreement with the binding affinities obtained in the SAR-study (see section 14.2.2), the efficacy of the *C*-terminal amide analog in reducing the opioid withdrawal signs was greater than SP(1-7).

14.6 ABILITY OF SP(1-7) AND ITS SYNTHETIC ANALOGS TO ALLEVIATE PAIN AND NEUROPATHIC PAIN

Three of the compounds showing the highest binding affinities, that is, SP(1-7), SP(1-7)-amide, and H-Phe-Phe-NH$_2$, were tested for their potential antinociceptive effect in both nondiabetic and diabetic mice after intrathecal administration (Figure 14.5) (Carlsson et al. 2010; Fransson, Carlsson et al. 2010; Nyberg et al. 2010; Ohsawa et al. 2011a, 2011b). Diabetic mice are shown to have a reduced pain threshold compared to nondiabetic mice and the reduction is thought to arise from hyperalgesia. Although morphine was unable to induce an antinociceptive effect

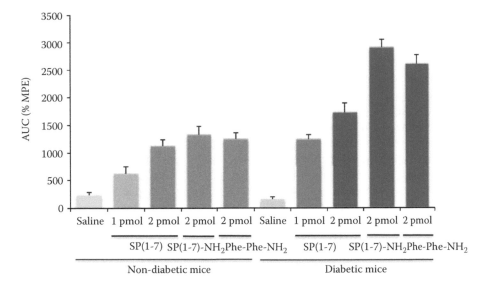

FIGURE 14.5 Antinociceptive effects of SP(1-7), SP(1-7)-amide and H-Phe-Phe-NH$_2$ in nondiabetic and diabetic mice. The antinociceptive effect was evaluated by the AUC calculated from the time-response curve of tail-flick latency. Each column represents the mean value with SEM ($n = 6$). (From Ohsawa, M.A., et al., *Mol. Pain*, 7, 85, 2011a; Ohsawa, M.A., et al., *Peptides*, 32(1), 93–98, 2011b. With permission.)

in diabetic mice, SP(1-7) showed a dose-dependent antinociceptive effect in both diabetic and nondiabetic mice (Carlsson et al. 2010). The effect was higher in diabetic mice, which suggests that the compound is more effective in neuropathic pain and that SP(1-7) ameliorates signs of hyperalgesia. Furthermore, the SP(1-7)-amide proved to be more efficient than the native heptapeptide (Figure 14.5). The dipeptide H-Phe-Phe-NH$_2$, which possessed the same binding affinity as SP(1-7) showed greater antinociceptive potency in diabetic mice than SP(1-7) (Fransson, Botros et al. 2010; Fransson, Carlsson et al. 2010; Nyberg et al. 2010; Ohsawa et al. 2011a).

14.7 ASPECTS ON THE MECHANISM OF SP(1-7)

In mouse and rat, three binding site populations have been identified for SP(1-7); one in the brain and two in the spinal cord (Igwe et al. 1990; Botros et al. 2006, 2008). Nevertheless, the exact mechanism by which SP(1-7) mediates its effects is still unknown. Early theories suggested that the heptapeptide operated via the opioid receptors, as the inhibitory effect of SP(1-7) on SP-induced behavior was shown to be reversed by the nonselective opioid receptor ligand, naloxone (Sakurada et al. 1988; Skilling, Smullin, and Larson 1990). However, none of the selective opioid receptor antagonists for the μ-, δ-, or κ-opioid receptors can reverse the SP(1-7) mediated effects, as illustrated in Figure 14.6 (Mousseau, Sun, and Larson 1992; Carlsson et al. 2010). Interestingly, naloxone itself cannot displace SP(1-7) from its binding site (Igwe et al. 1990; Botros et al. 2006). Recent results imply that the actions of SP(1-7) might be related to the naloxone-sensitive sigma receptor (σ_1-receptor) (Zhou et al. 2009; Carlsson et al. 2010; Ohsawa et al. 2011a, 2011b). Both the nociceptive effect induced in diabetic mice and the attenuating effect on morphine withdrawal by SP(1-7) and its analogs could be reversed by the σ_1-receptor agonist (+)-pentazocine. See Figure 14.6 for a summarized view of the SP(1-7) system.

The σ_1-receptor is widely expressed in the CNS and has been reported to modulate the release of several neurotransmitters, that is, serotonin, dopamine, noradrenaline, glutamate, and GABA (Diaz et al. 2009). Activation of the σ_1-receptor amplifies the N-methyl-D-aspartate (NMDA) receptor response, a consequence believed to be a crucial step in the biochemical events that lead to permanent changes in the brain and give rise to chronic pain conditions (Diaz et al. 2009). It is well-known that the NMDA receptor plays a key role in the development of hyperalgesia or allodynia (Sakurada et al. 2007). Moreover, the NMDA receptor is linked to an increased nitric oxide synthase (NOS) activity and NOS has been shown to be involved in thermal hyperalgesia (Meller and Gebhart 1993; Yamamoto and Shimoyama 1995; Sakurada et al. 2007).

The heptapeptide SP(1-7) and its synthetic analogs do not possess any affinities for the σ_1-receptor itself (unpublished results). More likely the SP(1-7) peptides influence the σ_1-receptor system indirectly, in a downstream manner, possibly by elevating the release of other neuroactive peptides/transmitter substances as illustrated in Figure 14.6. In fact, SP(1-7) has been shown to decrease the presynaptic release of the excitatory amino acid, neurotransmitter glutamate (Skilling et al. 1990), and reduce the production of nitric oxide metabolites in the dorsal spinal cord (Sakurada et al. 2007). Furthermore, it has been demonstrated that SP(1-7) modulates the dopamine system during morphine withdrawal (Zhou and Nyberg 2002; Zhou et al. 2004). Both increased dopamine release and alteration of the dopamine receptors have been observed after administration of SP(1-7) into ventral tegmental area of morphine-dependent rats, which might explain the attenuating effect of SP(1-7) on opioid withdrawal symptoms.

The neuropeptide EM-2 is, besides being a μ-opioid ligand, a relatively high-affinity ligand of the specific binding site of the SP(1-7), as discussed earlier. Apparently, the Tyr-Pro-sequence accounts for the μ-opioid affinity and the H-Phe-Phe-NH$_2$ sequence for the affinity to the SP(1-7) binding site (Okada et al. 2003; Kruszynski et al. 2005; Fichna et al. 2007, Fransson, Botros et al. 2010). It can be argued that the smaller size of H-Phe-Phe-NH$_2$ in comparison to the heptapeptide SP(1-7) might be accompanied with specificity problems. Moreover, H-Phe-Phe-NH$_2$ resembles ligands for the NK3 receptor (Boden et al. 1994, 1995). Hence, the possible binding of H-Phe-Phe-NH$_2$ to the human neurokinin receptors NK1 and NK3 was studied (Fransson, Botros et al. 2010). However, the dipeptide showed no affinity for any of the receptors, a fact, which further strengthens the hypothesis of an existing, not yet isolated, receptor specific for SP(1-7) and its synthetic analogs (Figure 14.6).

14.8 FUTURE PERSPECTIVE

The discovery of small peptides with high affinity to the SP(1-7) binding site and the intriguing results from *in vivo* assays reviewed herein have paved the way for

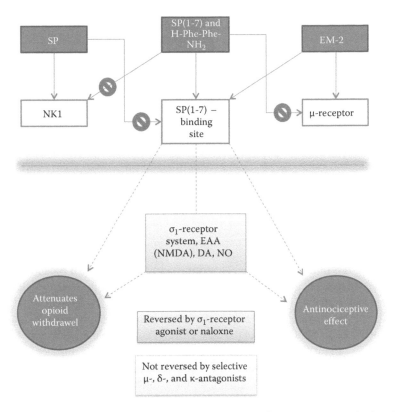

FIGURE 14.6 Summarized view on the interplay between SP(1-7)-system and other signaling pathways, and the corresponding effects.

more comprehensive research activity around the physiological function of neuropathic pain and opioid dependence related to the SP(1-7) system; something, which ultimately can have an impact on drug discovery in these two areas. One important aspect is that SP(1-7) derivatives do not seem to affect the same signal pathways as the opioids (Botros et al. 2006; Zhou et al. 2009), which is a problem with the drugs used today, that is, methadone. The SP(1-7) system might provide a different strategy to treat opiate abuse and to overcome the problem with accompanying addiction. Moreover, the SP(1-7) analogs seem to be more effective in treating neuropathic pain than morphine.

Indeed, dipeptides have previously served as the starting point for the elegant design of orally bioavailable peptidomimetics, as in the case of neurokinin (NK) 1-3 receptor antagonists (Boden et al. 1996) and the successful ACE inhibitors (Ondetti, Rubin, and Cushman 1977). The first ACE inhibitor, Captopril, was approved by FDA 1981 and has been followed by a large number of antihypertensive agents of the same class. Among others, ACE inhibitors and also penicillins utilize proton-coupled peptide transporters (PEPT1) for their passage through cell membranes (Brodin et al. 2002). PEPT1 transporters are normally responsible for the cellular uptake of naturally occurring di- and tripeptides. The potential of the dipeptide compounds already produced in this project and in ongoing optimization studies will depend on their stability, cell-penetrating ability, and on their bioavailability. Most likely, structural optimizations of the small SP(1-7) analogs are needed in order to develop a drug candidate fulfilling all the pharmacodynamic and pharmacokinetic demands of an orally available pharmaceutical. It should though be emphasized that there is a recent and renewed interest in peptides as drug-using alternative routes of administration. This can be seen as a consequence of both the current situation in the pharmaceutical industry where the numbers of approved drugs are decreasing and the massive interest and investment in the area of protein drugs (antibodies) (Nestor 2009; Vlieghe et al. 2010). The small SP(1-7) analogs identified so far in this project will surely be useful as research tools to disambiguate the intriguing mechanism and potential of the SP(1-7) system.

14.9 CONCLUSION

Rational design of drug-like molecules mimicking the actions of bioactive peptides can either be used as an alternative to the HTS techniques that are frequently used in the pharmaceutical industry today or can be used as a complement after high throughput peptide screening. Herein, the initial phase of rational peptide lead optimization has been exemplified in the case of SP(1-7). SAR elucidation and truncation of the parent heptapeptide in combination with identification of another tetrapeptide lead compound (EM-2) were crucial for the successful discovery of the dipeptide H-Phe-Phe-NH$_2$ (K_i = 1.5 nM), having equal affinity as the endogenous heptapeptide ligand SP(1-7) to the SP(1-7) binding site (Figure 14.4). Furthermore, consistent results from SP(1-7) and the identified SP(1-7) analogs on alleviation of opiate dependence and neuropathic pain in mice models have been demonstrated. In addition, the current knowledge about the interplay between SP(1-7) system and other signaling pathways has also been reviewed. Altogether, the data summarized herein constitute a promising platform for further studies on this exciting but yet not fully understood neuropeptide system.

REFERENCES

Ballantyne, J. C. and K. S. LaForge. 2007. Opioid dependence and addiction during opioid treatment of chronic pain. *Pain.* 129: 235–255.

Belanger, P. C. and C. Dufresne. 1986. Preparation of exo-6-benzyl-exo-2-(m-hydroxyphenyl)-1-dimethylaminomethylbicyclo[2.2.2]octane. A non-peptide mimic of enkephalins. *Can. J. Chem.* 64: 1514–1520.

Boden, P., J. M. Eden, J. Hodgson, et al. 1994. The rational development of small molecule tachykinin NK3 receptor selective antagonists—the utilization of a dipeptide chemical library in drug design. *Bioorg. Med. Chem. Lett.* 4: 1679–1684.

Boden, P., J. M. Eden, J. Hodgson, et al. 1995. The development of a novel series of non-peptide tachykinin NK3 receptor selective antagonists. *Bioorg. Med. Chem. Lett.* 5: 1773–1778.

Boden, P., J. M. Eden, J. Hodgson, et al. 1996. Use of a dipeptide chemical library in the development of non-peptide tachykinin NK3 receptor selective antagonists. *J. Med. Chem.* 39: 1664–1675.

Botros, M., M. Hallberg, T. Johansson, et al. 2006. Endomorphin-1 and endomorphin-2 differentially interact with specific binding sites for substance P (SP) aminoterminal SP1-7 in the rat spinal cord. *Peptides.* 27: 753–759.

Botros, M., T. Johansson, Q. Zhou, et al. 2008. Endomorphins interact with the substance P (SP) aminoterminal SP1-7 binding in the ventral tegmental area of the rat brain. *Peptides.* 29: 1820–1824.

Brodin, B., C. U. Nielsen, B. Steffansen, et al. 2002. Transport of peptidomimetic drugs by the intestinal di/tri-peptide transporter, PepT1. *Pharmacol. Toxicol.* 90: 285–296.

Carlsson, A., M. Ohsawa, M. Hallberg, et al. 2010. Substance P1-7 induces antihyperalgesia in diabetic mice through a mechanism involving the naloxone-sensitive sigma receptors. *Eur. J. Pharmacol.* 626: 250–255.

Contet, C., B. L. Kieffer, and K. Befort. 2004. Mu opioid receptor: a gateway to drug addiction. *Curr. Opin. Neurobiol.* 14: 370–378.

Culman, J. and T. Unger. 1995. Central tachykinins—mediators of defense reaction and stress reactions. *Can. J. Physiol. Pharm.* 73: 885–891.

Davies, M., S. Brophy, R. Williams, et al. 2006. The prevalence, severity, and impact of painful diabetic peripheral neuropathy in type 2 diabetes. *Diabetes Care.* 29: 1518–1522.

De Araujo, J. E., J. P. Huston, and M. L. Brandao. 2001. Opposite effects of substance P fragments C (anxiogenic) and N (anxiolytic) injected into dorsal periaqueductal gray. *Eur. J. Pharmacol.* 432: 43–51.

Diaz, J. L., D. Zamanillo, J. Corbera, et al. 2009. Selective sigma-1 receptor antagonists: emerging target for the treatment of neuropathic pain. *Cent. Nerv. Sys. Agents Med. Chem.* 9: 172–183.

Dworkin, R. H., M. Backonja, M. C. Rowbotham, et al. 2003. Advances in neuropathic pain—diagnosis, mechanisms, and treatment recommendations. *Arch. Neurol.* 60: 1524–1534.

Dyck, P. J., P. J. B. Dyck, J. A. Velosa, et al. 2000. Patterns of quantitative sensation testing of hypoesthesia and hyperalgesia are predictive of diabetic polyneuropathy—a study of three cohorts. *Diabetes Care.* 23: 510–517.

Ebner, K., P. Muigg, G. Singewald, et al. 2008. Substance P in stress and anxiety: NK-1 receptor antagonism interacts with key brain areas of the stress circuitry. *Ann. N. Y. Acad. Sci.* 1144: 61–73.

Feldman, E. L., J. W. Russell, K. A. Sullivan, et al. 1999. New insights into the pathogenesis of diabetic neuropathy. *Curr. opin. neurol.* 12: 553–563.

Fichna, J., A. Janecka, J. Costentin, et al. 2007. The endomorphin system and its evolving neurophysiological role. *Pharmacol. Rev.* 59: 88–123.

Fotsch, C., D. M. Smith, J. A. Adams, et al. 2003. Design of a new peptidomimetic agonist for the melanocortin receptors based on the solution structure of the peptide ligand, Ac-Nle-cyclo[Asp-Pro-dPhe-Arg-Trp-Lys]-NH2. *Bioorg. Med. Chem. Lett.* 13: 2337–2340.

Fransson, R., M. Botros, F. Nyberg, et al. 2008. Small peptides mimicking substance P (1-7) and encompassing a C-terminal amide functionality. *Neuropeptides.* 42: 31–37.

Fransson, R., M. Botros, C. Sköld, et al. 2010. Discovery of dipeptides with high affinity to the specific binding site for substance P1-7. *J. Med. Chem.* 53: 2383–2389.

Fransson, R., A. Carlsson, C. Sköld, et al. 2010. Design and synthesis of small substance P (1-7) mimetics. *International Narcotics Research Conference 2010*, July 11–16, Malmö, Sweden. Poster no. 6, Monday July 12.

Haanpaa, M. L., M. M. Backonja, M. I. Bennett, et al. 2009. Assessment of neuropathic pain in primary care. *Am. J. Med.* 122: S13–S21.

Hallberg, M. and F. Nyberg. 2003. Neuropeptide conversion to bioactive fragments—an important pathway in neuromodulation. *Curr. Protein Pept. Sci.* 4: 31–44.

Hanessian, S., G. McNaughton-Smith, H.-G. Lombart, et al. 1997. Design and synthesis of conformationally constrained amino acids as versatile scaffolds and peptide mimetics. *Tetrahedron.* 53: 12789–12854.

Hasenohrl, R. U., P. Gerhardt, and J. P. Huston. 1991. Naloxone blocks conditioned place preference induced by substance P and [pGlu6]-SP(6-11). *Regul. Pept.* 35: 177–187.

Hirschmann, R., K. C. Nicolaou, S. Pietranico, et al. 1992. Nonpeptidal peptidomimetics with beta -D-glucose scaffolding. A partial somatostatin agonist bearing a close structural relationship to a potent, selective substance P antagonist. *J. Am. Chem. Soc.* 114: 9217–9218.

Hirschmann, R., K. C. Nicolaou, S. Pietranico, et al. 1993. De novo design and synthesis of somatostatin non-peptide peptidomimetics utilizing b-D-glucose as a novel scaffolding. *J. Am. Chem. Soc.* 115: 12550–12568.

Hökfelt, T., T. Bartfai, and F. Bloom. 2003. Neuropeptides: opportunities for drug discovery. *Lancet neurology.* 2: 463–472.

Hökfelt, T., C. Broberger, Z. Q. D. Xu, et al. 2000. Neuropeptides—an overview. *Neuropharmacology.* 39: 1337–1356.

Horvath, G. 2000. Endomorphin-1 and endomorphin-2: pharmacology of the selective endogenous mu-opioid receptor agonists. *Pharmacol. Ther.* 88: 437–463.

Huston, J. P., R. U. Hasenoehrl, F. Boix, et al. 1993. Sequence-specific effects of neurokinin substance P on memory, reinforcement, and brain dopamine activity. *Psychopharmacology (Berl).* 112: 147–162.

Igwe, O. J., D. C. Kim, V. S. Seybold, et al. 1990. Specific binding of substance P aminoterminal heptapeptide [SP(1-7)] to mouse brain and spinal cord membranes. *J. Neurosci.* 10: 3653–3663.

Kakko, J., L. Gronbladh, D. Svanborg Kerstin, et al. 2007. A stepped care strategy using buprenorphine and methadone versus conventional methadone maintenance in heroin dependence: a randomized controlled trial. *Am. J. psychiatry.* 164: 797–803.

Kramer, M. S., N. Cutler, J. Feighner, et al. 1998. Distinct mechanism for antidepressant activity by blockade of central substance P receptors. *Science.* 281: 1640–1645.

Kreeger, J. S. and A. A. Larson. 1993. Substance P-(1-7), a substance P metabolite, inhibits withdrawal jumping in morphine-dependent mice. *Eur. J. Pharmacol.* 238: 111–115.

Kreek, M. J., L. Borg, E. Ducat, et al. 2010. Pharmacotherapy in the treatment of addiction: methadone. *J. Addict. Dis.* 29: 200–216.

Kruszynski, R., J. Fichna, J.-C. do-Rego, et al. 2005. Synthesis and biological activity of N-methylated analogs of endomorphin-2. *Bioorg. Med. Chem.* 13: 6713–6717.

Lee, C.-M., B. E. B. Sandberg, M. R. Hanley, et al. 1981. Purification and characterization of a membrane-bound substance P-degrading enzyme from human brain. *Eur. J. Biochem.* 114: 315–327.

Leung, D., G. Abbenante, and D. P. Fairlie. 2000. Protease inhibitors: current status and future prospects. *J. Med. Chem.* 43: 305–341.

Meller, S. T. and G. F. Gebhart. 1993. Nitric-oxide (No) and nociceptive processing in the spinal-cord. *Pain.* 52: 127–136.

Mousseau, D. D., X. F. Sun, and A. A. Larson. 1992. Identification of a novel receptor mediating substance-P-induced behavior in the mouse. *Eur. J. Pharmacol.* 217: 197–201.

Nestor, J. J. 2009. The medicinal chemistry of peptides. *Curr. Med. Chem.* 16: 4399–4418.

Nyberg, F., A. Hallberg, M. Hallberg, et al. 2010. Therapeutic methods and compositions employing peptide compounds. WO/2010/004535, 14 January.

Ohsawa, M., A. Carlsson, M. Asato, et al. 2011a. The dipeptide Phe-Phe amide attenuates signs of hyperalgesia, allodynia and nociception in diabetic mice using a mechanism involving the sigma receptor system. *Mol Pain.* 7: 85.

Ohsawa, M., A. Carlsson, M. Asato, et al. 2011b. The effect of substance P1-7 amide on nociceptive threshold in diabetic mice. *Peptides.* 32(1): 93–98.

Okada, Y., Y. Fujita, T. Motoyama, et al. 2003. Structural studies of [2',6'-dimethyl-L-tyrosine1]endomorphin-2 analogues: enhanced activity and cis orientation of the Dmt-Pro amide bond. *Bioorg. Med. Chem.* 11: 1983–1994.

Olson, G. L., D. R. Bolin, M. P. Bonner, et al. 1993. Concepts and progress in the development of peptide mimetics. *J. Med. Chem.* 36: 3039–3049.

Ondetti, M. A., B. Rubin, and D. W. Cushman. 1977. Design of specific inhibitors of angiotensin-converting enzyme: new class of orally active antihypertensive agents. *Science.* 196: 441–444.

Pasternak, G. W. and P. J. Wood. 1986. Multiple mu opiate receptors. *Life Sci.* 38: 1889–1898.

Rimon, R., P. Legreves, F. Nyberg, et al. 1984. Elevation of substance P-like peptides in the CSF of psychiatric-patients. *Biol. Psychiat.* 19: 509–516.

Ripka, A. S. and D. H. Rich. 1998. Peptidomimetic design. *Curr. Opin. Chem. Biol.* 2: 441–452.

Sakurada, C., C. Watanabe, and T. Sakurada. 2004. Occurrence of substance P(1-7) in the metabolism of substance P and its antinociceptive activity at the mouse spinal cord level. *Methods Find. Exp. Clin. Pharmacol.* 26: 171–176.

Sakurada, S., T. Hayashi, M. Yuhki, et al. 2001. Differential antinociceptive effects induced by intrathecally administered endomorphin-1 and endomorphin-2 in the mouse. *Eur. J. Pharmacol.* 427: 203–210.

Sakurada, T., T. Komatsu, H. Kuwahata, et al. 2007. Intrathecal substance P(1-7) prevents morphine-evoked spontaneous pain behavior via spinal NMDA-NO cascade. *Biochem. Pharmacol.* 74: 758–767.

Sakurada, T., H. Kuwahara, K. Takahashi, et al. 1988. Substance P(1-7) antagonizes substance P-induced aversive behavior in mice. *Neurosci. Lett.* 95: 281–285.

Sakurada, T., P. Le Greves, J. Stewart, et al. 1985. Measurement of substance P metabolites in rat CNS. *J. Neurochem.* 44: 718–722.

Skilling, S. R., D. H. Smullin, and A. A. Larson. 1990. Differential-effects of C-terminal and N-terminal substance-P Metabolites on the release of amino-acid neurotransmitters from the spinal-cord—potential role in nociception. *J. Neurosci.* 10: 1309–1318.

Treede, R. D., T. S. Jensen, J. N. Campbell, et al. 2008. Neuropathic pain—redefinition and a grading system for clinical and research purposes. *Neurology.* 70: 1630–1635.

Vagner, J., H. C. Qu, and V. J. Hruby. 2008. Peptidomimetics, a synthetic tool of drug discovery. *Curr. Opin. Chem. Biol.* 12: 292–296.

Vlieghe, P., V. Lisowski, J. Martinez, et al. 2010. Synthetic therapeutic peptides: science and market. *Drug Discov. Today.* 15: 40–56.

Wiktelius, D., Z. Khalil, and F. Nyberg. 2006. Modulation of peripheral inflammation by the substance P N-terminal metabolite substance P1-7. *Peptides.* 27: 1490–1497.

Yamamoto, T. and N. Shimoyama. 1995. Role of nitric-oxide in the development of thermal hyperesthesia induced by sciatic-nerve constriction injury in the rat. *Anesthesiology.* 82: 1266–1273.

Zadina, J. E. 2002. Isolation and distribution of endomorphins in the central nervous system. *Jap. J. Pharmacol.* 89: 203–208.

Zhou, Q., A. Carlsson, M. Botros, et al. 2009. The C-terminal amidated analogue of the Substance P (SP) fragment SP_{1-7} attenuates the expression of naloxone-precipitated withdrawal in morphine dependent rats. *Peptides.* 30: 2418–2422.

Zhou, Q., P.-A. Frandberg, A. M. S. Kindlundh, et al. 2003. Substance P(1-7) affects the expression of dopamine D2 receptor mRNA in male rat brain during morphine withdrawal. *Peptides.* 24: 147–153.

Zhou, Q., A. M. S. Kindlundh, M. Hallberg, et al. 2004. The substance P (SP) heptapeptide fragment SP1-7 alters the density of dopamine receptors in rat brain mesocorticolimbic structures during morphine withdrawal. *Peptides.* 25: 1951–1957.

Zhou, Q. and F. Nyberg. 2002. Injection of substance P (SP) N-terminal fragment SP1-7 into the ventral tegmental area modulates the levels of nucleus accumbens dopamine and dihydroxyphenylacetic acid in male rats during morphine withdrawal. *Neurosci. Lett.* 320: 117–120.

Zubrzycka, M. and A. Janecka. 2000. Substance P: transmitter of nociception (Minireview). *Endocr. Regul.* 34: 195–201.

15 Cognition-Enhancing Peptides and Peptidomimetics

Fred J. Nyberg and Mathias Hallberg

CONTENTS

15.1 INTRODUCTION

In recent decades, pharmacological treatments aimed at improving cognitive function across a wide range of brain disorders have been the center of attention for many investigators. In many cases, new drugs that potentially enhance cognitive capabilities have become established in clinical practice. In particular, the efficacy of cognitive-behavioral therapy can be improved by the use of cognition enhancers that augment the core learning processes of cognitive-behavioral therapy. This chapter provides a review of the development and use of peptides and peptidomimetics as cognition enhancers for the treatment of a variety of disorders associated with cognitive impairment.

Past and current research investigating the biological mechanisms underlying learning and memory has stimulated the discovery and development of many pharmacological therapies for cognitive dysfunction associated with neurological disorders (Roesler and Schröder 2011). Among a variety of agents used for this purpose are polypeptides, such as growth factors, and bioactive peptides acting on the central nervous system (CNS). These compounds can act as receptor ligands, channel blockers or activators, or enzyme inhibitors.

Although the extrapolation of results from preclinical to clinical research has important limitations, animal models have enabled the identification of selective molecular mechanisms that can be targeted by potential cognition enhancers. In this review, we address the experimental strategies and molecular targets used in the development of cognition enhancers, focusing on the preclinical effects of selected agents that modulate memory consolidation. This chapter does not provide a complete survey of studies on potential cognition enhancers. Instead, it discusses selected experimental approaches, neurochemical systems, and agents, which illustrate the wide range of mechanisms that can be targeted for the development of therapeutic approaches to the treatment of cognitive dysfunction.

15.2 SUBSTRATES OF COGNITION

The term *cognition* represents a scientific concept describing the abilities of humans and other mammals to process information; these abilities include perception, learning, remembering, judging, and problem solving. This topic is studied in various disciplines, such as psychology, philosophy, sociology, linguistics, and computer science, but is also studied in the areas of biomedicine and neuroscience. In some disciplines, such as psychology and sociology, the term cognition usually refers to an individual's psychological functioning or to the attitudes, attributes, and group dynamics of individuals and groups. However, in the areas of biomedicine and neuroscience, cognition usually refers to the brain networks and molecular mechanisms related to learning and memory.

Memory in the brain has been described as a multisystemed phenomenon; each system performs a different memory function that targets different neurological substrates. For instance, declarative memory may be described as the retention of conscious memories of facts and events. The medial temporal lobe and structures in the diencephalon are recognized as important brain areas in the establishment of new declarative memories and it has been suggested that these memory traces are built up in specific regions of the cerebral cortex. The frontal cortex and basal ganglia are also believed to be essential for the forms of declarative memory related to reasoning about memory content. Nondeclarative forms of memory (including skill learning, repetition priming, and classical conditioning) do not involve conscious recollection and are measured through changes in the way in which tasks are performed. These nondeclarative forms of memory are thought to involve the cerebral cortex, basal ganglia, and cerebellum.

At a cellular level, a form of neural synaptic plasticity believed to be involved in long-term memory (LTM) storage, called long-term potentiation (LTP), enhances signal transmission between adjacent neurons over a long period. LTP is induced by high-frequency stimulation of the synapse and is an important target for memory enhancement studies at this level. LTP shares many features with our understanding of LTM, making it an attractive candidate for an explanation of the cellular mechanism underlying learning and memory. For example, LTP and LTM are both triggered rapidly, and each seems to depend on the biosynthesis and formation of new proteins. These proteins are linked with associative memory and they are thought to last for several months. LTP is also believed to be associated with many types of learning, including both the simple classical conditioning that is seen in all animals and the more complex, higher-level cognition experienced by humans (Cooke and Bliss 2006).

Although LTP is not demonstrated in all brain regions, it has been clearly seen in the nucleus accumbens, prefrontal cortex, hippocampus, and amygdala, that is, regions involved in learning (Kenney and Gould 2008). A particular imprint of enhanced amygdala activity and enhanced amygdala–hippocampus connectivity that predicted long lasting, nontemporary memory alterations has been described in the brain (Edelson et al. 2011). It has been suggested that the hippocampus has a critical role in the transfer of short-term memories to LTMs (Santini, Muller, and Quirk 2001; Glannon 2006). Moreover, research involving neuropsychological patients and animal models has indicated that, in addition to playing a critical role in the formation of long-term memories, the hippocampus is critically involved in integrating and processing spatial and contextual information (Kim and Lee 2011).

15.3   MEMORY AND LEARNING

Memory and learning are important concepts to grasp in relation to cognition. They are closely connected and are often confused with each other. However, extensive studies clearly indicate that they should be considered as two distinct phenomena. Though memory depends on learning and learning depends on memory, learning is the acquisition of skill or knowledge, whereas memory is the expression of what has been acquired. As aforementioned, memory is not a single behavioral phenomenon, instead can be divided into a number of processes that are expressed in different combinations under different circumstances. Analyses of memory in healthy individuals and in patients with memory impairment have revealed various expressions of memory that can be assessed using specialized memory tests. Memory can be modified through reconsolidation and performance can change during extinction trials, whereas the original memory remains intact (Abel and Lattal 2001). Studies designed to clarify the molecular basis of these processes have identified a number of signaling molecules that are involved in various stages of memory formation (Govoni, et al. 2010; Lattal and Abel 2001; Ramirez-Amaya 2007). However, in some cases, the actual molecular pathways can be selectively recruited only during certain stages of memory formation.

15.4 MOLECULAR MECHANISMS RELATED TO MEMORY AND COGNITION

The most attractive theory of the cellular mechanism underlying higher cognitive functions, such as learning and memory, is activity-dependent synaptic plasticity, which has received attention for several decades since activity-induced LTP of hippocampal synapses was first discovered (Bliss and Lomo 1973; Teyler and DiScenna 1985). Over the years, a number of investigators have carried out studies on this issue and a solid basis for the importance of LTP in the mechanisms of laying down memories has been established (Abel and Nguyen 2008; Kenney and Gould 2008; Zhang and Poo 2010).

Studies of cognitive function at a molecular level suggest that memories are transformed from a labile to a more fixed state in a consolidation process that is dependent on protein synthesis (Lattal and Abel 2001; Hernandez and Abel 2008). Experimental data have demonstrated that protein synthesis occurs after learning how to perform a behavioral task, during consolidation of the learning. This includes Pavlovian conditioning, in which a conditioned stimulus is paired with an unconditioned stimulus; for example, a tone or a conditioning context might be combined with a footshock (Lattal and Abel 2001). Previous studies also suggest that retrieval of the original association during reexposure to a conditioned stimulus will induce another period of consolidation, in which protein synthesis is required for the original memory to be reconsolidated into a fixed state (Nadar, Schafe, and Le Doux 2000). Thus, data have shown that consolidated fear memories, when reactivated, return to a labile state that requires *de novo* protein synthesis for their reconsolidation (Nader, Schafe, and Le Doux 2000). This indicates that inhibition of protein synthesis after memory retrieval, like inhibition of protein synthesis after memory acquisition, will result in deficits in the consolidation of the original memory. The authors also found that consolidation of new fear memories involves both *de novo* mRNA synthesis and *de novo* protein synthesis in the lateral nucleus of the amygdala (Duvarci, Nader, and LeDoux 2008).

Another important aspect of cognitive function in the brain is related to astrocytes (Robertson 2002; Garcia-Ovejero et al. 2005; Aberg, Brywe, and Isgaard 2006; Pertusa et al. 2008; Pereira and Furlan 2010). The observation that astrocytes can participate as active elements in glutamatergic tripartite synapses (comprising two neurons and one astrocyte) has stimulated new studies of cognitive functioning in the human brain (Pereira and Furlan 2010). Models focusing on associative learning, sensory integration, conscious processing, and memory formation and retrieval have been established. Pereira and coworkers have modeled human cognitive functioning by using an ensemble of functional units connected by gap junctions, in turn connecting distributed astrocytes, which allow formation of cellular calcium waves that can mediate the processing of large-scale cognitive information (Pereira and Furlan 2010). Their model contains a chart of molecular mechanisms present in tripartite synapses that can be used to explain the physiological basis of cognitive functioning. At a molecular level, it has been shown that oestrogen enhances synaptic N-methyl-D-aspartyl (NMDA) receptor currents and the magnitude of LTP (Smith et al. 2009). Synaptic density, NMDA receptor function, and LTP at the hippocampal CA3-CA1 synapses are all thought to be associated with normal learning. Thus, it is likely that improved learning associated with agents, such as oestrogen, is the result of modulation of these parameters, leading to enhanced cognitive capacity (Smith et al. 2009).

The mechanisms underlying memory potentiation are thought to involve glutamate signaling (Abel and Lattal 2001). Glutamate binds to both NMDA and α-amino-3-hydroxy-5-methyl-4-isoxazole propionic acid (AMPA) receptors in the neuron membrane, leading to the opening of sodium and calcium channels into the cell. Calcium influx activates adenylate cyclase, which converts ATP to cAMP. Then the cAMP sequentially triggers activation of protein kinase A (PKA), mitogen-activated protein kinase/extracellular signal-regulated protein kinase (MAPK/ERK), and the cAMP response element-binding factor (CREB). CREB attaches to DNA, thus increasing production of protein for the construction of new synapses (Abel and Lattal 2001).

15.5 NEUROCOGNITIVE IMPAIRMENT

A pronounced decline in memory and cognitive capability is often seen in elderly individuals, sometimes across multiple performance domains. For many individuals, cognitive aging is particularly associated with impairments in remembering recent events and in learning complex associations (Wilson et al. 2006). However, a number of mental disorders and neurodegenerative and neurodevelopmental diseases also involve cognitive deficits (Han et al. 2011). Similarly, studies have shown that the abuse of alcohol, central stimulants, and opiates, such as heroin, destroys brain cells, reducing the attention span and negatively affecting memory (Nyberg 2009; Gould 2010). The molecular mechanisms behind memory decline and cognitive disorders have been the subject of many studies (e.g., Kadish et al. 2009). It appears that these disorders are connected with altered functioning of hippocampal circuits involved in γ-aminobutyric acid (GABA) and NMDA transmission (Rachidi and Lopes 2010). It also appears that age-dependent cognitive decline is linked to dysregulation of calcium homeostasis and that calcium-dependent signals may be key triggers of the molecular mechanisms underlying learning and memory (Oliveira and Bading 2011). Calcium signal-regulated transcription factors, namely, CREB, nuclear factors of activated T-cells (NFAT) and Dual Routing Engine Architecture in Multi-layer (DREAM) may be affected in these processes.

Recent advances in neurobiological science and the development of new technologies have provided a new understanding of the molecular basis of cognition. Changes in gene expression and protein synthesis interlace with the selection of synapses forming memory circuits. Regulation of protein translation and degradation as well as extracellular matrix interactions, second messenger signaling, and neurotransmitter receptor functioning are all components of synaptic remodeling that are essential for cognition (Bibb et al. 2010). Abnormalities or alterations in these components may thus result in impaired cognitive function.

15.6   ENHANCEMENT OF COGNITIVE FUNCTION

Cognition enhancement by pharmacological agents has been a topic of interest for many years. It is well-known that psychostimulants, such as caffeine and nicotine, can improve cognitive performance. Moreover, amphetamine-like drugs, such as methylphenidate, and many other medications are effective in the treatment of attention deficit disorders. However, because of their different etiologies, only a small proportion of learning and memory problems can be treated with central stimulants. There is an urgent need for more effective and precise therapeutic strategies to deal with most disorders associated with cognitive impairment. Cognitive deficiency can result from a countless number of divergent causes, and the list of genetic mutations associated with impaired cognitive function also continues to grow (Dearly et al. 2010). Therefore, it has been suggested that an efficient strategy for finding treatments for the various forms of cognitive disorder would be to target mechanisms that are known to mediate effects leading to improved cognitive function (Lee and Silva 2009). Consequently, it may be useful to consider the interrelated processes and factors that contribute to cognition.

Specific effectors in cognitive processes, such as NMDA receptors, are considered likely targets for such a strategy. The critical role of NMDA receptors in synaptic plasticity and memory has been extensively researched through both genetic and pharmacological manipulations (Lee and Silva 2009). These receptors are composed of an obligatory subunit, NR1, and other modulatory subunits, including NR2 (with A, B, C, and D subtypes) and NR3 (with A and B subtypes). The composition of the NMDA receptor subunits changes during normal development. For instance, the expression of NR2B slowly decreases during postnatal development. The prolonged NMDA receptor currents resulting from overexpression of NR2B lead to the enhancement of hippocampal CA1 LTP, a finding that is consistent with more robust levels of LTP found during the developmental stages, when higher levels of this receptor subunit are expressed.

15.7   COGNITION ENHANCERS

Modulatory influences on memory consolidation have been pharmacologically manipulated in attempts to find strategies to improve cognitive function (Lanni et al. 2008). Most research into cognitive enhancement using manipulation of the molecular mechanisms mediating memory formation has focused on drugs targeting neuronal receptors or their downstream protein kinase pathways. A number of different compounds, including a variety of neuropeptides and growth factors, have been addressed in this context. This section focuses on peptides and polypeptides previously known for their ability to promote cognition through interactions with neuronal pathways involved in the formation of memory.

15.7.1   NEUROPEPTIDES

Neuropeptides are an important group of signaling molecules in the brain. They are produced within specific neurons and, following release, they exert their effects by acting on pre- or postsynaptically located cell surface receptors, most of which belong to the G-protein family of receptors. Many neuropeptide systems regulate learning and memory, mainly by facilitating encoding, consolidation, or retrieval. Neuropeptides known to influence memory and cognition include adrenocorticotropic hormone (ACTH), cholecystokinin (CCK), corticotropin-releasing hormone (CRH), galanin, neuropeptide Y (NPY), the opioid peptides, oxytocin, somatostatin (SST), the tachykinin substance P (SP), vasoactive intestinal peptide (VIP), and vasopressin (Roesler and Schröder 2011).

A number of neuropeptides acting as agonists or antagonists at peptide receptors are currently available as recombinant or synthetic entities. They have been used as research tools or experimental drugs in a large number of studies on cognition enhancement. From the perspective of drug development, the use of peptidergic molecules, with their limited metabolic stability, as cognition enhancers is often aggravated by difficulties associated with suitable routes of administration. It has also been noted that the blood–brain barrier can be poorly permeable to peptides. However, these limitations can be circumvented by the development of intranasal formulations that allow efficient delivery to the brain.

15.7.1.1 Adrenocorticotropic Hormone and Corticotropin-Releasing Factor

It has been suggested that the hypothalamus–pituitary–adrenal (HPA) axis may be involved in the cognitive decline associated with aging in certain neurological patients (Ferrari and Magri 2008). Stress factors have also been shown to induce cognitive deficiency (Heffelfinger and Newcomer 2001). For instance, exposure of rodents to chronic stress and humans to psychosocial stress alters cognitive functioning and has been linked to the pathophysiology of mood disorders (Li et al. 2008). Moreover, modifiable psychological factors amplify or inhibit HPA axis activity in pharmacological activation paradigms, including CRH stimulation tests (Abelson et al. 2010).

However, ACTH and its fragments can also modify and improve cognitive impairment. Various dosages of an orally active analog of ACTH/melanocyte-stimulating hormone (MSH) 4–9 were administered to mentally retarded adults in a double-blind study (Sandman et al. 1980). During the first week, it was observed that behavior related to communication and sociability increased in clients receiving the ACTH analog. In the following week, the patients receiving the peptide appeared more productive and attentive to environmental events, whereas differences in sociability stabilized. Dosages of 5 and 10 mg per day enhanced productivity of tasks requiring precision and concentration, whereas higher doses (20 mg/day) depressed performance of all tasks. This finding encouraged the investigators to suggest that their ACTH/MSH analog should be considered as a potential pharmacological treatment for mentally retarded individuals (Sandman et al. 1980). Subsequent studies of the administration of the ACTH/MSH analog for other neurological indications did not result in any observable effect. In a later preclinical study on fowls subjected to reinforcing training, it was demonstrated that administration of ACTH induced both the development of the LTM stage and a preceding significant increase in the phosphorylation of the forebrain synaptosomal membrane protein GAP43 (Zhao et al. 1995). The authors concluded that their findings confirmed the view that a reinforcement-dependent neurohormone-mediated change to the phosphorylation of this synaptosomal membrane protein may be implicated in the triggering of LTM consolidation.

The retrieval of cognitive functioning is also believed to be modulated by peripheral ACTH, glucocorticoids, and catecholamines. These hormones probably act by activating β-noradrenergic receptor systems in the basolateral amygdala. Exposure to novel situations or systemic administration of antidepressant drugs prior to retention tests enhances memory retrieval, even for very remote memories. The effect of novelty is mediated by molecular mechanisms similar to those for memory retrieval itself.

15.7.1.2   Cholecystokinin

CCK, one of the most extensively characterized brain-gut peptides, has been the subject of numerous preclinical and clinical studies over the past 50 years. It has been shown to exert a variety of physiological actions, not only in the gastrointestinal tract but also in the CNS. The peptide occurs in several molecular forms of varying aminoacid length and is derived from the pre pro-CCK precursor (Rhefeld 1987). In the brain, the sulfated octapeptide (CCK-8) is the predominant form of active CCK peptide released from this prepropeptide. It mediates its effects through activation of two specific subtype receptors, previously known as CCK-A and CCK-B but now identified as CCK(1) and CCK(2); CCK(2) predominates in the brain. Both these receptors belong to the G-protein receptor class (Miller and Gao 2008). CCK is implicated in a variety of behavioral functions, such as satiety, anxiety, exploratory and locomotor activity, and learning and memory (Hadjiivanova, Belcheva, and Belcheva 2003). Studies investigating the involvement of CCK in memory function suggest that activation of CCK receptors plays a physiological role in the mediation of meal-induced enhancement of memory retention (Flood et al. 1987; Flood and Morley 1989). However, CCK is also involved in other types of memory. For example, the CCK(1) receptors appear to mediate the mnemonic effects of the peptide, whereas the CCK(2) receptors mediate its amnestic effects (Hadjiivanova et al. 2003). Furthermore, a CCK(2) receptor agonist (Boc-CCK-4) improved water-maze performance in rats with impaired dopaminergic neurotransmission (6-hydroxydopamine-lesioned) (Rex and Fink 2004).

A highly selective CCK(2) agonist, designated BC264, increased attention and/ or memory through a mechanism also involving dopaminergic pathways (Daugé and Léna 1998). However, this compound did not produce any anxiolytic-like effects, as normally seen following activation of the CCK(2) receptor. Subsequently, the existence of two CCK(2) binding sites, CCK(2)1 and CCK(2)2, was hypothesized. The authors suggested that the two subreceptors could correspond to different activation states of a single molecular entity (Daugé and Léna 1998). Though the CCK(2)1 receptor site is believed to account for the effects of anxiety, the improvements in attention and memory processes are thought to be mediated through the CCK(2)2 receptors, thus offering new options for research in the pharmacological therapy of attention and memory disorders (Daugé and Léna 1998).

The similar neuroanatomical distribution of CCK and opioid peptides in the limbic system is indicative of an opioid–CCK link in the modulation and expression of anxiety or stressor-related behavior. In fact, an antagonistic interaction of CCK and opioid peptides has been suggested in psychological disturbances and stress-induced analgesia (Hebb et al. 2005). There appears to be an intricate balance between the memory-enhancing and anxiety-provoking effects of CCK on one hand, and the amnesic and anxiolytic effects of opioid peptides on the other (Hebb et al. 2005).

The search for selective ligands of the CCK receptor over the past 10–15 years has evolved from the initial CCK structure-derived peptides toward peptidomimetic or nonpeptide agonists and antagonists with improved pharmacokinetic profiles. A broad selection of potent and selective CCK(1) and CCK(2) agonists and antagonists of diverse chemical structure has been produced from this research. These receptor ligands (which are mainly antagonists) have been discovered through optimization of various lead compounds, which were essentially designed on the basis of the structure of the *C*-terminal tetrapeptide CCK-4 or the nonpeptide natural compound asperlicin, or derived from random screening programs (Herranz 2003). However, no selective CCK agonist acting on the CCK(2) receptor has so far been developed for clinical use.

15.7.1.3 Galanin

The neuropeptide galanin is recognized for its multiple effects in both the central and the peripheral nervous systems. In the CNS, galanin and its receptors are located in brain pathways associated with learning and memory (Crawley 2010; Mitzukawa et al. 2010). It regulates numerous physiological and pathological processes through interactions with three G-protein-coupled receptors, GalR1, GalR2, and GalR3 (Mitzukawa et al. 2010). For example, it potently stimulates fat intake and impairs cognitive performance in animal models (Bedecs et al. 1995), and transgenic mice overexpressing galanin display deficits in some learning and memory tests (Crawley 2010).

The inhibitory role of galanin in cognitive processes and the fact that galanin is overexpressed in Alzheimer's disease (AD) suggests that galanin antagonists could offer a novel therapeutic approach to the treatment of memory loss in patients with AD (Crawley 2010). This hypothesis led many investigators to initiate the design and development of GalR antagonists for the treatment of AD.

The involvement of endogenous galanin in learning and memory was also indicated by the use of a synthesized high-affinity galanin antagonist, M35 [galanin(1-13)-bradykinin(2-9) amide]. Intracerebroventricular (i.c.v.) administration of M35 in an animal model significantly improved performance in the Morris water-maze (MWM) spacial learning test (Ogren et al. 1992). M35 binds preferentially to areas in the periventricular regions, including the hippocampus. The authors of this study concluded that galanin could be involved in the modulation of learning and memory and suggested that galanin receptor antagonists could provide a new direction for the treatment of individuals with cognitive disabilities, such as AD (Ogren et al. 1992).

Robinson (2004) points out that, though there are some consistent results regarding the behavioral effects of galanin using tests for learning and memory, extrapolation of these results must take into account confounding findings, a restricted range of tests, and the availability of several noncognitive tests for assessing the potential role of galanin in cognition that have not yet been thoroughly examined. It is hoped that future theory and experimental work will overcome these concerns.

The effects mediated by the galanin receptors are inhibitory. Galanin inhibits the release of anoxic glutamate in the hippocampus and coexists with acetylcholine in cholinergic neurons in the basal forebrain to modulate cholinergic activity in this brain area. Neurons in the cholinergic forebrain appear to play a significant role in learning and memory, as suggested by the loss of these neurons in patients with AD. The inhibitory effects of galanin on neurotransmitter release and the neuronal firing rate are thought to be effected through the inhibition of phosphatidyl inositol hydrolysis and adenylate cyclase (Wrenn and Crawley 2001).

Interestingly, a recent study designed to explore the neuroprotective role of galanin demonstrated that, in *in vitro*, galanin inhibited the neurotoxicity induced by the amyloid peptides, A-β(25-35) and A-β(1-42), in primary cultured rat hippocampal cells (Cheng and Yu 2010). In addition, the GalR2 and GalR3 receptor agonist, Gal(2-11), inhibited the neurotoxicity induced by A-β(25-35) in the cultured neuronal cells. Galanin also inhibited the activation of the apoptotic markers p53, Bax and caspase-3, and dysregulation of p53, Bax and MAP2, that were induced by A-β(25-35) in the cultured hippocampal cells and reversed the downregulation of Bcl-2 induced by A-β(25-35) in the cultured neurons. *In vivo*, in the MWM task, galanin blocked the spatial learning deficits induced in rats by intra-CA1 injection of A-β(25-35). The authors of this study were convinced that galanin has a neuroprotective role in nerve cells and in AD-induced learning and memory deficits (Cheng and Yu 2010).

A more recent review has offered further support for the role of galanin as a neuroprotective agent (Counts et al. 2010). It provided data supporting the hypothesis that overexpression of galanin preserves the neuronal function of the cholinergic basal forebrain, which could prevent or delay the onset of AD symptoms (Counts et al. 2010). Consequently, it appears that the therapeutic potential of GalR ligands in the treatment of AD is linked to their actions as agonists or partial agonists at the Gal receptor. Thus, the recently observed neuroprotective effect of galanin raises the possibility that pharmacological stimulation of galanin activity could result in neuroprotective effects, thus reducing the cognitive decline associated with AD. This concept differs from the traditional hypothesis that galanin inhibits neuronal function in relation to cognition, an effect that was thought to be corrected by treatment with GalR antagonists.

15.7.1.4   Neuropeptide Y

NPY is widely distributed throughout the CNS and has been attributed with an important role in the regulation of basic physiological functions, such as learning and memory. NPY is one of the most abundant peptides in the CNS. It is shown to mediate its physiological effects through at least four different receptors recognized as Y(1), Y(2), Y(4), and Y(5), the most abundant of which are the Y(1) and Y(2) receptors, which are densely expressed in the cortex, hippocampus, and amygdala (Lee and Herzog 2009; Morales-Medina, Dumont, and Quirion 2010).

NPY-like immunoreactivity and NPY receptors are found throughout the CNS but are mainly concentrated in the hippocampus. The hippocampal formation has been repeatedly implicated in the modulation of cognitive functioning but is also involved in the pathogenesis of stress and seizure (Redrobe et al. 1999; Eaton, Sallee, and Sah 2008; Morales-Medina et al. 2010). In fact, early studies suggested that NPY and its receptors may be directly implicated in several pathological disorders, including addictive diseases, depression, stress, and epilepsy/seizure (Redrobe et al. 1999; Gilpin, Misra, and Koob 2008; Thorsell 2008; Morales-Medina et al. 2010). Impaired central NPY signaling may, therefore, be involved in the pathophysiology of these disorders. In fact, analysis of plasma from psychiatric patients has provided some data on the relevance of NPY as a marker for sympathetic tone in certain conditions. Reports on cerebrospinal fluid (CSF) levels of NPY in subjects with depression (Heilig et al. 2004; Hou et al. 2006) suggest dysregulation of central NPY in this disorder; however, there is a need for more studies relating NPY levels in plasma and CSF to disorders involving cognitive decline (Eaton et al. 2008).

Data from studies on experimental animals also suggest a role for NPY in learning and memory (Thorsell et al. 2000). A study in mice designed to evaluate the effects of i.c.v. NPY indicated that it decreased the accuracy of short-term memory in a dose-dependent manner (Cleary et al. 1994). The decrease was significant with the two highest doses but, at lower doses, memory processes were facilitated. The effects of NPY on short-term memory were blocked by intraperitoneal naloxone, an opioid antagonist (Cleary et al. 1994).

Further, though learning and memory were not impaired in transgenic male and female mice lacking the NPY gene (Karl, Duffy, and Herzog 2008), spatial memory acquisition in the MWM was selectively impaired in young rats overexpressing hippocampal NPY (Thorsell et al. 2000). However, this effect was not observed in aged rats overexpressing NPY in the hippocampus (Carvajal et al. 2004). Thus, the cognitive deficits observed in young rats do not appear to occur in older animals, suggesting the existence of compensatory mechanisms leading to a reversal of the learning deficits noted in younger animals.

In conclusion, it appears that NPY and its receptors could be implicated in a range of diseases, including cognitive disorders. However, the development of NPY receptor ligands has been slow, with no clinically approved therapeutic agents currently available.

15.7.1.5 Opioid-Related Peptides—Dynorphin and Nociceptin

The classical endogenous opioid peptides, β-endorphin, the dynorphins, and the enkephalins, are derived from three genetically different precursors, proopiomelanocortin, prodynorphin, and proenkephalin. The opioid peptides have been studied with respect to their ability to influence cognitive functioning and are known to modulate neural transmission in the hippocampus (Simmons and Chavkins 1996). Their receptor activation profiles differ, with the dynorphins exhibiting preference for the κ-opioid peptide (KOP) receptor, the enkephalins preferring the δ-opioid peptide (DOP) receptor, and β-endorphin preferring the μ-opioid peptide (MOP) receptor. Peptides derived from proenkephalin act at both the DOP and MOP receptors to prevent GABA release from inhibitory interneurons, resulting in increased excitability of hippocampal, pyramidal, and dentate gyrus granule cells. Prodynorphin-derived peptides primarily act on presynaptically located KOP receptors to inhibit excitatory amino acid release from the perforant path and mossy fiber terminals.

Activation of the opioid receptors results in reduced membrane excitability via modulation of ion conductance, with resultant decreased voltage-dependent calcium influx and subsequent transmitter release. Synaptic plasticity in the hippocampus is also modulated by endogenous opioids. Though the enkephalins facilitate LTP, the dynorphins inhibit the induction of this type of neuroplasticity (Simmons and Chavkins 1996).

Several studies have indicated that the enkephalins could exert significant effects on learning and memory in rodents undertaking both negatively and positively motivated conditioning tasks (Martinez, Weinberger, and Schulteis 1988). It has also been suggested that the different functional roles and mechanisms of action of enkephalins within each lamina of the hippocampal formation could modulate hippocampal cell activity, both directly and indirectly (Commons and Milner 1995). Moreover, treatment of rats with the DOP receptor agonist [d-Ala2,d-Leu5]-enkephalin (DADLE) induced cognitive improvements, as recorded in the MWM (Wang et al. 2011).

In contrast to this action of DOP receptor agonists, agonist action at the KOP receptor results in cognitive impairment. For instance, learning and manipulation of spatial information (modeled in a modified water maze) were impaired when mice received an intrahippocampal injection (into the CA3 layer) of the KOP agonist U50,488H (Daumas et al. 2007). The effects of the agonist were decreased by the KOP receptor antagonist, norbinaltorphimine (norBNI). It was concluded that the acquisition and consolidation of contextual fear-related memory is impaired by overstimulation of CA3 KOP receptors and that even partial and topographically restricted activation of KOP receptors in CA3 causes impairment of hippocampus-dependent tasks (Daumas et al. 2007). This strengthens the notion of the functional relevance of the κ-opioid system in cognitive functioning.

The role of KOP receptor agonists in learning and memory is nonetheless somewhat controversial. The effects of dynorphin A (1-13) and the KOP agonist U50,488H on learning and memory impairment in mice and rats are difficult to interpret. In passive avoidance tests in rats, both dynorphin A (1-13) and U50,488H reversed the impairments to learning and memory induced by muscarinic and nicotinic cholinergic antagonists (Hiramatsu and Kameyama 1998). *In vivo* microdialysis showed that U50,488H completely blocked the decrease in hippocampal acetylcholine release induced by mecamylamine (an effect abolished by pretreatment with the KOP receptor antagonist norBNI) and also partially blocked the increase of acetylcholine induced by scopolamine. However, U50,488H did not influence the impairment of learning and memory induced by the NMDA-receptor antagonist, MK-801 (Hiramatsu and Kameyama 1998). From these results, it appears that KOP receptor agonists could abolish the decrease in acetylcholine release induced by cholinergic antagonists via the KOP receptor-mediated neuronal system and reverse the associated impairment of learning and memory.

In a later study performed in mice, stress-induced impairment of learning and memory tasks was prevented by pretreatment with norBNI (Carey et al. 2009). In the same behavioral test, norBNI reversed the decline in learning and memory induced by U50,488H. In addition, when transgenic mice lacking the prodynorphin gene were exposed to stress, they did not develop the normal learning and memory deficits. This observation strengthens the suggestion of the involvement of the KOP system in learning and memory (Carey et al. 2009).

Although it seems probable that the opioid receptors have a modulatory effect on cognitive functioning, very few attempts to design cognition-enhancing drugs targeting these receptors have been published to date.

However, drugs acting at the nociceptin (NOP) receptor are under development. Orphanin FQ/NOP (OFQ/N) is an opioid-related peptide studied for its influence on cognitive function (Civelli 2008). OFQ/N is the natural ligand of an orphan G-protein-coupled receptor. The OFQ/N receptor (NOP receptor) shares sequence similarities with the opioid receptors. This has encouraged numerous attempts to find functional similarities and differences between OFQ/N and the classical opioids. NOP is known to produce spinal analgesia but it also seems to antagonize the effects of opioids. Systemic administration of OFQ/N gives rise to a unique range of behavioral responses, including a variety of effects on pain processing (hyperalgesia, analgesia, and allodynia). The peptide also has anxiolytic activity, modulates several opioid-mediated effects, and influences learning and memory processes (Civelli 2008; Sandin, Ogren, and Terenius 2004; Ogren et al. 2010). The recent synthesis of the nonpeptide NOP receptor agonist, Ro-64-6198 (Hoffmann-La Roche), and the antagonist, J-113397 (Banyu Pharmaceutical investigators), has indicated that the NOP receptor is a feasible and interesting new target for drug design. NOP receptor agonists could have potential in the pharmacological management of anorexia, anxiety, cerebral ischemia, drug dependence, epilepsy, hypertension, and neuropathic pain. NOP receptor antagonists could be of interest in the improvement of memory function (Liu et al. 2007; Goeldner et al. 2010).

A recent study (Kuzmin et al. 2009) of spatial memory (in the water maze task) in NMRI mice and pronociceptin-knockout mice found that, though administration of OFQ/N, by i.c.v. and intrahippocampal (CA3) injection, dose-dependently impaired acquisition and retention activities in the maze, the effects were blocked by pretreatment (i.c.v.) with the NOP antagonist, [Nphe(1)]N/OFQ(1-13)-NH(2). The synthetic NOP receptor agonist, Ro-64-6198, also impaired learning dose-dependently but these effects were not modified by pretreatment with the NOP antagonist, naloxone benzoylhydrazone. Learning was improved in pronociceptin-knockout mice in retention trials and reversal training and their performance was not impaired by i.c.v. injection of OFQ/N. The investigators suggested that changes in postsynaptic NOP receptors can occur in these knockout mice and concluded that OFQ/N and NOP receptors could be important for hippocampus-dependent spatial learning and memory, most likely through their ability to modulate the activity and functioning of glutamatergic circuits (Kuzmin et al. 2009).

Thus, it appears that receptors for both opioid peptides and NOP could provide targets for peptides or peptidomimetics as receptor ligands for enhancing learning and memory. Both the endogenous opioids and NOP have modulatory effects on several CNS functions, including memory acquisition, stress, and movement.

15.7.1.6　Somatostatin

SST is a neuroactive peptide known for its modulatory effects on pancreatic exocrine and endocrine secretion (Kumar and Grant 2010). The SSTs act through six SST cell surface receptors (SSTRs), all of them members of the family of G-protein-coupled receptors. SST has two active forms, one composed of 28 amino acids and the other of 14 amino acids (SS14). SS14 has been implicated in various cognitive disorders and regulation of memory formation in the hippocampus and striatum. Four SS14 receptor subtypes (SSTR1-4) are expressed in the hippocampus, but their respective roles in memory processes have not yet been clarified. The role of SS14 in cognition enhancement is strengthened by the finding that SS14 levels were increased in mice after treatment with the known cognition enhancer and GABA (B) receptor antagonist, SGS742 (Sunyer et al. 2009). Several SST agonists, including SS14 and synthetic analogs, are known to affect memory formation.

However, all the SSTR agonists may not influence memory formation in the same way. For instance, though both SS14 and the SSTR4 agonist, L-803,087, dramatically impaired place memory formation in mice, SSTR1 (L-797,591), SSTR2 (L-779,976), and SSTR3 (L-796,778) agonists did not show any behavioral effects (Gastambide et al. 2009). Further, in contrast to SS14, L-803,087 enhanced cue-based memory formation. The authors suggested that hippocampal SSTR4 receptor subunits are selectively involved in the selection of memory strategies by switching from hippocampus-based multiple associations to simple dorsal striatum-based behavioral responses (Gastambide et al. 2009). It has also been suggested that SSTR4-mediated regulation of neuronal activity in the hippocampus depends on both competitive and cooperative interactions with the SSTR2 receptor subtype, although these interactions do not appear to be mediated by direct receptor coupling (Gastambide et al. 2010).

SST is expressed in a discrete population of interneurons in the dentate gyrus of the hippocampus (Kaneko, Maeda, and Satoh 1997). The soma of these interneurons is in the hilus and they project to the outer molecular layer onto the dendrites of dentate granule cells, adjacent to the perforant path input. SST-containing interneurons are very sensitive to excitotoxicity and are thus vulnerable to a variety of neurological disorders, such as epilepsy, AD, traumatic brain injury, and ischaemia (Tallent 2007). The cyclic AMP response element (CRE) site in the SST gene makes the gene activity dependent (i.e., it is turned on when neuronal activity is high). Thus, SST expression is increased by pathological conditions, such as seizures, and by natural stimulation, such as environmental enrichment. SST could, therefore, play an important role in cognition by modulating the response of neurons to synaptic input. In the dentate gyrus, SST and the related peptide cortistatin (CST) reduce the likelihood of generating LTP; in the case of SST, this is probably via inhibition of voltage-gated $Ca(2+)$ channels on dentate granule cell dendrites (Tallent 2007). Thus, the threshold of input required for acquisition of new memories is increased, filtering out irrelevant environmental cues. Transgenic overexpression of CST in the dentate gyrus leads to profound deficits in spatial learning and memory, validating its role in cognitive processing. A reduction in synaptic potentiation by SST and CST could also contribute to the well-characterized antiepileptic properties of these neuropeptides. Thus, SST and CST are important neuromodulators in the dentate gyrus, and disruption of this signaling system could have a major impact on hippocampal functioning (Kaneko et al. 1997).

15.7.1.7 Tachykinins—Substance P

The undecapeptide SP is widely distributed within the CNS but is also present in peripheral tissues. It belongs to the tachykinin family of neuropeptides and may activate a subset of neurokinin receptors (NK1-NK3) with binding preference for the NK1 receptor. SP modulates pain, inflammation, respiration, aggression, and anxiety. However, it is also involved in the enhancement of learning and memory (Tomaz and Nogueira 1997; Santangelo et al. 2001), possibly through its high affinity for the NK1 receptor (Liu et al. 2011) or via release of the *N*-terminal fragment SP(1-7) from the parent peptide (Tomaz and Nogueira 1997; Hallberg and Nyberg 2003).

When it was found that amnesia induced by the muscarinic antagonist scopolamine (SCP) or diazepam (DZP) was blocked by systemic and intraseptal administration of SP, SP(1-7), and choline chloride (Costa, Costa, and do Nascimento 2010), it was suggested that SP and cholinergic mechanisms could interact with GABAergic systems to modulate inhibitory avoidance retention. The authors also speculated that these effects could be mediated, at least in part, by interactions in the septohippocampal pathway (Costa et al. 2010).

In a double-blind, randomized cross-over study (Herpfer et al. 2007), 13 healthy young men received SP or placebo (NaCl) intravenously over 90 minutes for 2 days. Before and during the infusion, cognitive functioning was assessed by the Auditory-Verbal Learning Test (AVLT) and two subtests of the Test for Attentional Performance (TAP). The results indicated that administration of SP induced a disturbance in short-term memory. The authors interpreted this as possible evidence of a memory-disturbing effect of SP (Herpfer et al. 2007).

The effect of a prolyl endopeptidase (PEP) has also been examined in relation to the role of SP in spatial memory (Toide et al. 1997). PEP degrades SP from its *N*-terminal side. The potent PEP inhibitor, JTP-4819, ameliorated age-related impairment of spatial memory and partly reversed central cholinergic dysfunction in aged rats, possibly through the inhibition of PEP-mediated degradation of SP, arginine-vasopressin, and thyrotropin-releasing hormone (Toide et al. 1997). Interestingly, the PEP inhibitor also inhibits the degradation of SP(1-7) (Hallberg and Nyberg 2003).

Taken together, these data indicate that the role of SP in learning and memory is complex. There is evidence for both impairment of spatial performance and enhancement of learning and memory. However, the role of SP or other NK1 receptor ligands in cognition enhancement still seems to be an open question.

15.7.2 Growth Factors and Polypeptide Hormones

Research carried out during the past few years has provided strong support for the long-standing hypothesis that stabilization of both LTP and memory requires rapid reorganization of the actin cytoskeleton of the neuronal dendritic spine. LTP is mediated, in part, by the outgrowth of new dendritic spines or the enlargement of already existing spines to reinforce a particular neural pathway. Also, as the dendritic spines are plastic structures whose lifespan is influenced by input activity, spine dynamics may be essential for memory maintenance over a lifetime. These changes to the spine are influenced by several growth factors [e.g., the neurotrophins' insulin-like growth factor-1 (IGF-1) and growth hormone (GH)], which appear to induce actin polymerization and stable expansion of dendritic spines during LTP.

The neurotrophins represent a family of nerve growth factors consisting of nerve growth factor (NGF), brain-derived neurotrophic factor (BDNF), neurotrophin 3 (NT3), and neurotrophin 4 (NT4). These polypeptides bind to specific receptors that belong to the Trk family of tyrosine protein kinase receptors, where NGF recognizes trkA receptors, BDNF and NT4 specifically bind to and activate trkB receptors, and NT3 prefers trkC receptors. Furthermore, the effects of all the neurotrophins appear to be mediated through a low-affinity receptor, the p75 receptor, which is structurally unrelated to the trk receptors (Teng and Hempstead 2004).

In the context of age-related structural changes within the hippocampus, BDNF and its receptors are of particular interest.

15.7.2.1   Brain-Derived Neurotrophic Factor

BDNF is a polypeptide protein consisting of 252 amino acids. It is synthesized as a precursor with a signal peptide of 18 amino acids and a prosequence of 112 amino acids. Similar proteins isolated from various mammals have an almost identical amino acid sequence. The biological actions of BDNF, which is found in both the CNS and the periphery, are mediated through receptors that belong to the trk family. TrkB, the primary signal transduction receptor for BDNF, is a tyrosine kinase receptor (Lu et al. 2011).

BDNF has also emerged as an important regulator of synaptogenesis and the synaptic plasticity mechanisms underlying learning and memory in the adult CNS (Cunha, Brambilla, and Thomas 2010). The actions of this neurotrophin in the adult CNS have been extensively studied, probably because it has been shown to have a critical role in LTP. Thus, BDNF is a potent activator of the actin signaling cascades in adult dendritic spines. It has been shown to reverse potentiation in mutant mice with Huntington's disease, middle-aged rats, and a mouse model of Fragile-X syndrome (Lynch et al. 2008). A similar reversal of impairments to LTP was demonstrated in middle-aged rats by upregulating the production of BDNF production, following exposure to ampakines, a class of drugs that positively modulate AMPA-type glutamate receptors.

Research designed to test whether chronic elevation of BDNF enhances memory has revealed that activities or chemical agents that increase the CNS expression of this neurotrophin also improve learning and memory. For instance, it is well-known that physical exercise can enhance hippocampal-dependent forms of learning and memory in laboratory animals. This observation coincides with enhanced hippocampal neural plasticity, including increased expression of BDNF mRNA and protein, as well as increased neurogenesis and LTP (Hopkins, Nitecki, and Bucci 2011). Recently, the expression of both the gene transcript and protein of BDNF was upregulated by a novel tripeptide (Neuropep-1) in both cell lines and experimental rats (Shin, Kim, and Kim 2011). Tail vein injection of the tripeptide significantly upregulated BDNF expression, TrkB phosphorylation, and subsequent downstream signals, including activation of Akt, ERK, and cAMP response element binding in the rat hippocampus (Shin et al. 2011).

However, although BDNF has emerged as a possible broad-spectrum agent for the treatment of plasticity losses in rodent models of human conditions associated with memory and cognitive deficits, there are several concerns about the clinical use of this neurotropin. It has been noted that the ability of BDNF to regulate both excitatory and inhibitory synapses in the CNS could result in conflicting alterations in synaptic plasticity and memory formation. An obvious lack of detailed knowledge about the mechanisms by which BDNF influences higher cognitive functioning and complex behavior has worried some researchers regarding the possibility of devising BDNF-based therapeutics for human disorders of the CNS (Cunha et al. 2010). On the other hand, as BDNF supports the survival of basal forebrain and hippocampal cholinergic neurons, the possibility that it may be useful for the treatment of patients with neurological disorders, including cognitive dysfunction, should not be excluded (Nagahara and Tuszynski 2011). Alternatively, agents and conditions that induce expression of BDNF and the trkB receptor could also provide a possible therapeutic route.

15.7.2.2 Nerve Growth Factor

The neurotrophin NGF is a dimeric polypeptide, with 118 amino acids per protomer (McDonald et al. 1991). The active entity is released from pro-NGF by hydrolysis. The neurotrophins bind to and activate a class of receptors mentioned above, namely the Trk receptors. The NGF receptor TrkA, a tyrosine kinase receptor, is expressed primarily in cholinergic neurons known to be implicated in spatial learning and memory, whereas the NGF p75 receptor is expressed in both neuronal cells and glia. The effects of activating these receptors on learning, short-term memory, and LTM are not clearly understood. In a recent study, wild-type mice and transgenic mice overexpressing the amyloid precursor protein (APP mice) were given native NGF (a ligand of both the TrkA and p75 receptor) or selective agonists for the TrkA receptor (i.e., not binding to the p75 receptor) (Aboulkassim et al. 2011). Learning and short-term memory were significantly improved by the selective TrkA agonists in APP mice but were unaffected by any of the treatments in normal wild-type mice. However, one of the TrkA-selective agonists was found to cause persistent deficits in LTM in wild-type mice. The investigators concluded that selective activation of the TrkA receptor affects cognition differently in impaired APP mice compared to normal wild-type mice (Aboulkassim et al. 2011).

NGF is known to affect neuronal plasticity during mammalian development (Branchi et al. 2004); this effect is believed to be mediated through mechanisms involving glutamate transmission. Recent studies have suggested that NGF, along with many other neurotrophic factors, could regulate NMDA receptor gene expression, which in turn could stimulate LTP and subsequently improve learning and memory (Lessmann 1998).

Several compounds derived from plants enhance the expression of NGF with resultant enhanced learning and memory. Pycnogenol, which consists of flavonoids, mainly procyanidins and phenolic compounds, is a powerful antioxidant. It increased the NGF content in the hippocampus and cortex of orchidectomized rats, thus improving the spatial memory impairment (Hasegawa and Mochizuki 2009). Cistanches Herba also increased NGF release in the hippocampus and cortex of mice, significantly enhancing learning and memory, as shown by the passive avoidance test and novel object recognition test (Choi et al. 2011).

A dipeptide preparation, known as noopept (N-phenylacetyl-L-prolylglycine ethyl ester, GVS-111), appears to have nootropic and neuroprotective properties. In a study in rats, it depleted the antiamnesic effect induced by cholinoceptor antagonists and improved spatial preference (Radionova et al. 2008). Acute administration of noopept increased the expression of mRNA for both the neurotropic factors, NGF and BDNF, in the rat hippocampus (Ostrovskaya et al. 2008). The neurotophic effect was then potentiated by chronic treatment, without the development of tolerance (Ostrovskaya et al. 2008).

15.7.2.3 Insulin-Like Growth Factor-1

Insulin-like growth factor-1 (IGF-1) is a 70-amino acid peptide produced in the liver as a response to GH stimulation; it is widely distributed in the CNS and in various tissues and cells in the periphery (Okajima and Harada 2008). IGF-1 and its receptor IGF-1R play major roles in vertebrate growth and development. IGF-1 mediates the growth-promoting actions of GH and is essential for postnatal and adolescent growth.

Studies have demonstrated a neuroprotective effect of IGF-1 (Sharma et al. 1998, 2000) and the peptide was also found to increase excitatory synaptic transmission in the CA1 region of the hippocampus (Ramsey et al. 2005). Decreased serum levels of IGF-1 have been associated with memory loss on aging and the development of neurodegenerative disorders (Lanni et al. 2008). Exogenous administration of IGF-1 reversed the reduction in adult hippocampal neurogenesis and impairment of spatial learning in mutant mice that had low serum levels of IGF-1 (Trejo et al. 2008). Peripheral infusion of IGF-1 also selectively induced neurogenesis in the adult rat hippocampus (Aberg et al. 2000; Isgaard et al. 2007).

In another study, depletion of IGF-1 in transgenic mice, resulting in 80–85% reduction of IGF-1 levels in the circulation, had no effects on spatial learning and memory (MWM test) in young (6-month-old) mice (Svensson et al. 2006). However, the acquisition of the spatial task was slower in older (15- or 18-month-old) mutant mice than in controls, and impaired spatial working and reference memory were noted.

The mechanism by which IGF-1 enhances cognition has not been fully clarified. However, it may be linked with an interaction between IGF-1 and the NMDA receptor complex. The IGF-1-related reversal of the age-related decline in spatial working and reference memory and impairment of learning and reference memory by antagonism of IGF-1 in the brains of young animals appears to involve the NMDA receptor (Sonntag et al. 2000), particularly the NMDA receptor subunits, NR2A and NR2B.

In confirmation of this hypothesis, after 10 daily subcutaneous injections of IGF-1, the hippocampal mRNA levels of the NR2B subunit increased in young (11 weeks) but not in older (14–16 months) rats and the expression of NR2A mRNA was decreased in both groups (Le Grevés et al. 2002). This significant increase in the NR2B:NR2A ratio at transcription level is thought to indicate potential for synaptic plasticity, as the NR2B subtype appears to be essential for spatial learning and LTP (Tang et al. 1999). The IGF-1 receptor gene transcript was decreased in the older rats after IGF-1 treatment. These results suggest that IGF-1 mediates its effects on the hippocampus via the NMDA receptor complex. In many respects, these effects of IGF-1 are identical to those shown for GH (Le Grevés et al. 2002).

As indicated above, animal models have suggested that treatment with IGF-1 could improve various neurological disorders associated with cognitive deficiency. Based on this preclinical evidence, many investigators have proposed that IGF-1 should be tested in clinical trials of dysfunctional cognition in humans. However, the only clinical approaches using recombinant human IGF-1 (rhIGF-1) treatment involve children with short stature as a result of severe primary IGF deficiency. A recent study designed to assess the safety and efficacy of rhIGF-1 treatment in short children with low IGF-1 levels found age- and dose-dependent increases in height velocity during the first year of treatment (Midyett et al. 2010). Adverse effects during treatment (headache, vomiting, and hypoglycaemia) were generally transient and easily managed. The efficacy and safety of IGF-1 therapy in these patients has been demonstrated, and investigations are in progress to determine optimal dosing (Bright, Mendoza, and Rosenfeld 2010), but the availability of IGF-1 therapy remains limited to children with severe primary IGF-1 deficiency or growth hormone resistance. No attempts to treat patients with cognitive deficits with rhIGF-1 have been described to date.

15.7.2.4 Growth Hormone

GH is a major hormone that is produced in the anterior pituitary cells. It is a polypeptide and consists of a primary sequence of 191 amino acids. The hormone is mainly recognized for its effects on growth and metabolism. However, early studies (for review, see Nyberg 2000) demonstrated that GH can target many areas of the CNS, and that a deficiency of this hormone has been associated with cognitive impairment, memory loss, and diminished overall well-being (Bengtsson et al. 1993; Burman and Deijen 1998). GH replacement therapy given to GH-deficient patients improves these effects (Burman and Deijen 1998; Wass and Reddy 2010).

In aged rats, the hormone prevents neuronal loss in the hippocampus, indicating a neuroprotective effect (Azcoitia et al. 2005). The levels of circulating GH and the density of GH-binding sites decrease with age in several areas of the human and rat brain, including the hippocampus (Lai et al. 1993; Zhai et al. 1994; van Dam et al. 2000). GH also enhances the expression of the gene transcript for NR2B in the rat hippocampus (Le Grevés et al. 2002). Overexpression of NR2B has been associated with age-dependent increases in cognitive capabilities (Tang et al. 1999; Cui et al. 2011). In addition, peripheral administration of GH to hypophysectomized rats improved spatial performance (Le Grevés et al. 2006, 2011) and increased the expression of the hippocampal gene transcript of both NMDA receptor subunits as well as postsynaptic density protein 95 (Le Grevés et al. 2006). These results suggest a link between decreased levels of GH in the elderly and impairment of cognitive functioning and offer a clear indication that GH replacement could improve memory and cognitive capabilities (Le Grevés et al. 2005).

Several neurological complications are associated with alterations to the somatotrophic axis, affecting levels of both IGF-1 and GH. For instance, traumatic brain injury has recently been recognized as a risk factor for cognitive impairment and hypopituitarism, most frequently with GH deficiency (GHD). In fact, trauma to the spinal cord decreases GH levels in both humans (Nyberg 2000) and rats (Nyberg and Sharma 2002). However, it is also shown that GH administration may restore trauma-induced brain damages to the spinal cord (Nyberg and Sharma 2002). Also, the hormone has been shown to reverse opioid-induced apoptosis in murine hippocampal cells (Svensson et al. 2008).

In a clinical trial, six GHD patients with traumatic brain injuries received GH replacement therapy for 6 months (Maric et al. 2010). Not only did their cognitive abilities, including verbal and nonverbal memory, improve, but their psychiatric functioning also improved. Twelve months after GH therapy had been discontinued, verbal and nonverbal memory deteriorated in three of these patients (Maric et al. 2010). In another study, 11 patients with brain trauma were confirmed to have GHD. These patients received GH or vehicle, plus daily cognitive rehabilitation therapy for three months. GH-recipients had significantly greater improvements than controls in the similarities and vocabulary sections of the verbal IQ test and in total IQ. Thus, GH administration significantly improved cognitive rehabilitation of GHD patients and, as the plasma levels of IGF-1 were similar in both groups at the end of the treatment period, it was believed that GH was responsible for the observed improvement (Reimunde et al. 2011).

Although the overall intellectual ability of children with GHD is comparable to that of a normal reference population, some areas such the motor-component scale and performance IQ are below the mean value. GH replacement in these individuals resulted in significant improvement (Puga González et al. 2010). It has previously been shown that GH treatment during infancy and childhood normalizes growth velocity and improves fine motor skill performance in individuals with Downs Syndrome (Annerén et al. 2000). In a recent study designed to investigate the late effects of early GH treatment on growth and psychomotor development in adolescents with Down syndrome (Myrelid et al. 2010), 12 subjects, who had received GH for 3 years from the age of 6–9 months in an earlier trial, and 10 controls were followed for 15 years. The adolescents previously treated with GH had cognitive function scores above those of the controls in all tested subcategories of the Leiter International Performance Scale-Revised and the Wechsler Intelligence Scale for Children, but there were no differences in the Brief IQ scores. The GH-treated subjects also performed better than the controls in age-adjusted motor performance subtests. These findings indicate that children with Downs Syndrome may benefit from early GH treatment (Myrelid et al. 2010).

It is evident that GH replacement, along with psychomotor and cognitive stimulation, can be useful for appropriate neurodevelopment of children with GHD. Treatment of these patients with GH is associated with improvements in both memory and attention. Improvements in verbal memory in patients with childhood-onset GHD have also been confirmed. However, some questions concerning the possibility of extending the areas of indication and better defining the treatment of GHD in different age groups remain unanswered.

15.8   CONCLUDING REMARKS

Research carried out during recent decades has uncovered a number of new aspects of molecular and cellular mechanisms involved in memory formation. This, in turn, has provided new opportunities for the development of cognition-enhancing drugs. This chapter has reviewed the effects of various peptides and polypeptides on memory enhancement and has discussed their implications for the development of cognition enhancers. Although a large number of neuroactive peptides and hormones has been implicated in cognitive functioning, only a limited number of these compounds has been developed for clinical use. In many cases, it appears that smaller molecules that affect these peptides or their receptors are more likely to become drugs for clinical use.

ACKNOWLEDGMENTS

This study was supported by The Swedish Research Council (Grant 9459), the Berzelii Center of neurodiagnostics, and the Swedish Council for Working Life and Social Research.

REFERENCES

Abel, T., Lattal, K.M. 2001. Molecular mechanisms of memory acquisition, consolidation and retrieval. *Curr Opin Neurobiol*, 11:180–187.

Abel, T., Nguyen, P.V. 2008. Regulation of hippocampus-dependent memory by cyclic AMP-dependent protein kinase. *Prog Brain Res*, 169:97–115.

Abelson, J.L., Khan, S., Young, E.A., Liberzon, I. 2010. Cognitive modulation of endocrine responses to CRH stimulation in healthy subjects. *Psychoneuroendocrinology*, 35:451–459.

Aberg, M.A., Aberg, N.D., Hedbäcker, H., Oscarsson, J., Eriksson, P.S. 2000. Peripheral infusion of IGF-I selectively induces neurogenesis in the adult rat hippocampus. *J Neurosci*, 20:2896–2903.

Aberg, N.D., Brywe, K.G., Isgaard, J. 2006. Aspects of growth hormone and insulin-like growth factor-I related to neuroprotection, regeneration, and functional plasticity in the adult brain. *Scientific World J*, 6:53–80.

Aboulkassim, T., Tong, X.K., Chung Tse, Y., Wong, T.P., Woo, S.B., Neet, K.E., Brahimi, F., Hamel, E., Saragovi, H.U. 2011. Ligand-dependent TrkA activity in brain differentially affects spatial learning and long-term memory. *Mol Pharmacol*, 80:498–508.

Annerén, G., Tuvemo, T., Gustafsson, J. 2000. Growth hormone therapy in young children with Down syndrome and a clinical comparison of Down and Prader-Willi syndromes. *Growth Horm IGF Res*,10 Suppl B:S87–91.

Azcoitia, I., Perez-Martin, M., Salazar, V., Castillo, C., Ariznavarreta, C., Garcia-Segura, L.M., Tresguerres, J.A. 2005. Growth hormone prevents neuronal loss in the aged rat hippocampus. *Neurobiol Aging*, 26:697–703.

Bedecs, K., Berthold, M., Bartfai, T. 1995. Galanin—10 years with a neuroendocrine peptide. *Int J Biochem Cell Biol*, 27:337–349.

Bengtsson, B.A., Edén, S., Lönn, L., Kvist, H., Stokland, A., Lindstedt, G., Bosaeus, I., Tölli, J., Sjöström, L., Isaksson, O.G. 1993. Treatment of adults with growth hormone (GH) deficiency with recombinant human GH. *J Clin Endocrinol Metab*, 76:309–317.

Bibb, J.A., Mayford, M.R., Tsien, J.Z., Alberini, C.M. 2010. Cognition enhancement strategies. *J Neurosci*, 30:14987–14992.

Bliss, T.V., Lomo, T. 1973. Long-lasting potentiation of synaptic transmission in the dentate area of the anaesthetized rabbit following stimulation of the perforant path. *J Physiol*, 232:331–356.

Branchi, I., Francia, N., Alleva, E. 2004. Epigenetic control of neurobehavioural plasticity: the role of neurotrophins. *Behav Pharmacol*, 15:353–362.

Bright, G.M., Mendoza, J.R., Rosenfeld, R.G. 2010. Recombinant human insulin-like growth factor-1 treatment: ready for primetime. *Endocrinol Metab Clin North Am*, 38:625–638.

Burman, P., Deijen, J.B. 1998. Quality of life and cognitive function in patients with pituitary insufficiency. *Psychother Psychosom*, 67:154–167.

Carey, A.N., Lyons, A.M., Shay, C.F., Dunton, O., McLaughlin, J.P. 2009. Endogenous kappa opioid activation mediates stress-induced deficits in learning and memory. *J Neurosci*, 29:4293–4300.

Carvajal, C.C., Vercauteren, F., Dumont, Y., Michalkiewicz, M., Quirion, R. 2004. Aged neuropeptide Y transgenic rats are resistant to acute stress but maintain spatial and non-spatial learning. *Behav Brain Res*, 153:471–480.

Cheng, Y., Yu, L.C. 2010. Galanin protects amyloid-beta-induced neurotoxicity on primary cultured hippocampal neurons of rats. *J Alzheimers Dis*, 20:1143–1157.

Choi, J.G., Moon, M., Jeong, H.U., Kim, M.C., Kim, S.Y., Oh, M.S. 2011. Cistanches Herba enhances learning and memory by inducing nerve growth factor. *Behav Brain Res*, 216:652–658.

Civelli, O. 2008. The orphanin FQ/nociceptin (OFQ/N) system. *Results Probl Cell Differ*, 46:1–25.

Cleary, J., Semotuk, M., Levine, A.S. 1994. Effects of neuropeptide Y on short-term memory. *Brain Res*, 653:210–214.

Commons, K.G., Milner, T.A. 1995. Ultrastructural heterogeneity of enkephalin-containing terminals in the rat hippocampal formation. *J Comp Neurol*, 358:324–342.

Cooke, S.F., Bliss, T.V. 2006. Plasticity in the human central nervous system. *Brain*, 129 Pt 7:1659–1673.

Costa, J.C., Costa, K.M., do Nascimento, J.L. 2010. Scopolamine- and diazepam-induced amnesia are blocked by systemic and intraseptal administration of substance P and choline chloride. *Peptides*, 31:1756–1760.

Counts, S.E., Perez, S.E., Ginsberg, S.D., Mufson, E.J. 2010. Neuroprotective role for galanin in Alzheimer's disease. *EXS*, 102:143–162.

Crawley, J.N. 2010. Galanin impairs cognitive abilities in rodents: relevance to Alzheimer's disease. *EXS*, 102:133–141.

Cui, Y., Jin, J., Zhang, X., Xu, H., Yang, L., Du, D., Zeng, Q., Tsien, J.Z., Yu, H., Cao, X. 2011. Forebrain NR2B overexpression facilitating the prefrontal cortex long-term potentiation and enhancing working memory function in mice. *PLoS One*, 6:e20312.

Cunha, C., Brambilla, R., Thomas, K.L. 2010. A simple role for BDNF in learning and memory? *Front Mol Neurosci*, 3:1.

Daugé, V., Léna, I. 1998. CCK in anxiety and cognitive processes. *Neurosci Biobehav Rev*, 22:815–825.

Daumas, S., Betourne, A., Halley, H., Wolfer, D.P., Lipp, H.P., Lassalle, J.M., Francés, B. 2007. Transient activation of the CA3 Kappa opioid system in the dorsal hippocampus modulates complex memory processing in mice. *Neurobiol Learn Mem*, 88:94–103.

Duvarci, S., Nader, K., LeDoux, J.E. 2008. De novo mRNA synthesis is required for both consolidation and reconsolidation of fear memories in the amygdala. *Learn Mem*, 15:747–755.

Eaton, K., Sallee, F.R., Sah, R. 2008. Relevance of neuropeptide Y (NPY) in psychiatry. *Curr Top Med Chem*, 7:1645–1659.

Edelson, M., Sharot, T., Dolan, R.J., Dudai, Y. 2011. Following the crowd: brain substrates of long-term memory conformity. *Science*, 333:108–111.

Ferrari, E., Magri, F. 2008. Role of neuroendocrine pathways in cognitive decline during aging. *Ageing Res Rev*, 7:225–233.

Flood, J.F., Morley, J.E. 1989. Cholecystokinin receptors mediate enhanced memory retention produced by feeding and gastrointestinal peptides. *Peptides*, 10:809–813.

Flood, J.F., Smith, G.E., Morley, J.E. 1987. Modulation of memory processing by cholecystokinin: dependence on the vagus nerve. *Science*, 236:832–834.

Garcia-Ovejero, D., Azcoitia, I., Doncarlos, L.L., Melcangi, R.C., Garcia-Segura, L.M. 2005. Glia-neuron crosstalk in the neuroprotective mechanisms of sex steroid hormones. *Brain Res Brain Res Rev*, 48:273–286.

Gastambide, F., Lepousez, G., Viollet, C., Loudes, C., Epelbaum, J., Guillou, J.L. 2010. Cooperation between hippocampal somatostatin receptor subtypes 4 and 2: functional relevance in interactive memory systems. *Hippocampus*, 20:745–757.

Gastambide, F., Viollet, C., Lepousez, G., Epelbaum, J., Guillou, J.L. 2009. Hippocampal SSTR4 somatostatin receptors control the selection of memory strategies. *Psychopharmacology (Berl)*, 202:153–163.

Gilpin, N.W., Misra, K., Koob, G.F. 2008. Neuropeptide Y in the central nucleus of the amygdala suppresses dependence-induced increases in alcohol drinking. *Pharmacol Biochem Behav*, 90:475–480.

Glannon, W. 2006. Psychopharmacology and memory. *J Med Ethics*, 32:74–78.

Goeldner, C., Reiss, D., Kieffer, B.L., Ouagazzal, A.M. 2010. Endogenous nociceptin/ orphanin-FQ in the dorsal hippocampus facilitates despair-related behavior. *Hippocampus*, 20:911–916.

Gould, T.J. 2010. Addiction and cognition. *Addict Sci Clin Pract*, 5:4–14.

Govoni, S., Amadio, M., Battaini, F., Pascale, A. 2010. Senescence of the brain: focus on cognitive kinases. *Curr Pharm Des*, 16:660–671.

Hadjiivanova, C., Belcheva, S., Belcheva, I. 2003. Cholecystokinin and learning and memory processes. *Acta Physiol Pharmacol Bulg*, 27:83–88.

Hallberg, M., Nyberg, F. 2003. Neuropeptide conversion to bioactive fragments—an important pathway in neuromodulation. *Curr Protein Pept Sci*, 4:31–44.

Han, G., Klimes-Dougan, B., Jepsen, S., Ballard, K., Nelson, M., Houri, A., Kumra, S., Cullen, K. 2011. Selective neurocognitive impairments in adolescents with major depressive disorder. *J Adolesc*, 35(1):11–20.

Hasegawa, N., Mochizuki, M. 2009. Improved effect of Pycnogenol on impaired spatial memory function in partial androgen deficiency rat model. *Phytother Res*, 23:840–843.

Hebb, A.L., Poulin, J.F., Roach, S.P., Zacharko, R.M., Drolet, G. 2005. Cholecystokinin and endogenous opioid peptides: interactive influence on pain, cognition, and emotion. *Prog Neuropsychopharmacol Biol Psychiatry*, 29:1225–1238.

Heffelfinger, A.K., Newcomer, J.W. 2001. Glucocorticoid effects on memory function over the human life span. *Dev Psychopathol*, 13:491–513.

Herranz. 2003. Cholecystokinin antagonists: pharmacological and therapeutic potential. *Med Res Rev*, 23:559–605.

Heilig, M., Zachrisson, O., Thorsell, A., Ehnvall, A., Mottagui-Tabar, S., Sjögren, M., Asberg, M., Ekman, R., Wahlestedt, C., Agren, H. 2004. Decreased cerebrospinal fluid neuropeptide Y (NPY) in patients with treatment refractory unipolar major depression: preliminary evidence for association with preproNPY gene polymorphism. *J Psychiatr Res*, 38:113–121.

Hernandez, P.J., Abel, T. 2008. The role of protein synthesis in memory consolidation: progress amid decades of debate. *Neurobiol Learn Mem*, 89:293–311.

Herpfer, I., Katzev, M., Feige, B., Fiebich, B.L., Voderholzer, U., Lieb, K. 2007. Effects of substance P on memory and mood in healthy male subjects. *Hum Psychopharmacol*, 22:567–573.

Hiramatsu, M., Kameyama, T. 1998. Roles of kappa-opioid receptor agonists in learning and memory impairment in animal models. *Methods Find Exp Clin Pharmacol*, 20:595–559.

Hopkins, M.E., Nitecki, R., Bucci, D.J. 2011. Physical exercise during adolescence versus adulthood: differential effects on object recognition memory and brain-derived neurotrophic factor levels. *Neuroscience*, 194:84–94.

Hou, C., Jia, F., Liu, Y., Li, L. 2006. CSF serotonin, 5–hydroxyindolacetic acid and neuropeptide Y levels in severe major depressive disorder. *Brain Res*, 1095:154–158.

Isgaard, J., Aberg, D., Nilsson, M. 2007. Protective and regenerative effects of the GH/IGF-I axis on the brain. *Minerva Endocrinol*, 32:103–113.

Kadish, I., Thibault, O., Blalock, E.M., Chen, K.C., Gant, J.C., Porter, N.M., Landfield, P.W. 2009. Hippocampal and cognitive aging across the lifespan: a bioenergetic shift precedes and increased cholesterol trafficking parallels memory impairment. *J Neurosci*, 29:1805–1816.

Kaneko, S., Maeda, T., Satoh, M. 1997. Cognitive enhancers and hippocampal long-term potentiation in vitro. *Behav Brain Res*, 83:45–49.

Karl, T., Duffy, L., Herzog, H. 2008. Behavioural profile of a new mouse model for NPY deficiency. *Eur J Neurosci*, 28:173–180.

Kenney, J.W., Gould, T.J. 2008. Modulation of hippocampus-dependent learning and synaptic plasticity by nicotine. *Mol Neurobiol*, 38:101–121.

Kim, J., Lee, I. 2011. Hippocampus is necessary for spatial discrimination using distal cue-configuration. *Hippocampus*, 21:609–621.

Kumar, U., Grant, M. 2010.Somatostatin and somatostatin receptors. *Results Probl Cell Differ*, 50:137–184.

Kuzmin, A., Madjid, N., Johansson, B., Terenius, L., Ogren, S.O. 2009. The nociceptin system and hippocampal cognition in mice: a pharmacological and genetic analysis. *Brain Res*, 1305 Suppl:S7–S19.

Lai, Z., Roos, P., Zhai, O., Olsson, Y., Fhölenhag, K., Larsson, C., Nyberg, F. 1993. Age-related reduction of human growth hormone-binding sites in the human brain. *Brain Res*, 621(2):260–266.

Lanni, C., Lenzken, S.C., Pascale, A., Del Vecchio, I., Racchi, M., Pistoia, F., Govoni, S. 2008. Cognition enhancers between treating and doping the mind. *Pharmacol Res*, 57:196–213.

Lattal, K.M., Abel, T. 2001. Different requirements for protein synthesis in acquisition and extinction of spatial preferences and context-evoked fear. *J Neurosci*, 21:5773–5780.

Lee, N.J., Herzog, H. 2009. NPY regulation of bone remodelling. *Neuropeptides*, 43:457–463.

Lee, Y.S., Silva, A.J. 2009. The molecular and cellular biology of enhanced cognition. *Nat Rev Neurosci*, 10:126–140.

Le Grevès, M., Enhamre, E., Zhou, Q., Fhölenhag, K., Berg, M., Meyerson, B., Nyberg, F. 2011. Growth hormone enhances cognitive functions in hypophys-ectomized male rats. *Am J Neuroprot and Neuroregen* 3:53–58.

Le Grevès, M., Le Grevès, P., Nyberg, F. 2005. Age-related effects of IGF-1 on the NMDA-, GH- and IGF-1-receptor mRNA transcripts in the rat hippocampus. *Brain Res Bull*, 65:369–374.

Le Grevès, M., Steensland, P., Le Grevès, P., Nyberg, F. 2002. Growth hormone induces age-dependent alteration in the expression of hippocampal growth hormone receptor and *N*-methyl-D-aspartate receptor subunits gene transcripts in male rats. *Proc Natl Acad Sci U.S.A.*, 99:7119–7123.

Le Grevès, M., Zhou, Q., Berg, M., Le Grevès, P., Fhölenhag, K., Meyerson, B., Nyberg, F. 2006. Growth hormone replacement in hypophysectomized rats affects spatial performance and hippocampal levels of NMDA receptor subunit and PSD-95 gene transcript levels. *Exp Brain Res*, 173:267–273.

Lessmann, V. 1998. Neurotrophin-dependent modulation of glutamatergic synaptic transmission in the mammalian CNS. *Gen Pharmacol*, 31(5):667–674.

Li, S., Wang, C., Wang, W., Dong, H., Hou, P., Tang, Y. 2008. Chronic mild stress impairs cognition in mice: from brain homeostasis to behavior. *Life Sci*, 82:934–942.

Liu, E.H., Lee, T.L., Nishiuchi, Y., Kimura, T., Tachibana, S. 2007. Nocistatin and its derivatives antagonize the impairment of short-term acquisition induced by nociceptin. *Neurosci Lett*, 416:155–159.

Liu, X.M., Shu, S.Y., Zeng, C.C., Cai, Y.F., Zhang, K.H., Wang, C.X., Fang, J. 2011. The role of substance P in the marginal division of the neostriatum in learning and memory is mediated through the neurokinin 1 receptor in rats. *Neurochem Res*, 36:1896–1902.

Lu, Y., Ji, Y., Ganesan, S., Schloesser, R., Martinowich, K., Sun, M., Mei, F., Chao, M.V., Lu, B. 2011. TrkB as a potential synaptic and behavioral tag. *J Neurosci*, 31:11762–1171.

Lynch, G., Rex, C.S., Chen, L.Y., Gall, C.M. 2008. The substrates of memory: defects, treatments, and enhancement. *Eur J Pharmacol*, 585:2–13.

Maric, N.P., Doknic, M., Pavlovic, D., Pekic, S., Stojanovic, M., Jasovic-Gasic, M., Popovic V. 2010. Psychiatric and neuropsychological changes in growth hormone-deficient patients after traumatic brain injury in response to growth hormone therapy. *J Endocrinol Invest*, 33:770.

Martinez, J.L. Jr, Weinberger, S.B., Schulteis, G. 1988. Enkephalins and learning and memory: a review of evidence for a site of action outside the blood-brain barrier. *Behav Neural Biol*, 49:192–221.

McDonald, N.Q., Lapatto, R., Murray-Rust, J., Gunning, J., Wlodawer, A., Blundell, T.L. 1991. New protein fold revealed by a 2.3-A resolution crystal structure of nerve growth factor. *Nature*, 354:411–414.

Miller, L.J., Gao, F. 2008. Structural basis of cholecystokinin receptor binding and regulation. *Pharmacol Ther*, 119:83–95.

Mitsukawa, K., Lu, X., Bartfai, T. 2010. Galanin, galanin receptors, and drug targets. *EXS*, 102:7–23.

Midyett, L.K., Rogol, A.D., Van Meter, Q.L., Frane, J., Bright, G.M., MS301 Study Group. 2010. Recombinant insulin-like growth factor (IGF)-I treatment in short children with low IGF-I levels: first-year results from a randomized clinical trial. *J Clin Endocrinol Metab*, 95:611–619.

Morales-Medina, J.C., Dumont, Y., Quirion, R. 2010. A possible role of neuropeptide Y in depression and stress. *Brain Res*, 1314:194–205.

Myrelid, A., Bergman, S., Elfvik, Strömberg, M., Jonsson, B., Nyberg, F., Gustafsson, J., Annerén, G. 2010. Late effects of early growth hormone treatment in Down syndrome. *Acta Paediatr*, 99:763–699.

Nader, K., Schafe, G.E., Le Doux, J.E. 2000. Fear memories require protein synthesis in the amygdala for reconsolidation after retrieval. *Nature*, 406:722–726.

Nagahara, A.H., Tuszynski, M.H. 2011. Potential therapeutic uses of BDNF in neurological and psychiatric disorders. *Nat Rev Drug Discov*, 10:209–219.

Nyberg F. 2000. Growth hormone in the brain: characteristics of specific brain targets for the hormone and their functional significance. *Front Neuroendocrinol*, 21:330–348.

Nyberg, F. 2009. The role of the somatotrophic axis in neuroprotection and neuroregeneration of the addictive brain. *Int Rev Neurobiol*, 88:399–427.

Nyberg, F., Sharma, H.S. 2002. Repeated topical application of growth hormone attenuates blood-spinal cord barrier permeability and edema formation following spinal cord injury: an experimental study in the rat using Evans blue, ([125])I-sodium and lanthanum tracers. *Amino Acids*, 23:231–239.

Ogren, S.O., Hökfelt, T., Kask, K., Langel, U., Bartfai, T. 1992. Evidence for a role of the neuropeptide galanin in spatial learning. *Neuroscience*, 51:1–5.

Ogren S.O., Kuteeva E., Elvander-Tottie E., Hökfelt T. 2010. Neuropeptides in learning and memory processes with focus on galanin. *Eur J Pharmacol*, 626:9–17.

Okajima, K., Harada, N. 2008. Promotion of insulin-like growth factor-I production by sensory neuron stimulation; molecular mechanism(s) and therapeutic implications. *Curr Med Chem*, 15:3095–3112.

Oliveira, A.M., Bading, H. 2011. Calcium signaling in cognition and aging-dependent cognitive decline. *Biofactors*, 37:168–174.

Ostrovskaya, R.U., Gudasheva, T.A., Zaplina, A.P., Vahitova, J.V., Salimgareeva, M.H., Jamidanov, R.S., Seredenin, S.B. 2008. Noopept stimulates the expression of NGF and BDNF in rat hippocampus. *Bull Exp Biol Med*, 146:334–337.

Pereira, A. Jr., Furlan, F.A. 2010. Astrocytes and human cognition: modeling information integration and modulation of neuronal activity. *Prog Neurobiol*, 92:405–440.

Pertusa, M., García-Matas, S., Mammeri, H., Adell, A., Rodrigo, T., Mallet, J., Cristòfol, R., Sarkis, C., Sanfeliu, C. 2008. Expression of GDNF transgene in astrocytes improves cognitive deficits in aged rats. *Neurobiol Aging*, 29:1366–1379.

Puga González, B., Ferrández Longás, A., Oyarzábal, M., Nosas, R., Grupo Colaborativo Español. 2010. The effects of growth hormone deficiency and growth hormone replacement therapy on intellectual ability, personality and adjustment in children. *Pediatr Endocrinol Rev*, 7:328–338.

Rachidi, M., Lopes, C. 2010. Molecular and cellular mechanisms elucidating neurocognitive basis of functional impairments associated with intellectual disability in Down syndrome. *Am J Intellect Dev Disabil*, 115:83–112.

Ramirez-Amaya, V. 2007. Molecular mechanisms of synaptic plasticity underlying long-term memory formation. In: Bermúdez-Rattoni F, editor. *Neural Plasticity and Memory: From Genes to Brain Imaging*. Boca Raton, FL: CRC Press.

Ramsey, M.M., Adams, M.M., Ariwodola, O.J., Sonntag, W.E., Weiner, J.L. 2005. Functional characterization of des-IGF-1 action at excitatory synapses in the CA1 region of rat hippocampus. *J Neurophysiol*, 94:247–254.

Redrobe, J.P., Dumont, Y., St-Pierre, J.A., Quirion, R. 1999. Multiple receptors for neuropeptide Y in the hippocampus: putative roles in seizures and cognition. *Brain Res*, 848:153–166.

Rhefeld, J.F. 1987. Preprocholecystokinin processing in the normal human anterior pituitary. *Proc Natl Acad Sci U.S.A.*, 84:3019–3023.

Reimunde, P., Quintana, A., Castañón, B., Casteleiro, N., Vilarnovo, Z., Otero, A., Devesa, A., Otero-Cepeda, X.L., Devesa, J. 2011. Effects of growth hormone (GH) replacement and cognitive rehabilitation in patients with cognitive disorders after traumatic brain injury. *Brain Inj*, 25:65–73.

Rex, A., Fink, H. 2004. Cholecystokinin tetrapeptide improves water maze performance of neonatally 6-hydroxydopamine-lesioned young rats. *Pharmacol Biochem Behav*, 79:109–117.

Robertson, J.M. 2002. The astrocentric hypothesis: proposed role of astrocytes in consciousness and memory formation. *J Physiol Paris*, 96:251–255.

Robinson, J.K. 2004. Galanin and cognition. *Behav Cogn Neurosci Rev*, 3:222–242.

Roesler, R., Schröder, N. 2011. Cognitive enhancers: focus on modulatory signaling influencing memory consolidation. *Pharmacol Biochem Behav*, 99:155–163.

Sandin, J., Ogren, S.O., Terenius, L. 2004. Nociceptin/orphanin FQ modulates spatial learning via ORL-1 receptors in the dorsal hippocampus of the rat. *Brain Res*, 997:222–233.

Sandman, C.A., Walker, B.B., Lawton, C.A. 1980. An analog of MSH/ACTH 4–9 enhances interpersonal and environmental awareness in mentally retarded adults. *Peptides*, 1:109–114.

Santangelo, E.M., Morato, S., Mattioli, R. 2001. Facilitatory effect of substance P on learning and memory in the inhibitory avoidance test for goldfish. *Neurosci Lett*, 303:137–139.

Santini, E., Muller, R.U., Quirk, G.J. 2001. Consolidation of extinction learning involves transfer from NMDA-independent to NMDA-dependent memory. *J Neurosci*, 21:9009–9017.

Sharma, H.S., Nyberg, F., Westman, J., Alm, P., Gordh, T., Lindholm, D. 1998. Brain derived neurotrophic factor and insulin like growth factor-1 attenuate upregulation of nitric oxide synthase and cell injury following trauma to the spinal cord. An immunohistochemical study in the rat. *Amino Acids*, 14:121–129.

Sharma, H.S., Nyberg, F., Gordh, T., Alm, P., Westman, J. 2000. Neurotrophic factors influence upregulation of constitutive isoform of heme oxygenase and cellular stress response in the spinal cord following trauma. An experimental study using immunohistochemistry in the rat. *Amino Acids*, 19:351–361.

Shin, M.K., Kim, H.G., Kim, K.L. 2011. A novel trimeric peptide, Neuropep-1-stimulating brain-derived neurotrophic factor expression in rat brain improves spatial learning and memory as measured by the Y-maze and Morris water maze. *J Neurochem*, 116:205–216.

Simmons, M.L., Chavkin, C. 1996. Endogenous opioid regulation of hippocampal function. *Int Rev Neurobiol*, 39:145–196.

Smith, G.B., Heynen, A.J., Bear, M.F. 2009. Bidirectional synaptic mechanisms of ocular dominance plasticity in visual cortex. *Philos Trans R Soc Lond B Biol Sci*, 364:357–367.

Sonntag, W.E., Lynch, C., Thornton, P., Khan, A., Bennett, S., Ingram, R. 2000. The effects of growth hormone and IGF-1 deficiency on cerebrovascular and brain ageing. *J Anat*, 197 Pt 4:575–585.

Sunyer, B., Shim, K.S., An, G., Höger, H., Lubec, G. 2009. Hippocampal levels of phosphorylated protein kinase A (phosphor-S96) are linked to spatial memory enhancement by SGS742. *Hippocampus*, 19:90–98.

Svensson, A.L., Bucht, N., Hallberg, M., Nyberg, F. 2008. Reversal of opiate-induced apoptosis by human recombinant growth hormone in murine foetus primary hippocampal neuronal cell cultures. *Proc Natl Acad Sci U.S.A.*, 105:7304–7308.

Svensson, J., Diez, M., Engel, J., Wass, C., Tivesten, A., Jansson, J.O., Isaksson, O., Archer, T., Hökfelt, T., Ohlsson, C. 2006. Endocrine, liver-derived IGF-I is of importance for spatial learning and memory in old mice. *J Endocrinol*, 189:617–627.

Tallent, M.K. Somatostatin in the dentate gyrus. *Prog Brain Res*, 163:265–284.

Tang, Y.P., Shimizu, E., Dube, G.R., Rampon, C., Kerchner, G.A., Zhuo, M., Liu, G., Tsien, J.Z. 1999. Genetic enhancement of learning and memory in mice. *Nature*, 401:63–69.

Teng, K.K., Hempstead, B.L. 2004. Neurotrophins and their receptors: signaling trios in complex biological systems. *Cell Mol Life Sci*, 61:35–48.

Teyler, T.J., DiScenna, P. 1985. The role of hippocampus in memory: a hypothesis. *Neurosci Biobehav Rev*, 9:377–389.

Thorsell, A. 2008. Central neuropeptide Y in anxiety- and stress-related behavior and in ethanol intake. *Ann N Y Acad Sci*, 1148:136–140.

Thorsell, A., Michalkiewicz, M., Dumont, Y., Quirion, R., Caberlotto, L., Rimondini, R., Mathé, A.A., Heilig, M. 2000. Behavioral insensitivity to restraint stress, absent fear suppression of behavior and impaired spatial learning in transgenic rats with hippocampal neuropeptide Y overexpression. *Proc Natl Acad Sci U.S.A.*, 97:12852–12857.

Toide, K., Shinoda, M., Fujiwara, T., Iwamoto, Y. 1997. Effect of a novel prolyl endopeptidase inhibitor, JTP-4819, on spatial memory and central cholinergic neurons in aged rats. *Pharmacol Biochem Behav*, 56:427–434.

Tomaz, C., Nogueira, P.J. 1997. Facilitation of memory by peripheral administration of substance P. *Behav Brain Res*, 83:143–145.

Trejo, J.L., Llorens-Martín, M.V., Torres-Alemán, I. 2008. The effects of exercise on spatial learning and anxiety-like behavior are mediated by an IGF-I-dependent mechanism related to hippocampal neurogenesis. *Mol Cell Neurosci*, 37:402–141.

van Dam, P.S., Aleman, A., de Vries ,W.R., Deijen, J.B., van der Veen, E.A., de Haan, E.H., Koppeschaar, H.P. 2000. Growth hormone, insulin-like growth factor I and cognitive function in adults. *Growth Horm IGF Res*, 10 Suppl B:S69–73.

Wang, S., Duan, Y., Su, D., Li, W., Tan, J., Yang, D., Wang, W., Zhao, Z., Wang, X. 2011. Delta opioid peptide [D-Ala2, D-Leu5] enkephalin (DADLE) triggers postconditioning against transient forebrain ischemia. *Eur J Pharmacol*, 658:140–144.

Wass, J.A., Reddy, R. 2010. Growth hormone and memory. *J Endocrinol*, 207:125–126.

Wilson, I.A., Gallagher, M., Eichenbaum, H., Tanila, H. 2006. Neurocognitive aging: prior memories hinder new hippocampal encoding. *Trends Neurosci*, 29:662–670.

Wrenn, C.C., Crawley, J.N. 2001. Pharmacological evidence supporting a role for galanin in cognition and affect. *Prog Neuropsychopharmacol Biol Psychiatry*, 25:283–299.

Zhai, Q., Lai, Z., Roos, P., Nyberg, F. 1994. Characterization of growth hormone binding sites in rat brain. *Acta Paediatr Suppl*, 406:92–95.

Zhang, X., Poo, M.M. Progress in neural plasticity. 2010. *Sci China Life Sci*, 53:322–329.

Zhao, W., Sedman, G., Gibbs, M., Ng, K.T. 1995. Phosphorylation changes following weakly reinforced learning and ACTH-induced memory consolidation for a weak learning experience. *Brain Res Bull*, 36:161–168.

Index

T - #0392 - 071024 - C320 - 234/156/14 - PB - 9780367381233 - Gloss Lamination